THE NEW ELEMENTS OF MATHEMATICS

The Peirce Homestead, Arisbe.

(From the Charles S. Peirce Collection in the Houghton Library, Harvard University.)

THE NEW ELEMENTS OF MATHEMATICS

by

CHARLES S. PEIRCE

Edited by

CAROLYN EISELE

VOLUME III/2

MATHEMATICAL MISCELLANEA

1976

MOUTON PUBLISHERS
THE HAGUE – PARIS

HUMANITIES PRESS
ATLANTIC HIGHLANDS N.J.

ISBN 90 279 3035 X

American edition published by Humanities Press Inc. – Atlantic Highlands, N.J.

Library of Congress Cataloging in Publication Data

Peirce, Charles S.
Mathematical Miscellanea.

(His The new elements of mathematics; v. 3)
Includes index.
1. Mathematical Miscellanea. I. Title.
QA39.2.P42 vol. 3 [QA531] 510'.8s [516'.24] 76-15328
ISBN 0-391-00641-X

Printed in the Netherlands

18

MATHEMATICAL CORRESPONDENCE*

* Bracketed dates supplied by Max Fisch.

A. GEORG CANTOR (L 73)

Milford Pa 1900 Dec. 21

Dear Sir:

Before reading your wonderfully beautiful and profound studies (in fact, I have only just read your memoir in the Math. Annalen XLVI and XLIX, and have not had time to digest them, and have only looked a little at the memoir in Vol. XXI, because I could not catch the idea of your ordinal numbers) I was led, in studying the *Logic of Relatives*, to some slight considerations concerning *multitudes*, as I have been in the habit of calling your *Mächtigkeiten*; and have reached a few conclusions and points of view, which I do not find set forth in your papers. They have been partly printed, and I take the liberty of submitting to your judgment some of my papers, which I will send you by mail. Other results have not been printed; but I will here explain my thoughts.

I will first define a *collection*. Whether or not this is precisely what you call a *Menge*, is not clear to me. In order to define it I must first define what I mean by *individual*, *coördinate*, and *independent*. *Individual* (*Individuum*) I use in the logical sense to mean that which has no generality, but is fully determinate (*bestimmt*). Thus, the German people is an individual. So a relation which subsists between two individuals and no others is an *individual* relation. I call subjects *coördinate*, if there is nothing in their essential nature to prevent one from taking any character, or predicate, which the other can take. Thus, any two points are coordinate. But a material thing and a relation are not coordinate; for there is no sense in saying that a relation is, for example, blue, while nothing logically prevents a material thing from being blue. I call subjects independent if it is logically possible for one to have any character while another has any character, provided these two characters are not themselves, either directly or indirectly, in any logical relation which prevents one from being affirmed (or denied) while the other is affirmed or while it is denied. Thus, it does not prevent *A* from being independent of *B*, that *A* cannot be greater than *B*, while *B* is not less than *A*. Nor does it prevent *A* and *B* from being independent that both cannot be the sole thing adored by *C*.

I now define a *collection* as an individual which possesses no character whatever except such as consists in one or other or both of two kinds of fact, namely, 1st, that certain coördinate and independent individuals (if they are more than one), called the members (*Elemente*) of the collection, possess a certain character (absolute or relative) which, if of this derivative nature, is not derived in precisely the same way, and second, that certain other coordinate and independent individuals (coordinate and independent of all the first ones as well as of each other) do not possess this non-derivative character. Thus, if it is true that the German people is at war, that consists in the fact that every member of the German people is subject to an Emperor who has sent his soldiers to do certain things.

We may discriminate between the Earth as possessed of certain non-derivative characters, and the Earth considered as the *only* body to which all tangible objects will fall. The latter is the collection which contains just one single member. Moreover, *nothing* may be considered as a collection which has no character except what consists in the fact that everything wants some character. This view is convenient because it makes the number of possible collections contained in a collection of N ultimate members, equal to 2^N.

Now I say that, no matter how great or how small a collection, say the *S*s, or, as you would write it, $\{S\}$, may be, there is a collection having as its members the different possible collections of *S*s (which collections I will call the *T*s), which latter collection is greater than the collection of *S*s. To prove this, I must prove that there is such a collection, and that, if there is such a collection, there is some relation, *r*, in which every *S* stands to some *T* to which no other *S* stands in the same relation, *r*, and also that there is no relation, ϱ, in which every *T* stands to an *S* to which no other *T* stands in the relation ϱ.

Postponing for a moment the proof that there is such a collection of *T*s, I first prove that, granting that is, it is greater than the collection of *S*s, that is, that a relation such as *r* exists, but no relation such as ϱ. That there is a relation such as *r* is obvious since "being a member of" is such a relation. For every *S* is the sole member of that *T*, which is the collection which includes that *S*, and excludes all others. So is "not being a member of" such a relation.

Now I will prove that there is no relation such as ϱ. For let *R* be any relation whatsoever, I will describe a *T* which is either not in the relation *R* to any *S*, or else is not the *only* *T* in the relation *R* to any *S*. For this purpose I will divide the *S*s into four classes, as follows:

1st, those *S*s (if there be any) to which no *T* is in the relation *R*;

2nd, those *S*s (if there be any) to which more than one *T* are in the relation *R*;

3rd, those *S*s (if there be any) each of which is contained in a *T* which is the only *T* in the relation *R* to it;

4th, those *S*s (if there be any), to each of which one sole *T* in which it is not contained as a member stands in the relation *R*.

Now, let us take any collection of *S*s which contains all of the 4th class and none of the third. Whether or not it contains any or none of the first two classes makes no difference. Considering that collection of *S*s as a *T*, I say that it cannot be the sole *T* in the relation *R* to any member of any of the four classes.

Not to an *S* of the first class, because *no T* is in the relation *R* to them;

Not to an *S* of the second class, because *no T* is the sole one in the relation *R* to any of them;

Not to an *S* of the third class, because these are all contained in the *T*s which are *R* to them while the *T* described contains none of this class and is therefore not *R* to any of them;

Not to an *S* of the fourth class, because none of these is contained in a *T* which is *R* to it, while the *T* described contains all the *S*s of this class, and is therefore not *R* to any of them.

It is thus proved that if there is any collection of all possible collections of *S*s, it is greater than the collection of *S*s.

[FROM ANOTHER DRAFT]

I hesitate to ask you, as a mathematician, to give your time, so precious to the cause of reason, to reading how I, as a logician who strives to look beneath the mere formalisms to the essential affinities of conceptions, would apply the definition of a collection to the various kinds of classes. But then, a student of such subjects as yours is much more a logician than a mathematician; that is to say, however essential to your work it is to deduce the consequences of exact hypotheses (which is mathematical business) it undoubtedly is, yet still more it belongs to you to *frame* those exact hypotheses by the analysis of unclear notions; and that makes you a logician of our school of Exact Logic. We claim you as one of our greatest objects of pride.

A collection has been defined as an Individual, and its members (*Elemente*) are individuals. Now I wish to penetrate below the superficial formalistic definition of an individual. A definition professes to show the relations of the concept (*Begriff*) defined to other conceptions. But the formal definition is content when it has prevented any mistake in the use of a word. It does not always bring out the most significant relations of the *definitum*. This I wish to do. For this purpose, I will imitate Euclid who first states his theorem *in abstracto*, and then constructs a diagram to render his meaning mathematically convenient.

No man can communicate any information to another without referring to some *experiences* known to be shared by him and the person whom he addresses. When a man forms a judgment (*Urtheil*) in his own mind, he addresses his future self (perhaps of the next minute) exactly as he would address another man, and subject to essentially the same conditions. "*Experiences*," I have said. Experience is what? It is the cognitive resultant of the *compulsions* to which our thought has been subject during our lives. Compulsion implies force, action and reaction, struggle. Struggle is something *hîc et nunc*, without generality. It is, indeed, incapable of being generalized. For a generalized force is a mere law. A law remains an idle formula until an occasion for its application blindly and irrationally insists upon existence. The conception of individuality owes its origin to the sense of resistance to an effort; and however abstract a form it may assume, it always retains that reference.

There is my theorem. As it stands, it is not mathematically intelligible. I will endeavor to make it so. A man walking along a lonely road sees at a distance a house on fire. He trudges on for twenty kilometres, when he meets another man. He says "There is a fire." The second man does not derive much information from this; so he asks "Where?" The first man stretches out his arm and points in the direction of the fire, and says "There, twenty kilometres back." The second man has now to look at the arm, which compels his eyes to look at the country in a certain direction, to recall what he remembers to have experienced when he was formerly in that part of the country, to remember that persons from the *Eichungsamt* have made certain experiments which have led them to declare certain rods to be *metres*, and that other persons have laid down those rods and measured a kilometer which he has often walked over, to estimate how much effort it would be to walk that twenty times over, to remember where he was when he had walked in the direction pointed out until he had made about that amount of effort, and to remember what houses he saw when he was there. All these things, — or their equivalents, — must come before his mind before that first man's proposition "There is a fire" can convey any information. All those memories are resultants of actions and reactions and their compulsions on his mind, each of which actually happened to him at moments of his past existence.

The word kilometre was used. That refers to an individual bar kept in Sêvres, in the Park of St.-Cloud. Strictly speaking, there are many metres, any finite multitude of *possible* metres. Strictly speaking, the bar in the *Pavillon de Breteuil* is not an individual; for it is *sometimes* one length and sometimes another; so that it does not *universally* possess any exact length. In comparisons, the temperature is observed, and allowed for. But experience compels us to believe that, those corrections applied, the ideal corrected bar is *virtually* an individual, so far as our use of the predicate of length gives it a meaning. "Virtually," that is, that there is something which compels the facts of observation and the deductions from them to be such as they would be if there were an individual there.

Thus the only perfect individual is the effort and resistance that happens *hîc et nunc*. Every thing else that I call so is only a *virtual* individual, — even the very inkstand I see before me on the table. The reaction element of the percept is perfectly individual; but all else that goes to make up the percept of the inkstand is only virtually individual.

I see an inkstand and I imagine another standing beside it. The latter is very faint, but were it as vivid as it would be to some painters, were it as vivid as the one I see, it would still be a mere fancy, possessing no self-identity, since I can at will make it tall or short, large or small, clean or dirty. It thus has no individuality. For an individual has self-identity. But perhaps the inkstand I think I see is only a hallucination. I call in my wife. She sees it too, and cannot unsee it. Perhaps, however, we are mad together. To test this, I set up a camera before it, and I *predict* that if it is a real inkstand the camera will show it. My prediction is verified; and the *insistency* of this image in retaining definite characters, regardless of who thinks of it, and for all purposes, compels me to think of it as virtually individual. Here again it is the compulsive force of the percept that makes me call it a real individual inkstand.

Let us pass to a case in which individuality is still weaker. I imagine an equilateral triangle. It has three angles. I imagine it to rotate too rapidly for the eye to follow about its centre and in its own plane. When it comes to rest, have the three angles preserved each its own identity? Undoubtedly; for they might have had the letters *A*, *B*, *C*, attached to them severally, so that these letters would be carried round with them. ...

Milford Pa 1900 Dec 23

Dear Sir:

I yesterday succeeded in borrowing Volumes XLVI and XLIX of the *Mathematische Annalen* containing your wonderfully beautiful and masterly memoir, which I have read but have not had time to digest. I am nothing but a farmer living in the wildest part of the Eastern States; although our National Academy of Sciences has most indulgently honored me with one of its chairs. So my isolation accounts for my not having read that great work before. I had already read the translations in Vol. II of the *Acta Mathematica*; but could not understand that from Vol. XXI of the *Annalen*, because I failed to grasp the idea of an ordinal number. Even before I knew any of your papers, I had been led, in 1881, from the study of the Logic of Relatives, to a few ideas about numbers, and had particularly seen that finite classes differ from infinite ones in that a certain form of reasoning is valid of the former, that is not valid of the latter. It is De Morgan's *Syllogism of Transposed Quantity* of which this is an example:

Every Hottentot kills a Hottentot
No Hottentot is killed by more than one Hottentot
∴ Every Hottentot is killed by a Hottentot.

Every property which distinguishes a finite from an infinite class can be deduced from that. I also saw that what you call a class of *Mächtigkeit* $\aleph_0$ is distinguished from other infinite classes in that the *Fermatian Inference* (very improperly called *vollständige Induktion*, in German) is applicable to the former and not to the latter; and that generally, *to any smaller class some mode of reasoning is applicable which is not applicable to any greater one.* For greater classes allow greater possibilities, — a very significant fact. Later, after reading your papers in the *Acta Mathematica*, except the one I could not understand, I was led to some further considerations, which I shall venture to submit to your judgment. They have been printed in part only. One of these is a proof (without resort to ordinal numbers which were a closed book to me) that two classes could not be each greater than the other. I will mail you a number of the *Monist* containing this proof; and since I there use my two algebras for the logic of relatives, I will mail also some papers about that and a book where in "Note B" the notation is briefly explained.

I have also reached another result which may very likely be well-known to you but which I do not find that you state and which leads directly to ideas which seem quite opposed to yours about continuity, and which appear to me extremely important not only for the theory of logic, in general, but also as leading to a strictly rigid demonstrative method in Topical Geometry (*Topologie*) and I hope may lead to some practical way of dealing with that puzzling subject, — and show, for example, how to treat the problem of coloring a map.

In order to explain the little matter to which I allude, I will begin with a few definitions. Some of the first ones seem, at first sight, quite foreign to the subject; but their connection with it will appear as I proceed further.

By a *true* proposition (if there be any such thing) I mean a proposition which at some time, past or future, emerges into thought, and has the following three characters:

1st, no direct effort of yours, mine, or anybody's, can reverse it permanently, or even permanently prevent its asserting itself;

2nd, no reasoning or discussion can permanently prevent its asserting itself;

3rd, any prediction based on the proposition, as to what ought to present itself in experience under certain conditions, will be fulfilled when those conditions are satisfied.

By a *reality*, I mean anything represented in a true proposition.

By a *positive* reality or truth, I mean one to which all three of the above criteria can be applied, — of course imperfectly, since we can never carry them out to the end.

By an *ideal* reality or truth, I mean one to which the first two criteria can be applied imperfectly, but the third not at all, since the proposition does not imply that any particular state of things will ever appear in experience. Such is a truth of pure mathematics.

By an *ultimate* reality or truth, I mean one to which the first criterion can be in some measure applied, but which can never be overthrown or rendered clearer by any reasoning, and upon which alone no predictions can be based. Thus, if you are kicked by a horse, the fact of the pain is beyond all discussion, and far less can it be shaken or established by any experimentation.

By an *individual* I mean a subject of which every predicate is either universally true or universally false.

By a *primitive individual* (πρώτη οὐσία), I mean an individual having characters which it might, without contradiction, be supposed to possess although there were no regularities among other things, except certain regularities presupposed to render knowledge possible. Thus, the fact that a stone is hard is explained by certain things being true of its molecules generally, which constitute a regularity among them. Still, it is conceivable that it might be hard even if it did not consist of molecules. So a point has its position, so long as the general regularities of space exist, no matter how it may be with other points.

By a *derivative individual*, I mean an individual which has no characters except such as truly consist in certain regularities among other things, beyond such regularities as are requisite to cognition. Thus, the German People is an individual of which nothing whatever is true except in virtue of facts concerning individual men, their country, etc. So a line, whether it be regarded as an aggregation of points or not, has no character whatever except what consists in relations between its parts and other parts of space. Thus, if it is black, to say it is so only means that all its parts are black. If it is the boundary between two areas, that means that there are certain relations between its parts and those of the areas, — a movable area cannot move from any part of one of the bounded areas to any part of the other without at some time occupying some part of the line. True, we may abstract from certain circumstances and conceive of rays as primitive individuals and points as derivative; but the *truth* is that the line consists of points, not the points of lines. On an extreme nominalistic theory, there is no *true* order of dependence; and in that case, the distinction between primitive and derivative individuals is relative to a particular aspect of the facts.

By subjects *of the same category* I mean two or more subjects each of which may, without contradiction, be supposed to be modified, without altering its general nature, so as to take any character that any of the others can take. Thus, two points are of the same category. But a movable particle and its position are not; for a position cannot burst into two, or move so as to lose its connection with neighboring positions, as a particle can, without ceasing to be a position. Of course the "category," may be more or less widely conceived.

By saying that subjects are *more or less independent*, I mean, that they can be supposed, without absurdity, to interchange their characters, more or less, without losing their identities. Thus, two atoms can be supposed, in the course of time, to interchange every instantaneous character; but two colors cannot interchange many of their characters without becoming different colors.

By a *collection*, I mean a derivative individual which possesses no characters except such as consist in certain characters of certain individuals, distinct from, of the same category with, and more or less independent of, one another, and that in one or other or both of the following ways:

1st, the characters of the collection may consist in the possession of certain characters by certain relatively primitive individuals, termed its *ineunts* or *members* (*Elemente*).

2nd, the characters of the collection may consist in the possession of certain characters by others of the individuals of the same category as its ineunts, these being termed its *exeunts*.

For example, the letters *A* and *B* are objects of the same category, distinct from one another and more or less independent of one another. Of them, there are four collections, viz:

1st, the collection which has no ineunts, and *A* and *B* as exeunts;

2nd, the collection which has *A* as ineunt and *B* as exeunt;

3rd, the collection which has *B* as ineunt and *A* as exeunt;

4th, the collection which has *A* and *B* as ineunts and no exeunt.

The collection which has *A* as ineunt is no more to be confounded with *A* than with *B*. The letter *A* is one thing; it is not a collection at all. The collection which contains *A* and excludes *B* is not a letter; it is an *ens rationis*. It is much more like the collection which includes *B* and excludes *A*, than it is to anything else. The null-collection which has no ineunts is distinguishable, but barely different, from the all-collection which has no exeunts. It is utterly unlike *nothing*. Nor is the all-collection to be confounded with the logical aggregate of *A* and *B*. Indeed, all the old logic[s] caution against this error. Of the above four collections, there are again 16 collections, as follows:

I, the collection which has no ineunts and the four as exeunts;

II, III, IV, V, the collections having single ineunts and three exeunts;

VI, VII, VIII, IX, X, XI, the collections having two ineunts and two exeunts;

XII, XIII, XIV, XV, the collections having three ineunts and single exeunts;

XVI, the collection having four ineunts and no exeunt.

Of these again there are 65536 collections. The collection which has no ineunt and four exeunts is, by no means, to be confounded with the collection which has for its sole ineunt the null-collection of *A* and *B*, and the other three for its exeunts. Still less, is this collection to be confounded with the collection which has no ineunt and *A* and *B* as exeunts; for a *collection*, as I have defined it, depends just as much on its exeunts as on its ineunts. Now the three collections of A, of B, and of AB, are not even of the same category as the letters A and B.

It is not necessary, in speaking of a *collection*, to have a perfectly definite idea of both ineunts and exeunts. I may say that three dots on a sheet of paper from a collection, meaning by the exeunts any other that could conveniently be made at the same time. But, since a collection is not well defined until what it excludes is defined, it follows from the above definition of truth, that a collection does not really exist, even ideally, until the complementary collection exists. Hence, it is incorrect to speak of three possible objects as a collection, meaning to exclude every other object which could exist without contradiction. For those possibilities (unless some explicit conditions are laid down) are not distinct from one another. No actual definitions, no possible definitions, no forces, no facts of existence, sharply separate one from another. I do not deny that there is such an idea as 'all angels' taken, not *distributive*, as the logics say, but *collective*. I only say, that in the sense above defined, they are not a collection. For Aquinas was quite right in saying that ten thousand of them could dance on the point of a needle, because they had no "matter"; for "matter" was defined as that which confers existence. Mere possibilities have no distinct identity.

I must beg your pardon for asking you to read this too verbose letter; but you will see presently that all this has a bearing on the subject to which you have contributed so much.

I use your definition of *greater*, *less*, and *equal*, as applied to collections, merely stating it in slightly different form.

Namely it is assumed that, given any two collections, the As, or $\{A\}$, and the Bs, or $\{B\}$, whatsoever, there is either some relation r such that every A stands in the relation r to some B to which no other A stands in the relation r, or every B stands in the relation r to some A to which no other B stands in the relation r. This I write thus:

$$\Sigma_r \Pi_i \Sigma_k \Pi_j \Pi_p \Sigma_s \Pi_q (\bar{A}_i \Psi r_{ik} \cdot B_k) \cdot (\bar{A}_j \Psi l_{ij} \Psi \bar{r}_{jk}) \Psi (\bar{B}_p \Psi r_{ps} \cdot A_s) \cdot (\bar{B}_q \Psi l_{pq} \Psi \bar{r}_{qs})^1$$

Then the collection $\{A\}$ is *greater* than the collection $\{B\}$
and the collection $\{B\}$ is *less* than the collection $\{A\}$
if, and only if, no relation, ϱ, can be found such that every A is in the relation, ϱ, to a B to which no other A is in the relation ϱ.

If neither of two collections is greater than the other, they are *equal.*

[1] The capital psi is being used by the printer to represent Peirce's summation symbol Ψ. It is an approximation to Peirce's notation for "or". See pages 66 and 907.

By the *multitude* of a collection, I mean that character of it by virtue of which it is greater than some collections and less than others. (*Mächtigkeit*). By the *arithm* of a collection, I mean that one of a series of vocables or signs which is applied as an adjective to a collection to express its ordinal place in the scale of multitudes. (*Cardinalzahl.*)

Now the first step toward the result to which I beg leave to submit to your kind consideration is to show that, no matter how small or how great a collection $\{A\}$ may be, the multitude of the collection of different possible collections of ineunts of $\{A\}$ is greater than the multitude of $\{A\}$. That is to say, calling the *B*s the different collections whose ineunts are ineunts of *A*, no relation ϱ can be found such that every *B* is in the relation ϱ to an *A* to which no other *B* is in the relation ϱ. For, given any relation, ϱ, whatever, I will describe a *B* which does not stand in the relation ϱ to any *A* to which no other *B* stands in the same relation. Namely, I first divide the *A*s into four classes, as follows:

1st, those (if such there be) to which no *B* is in the relation ϱ;

2nd, those (if such there be) to each of which more than one *B* is in the relation ϱ;

3rd, those (if such there be) each of which is an ineunt of that *B* which is ϱ to it, and to no other *A*;

4th, those (if such there be) each of which is an exeunt of the *B* which alone of the *B*s stands in the relation ϱ to it.

Now consider any *B*, say *B*, which has all the *A*s of the fourth class as ineunts and all those of the third class as exeunts, regardless of whether it has the members of the first two classes as ineunts or exeunts. There is such a *B*; for this describes a possible collection of *A*s. Now *B* is the *sole B* in the relation ϱ

to any *A* of the first class, since *no B* is ϱ to any *A* of this class; nor to any *A* of the second class, since *no B* is the *sole B* that is ϱ to any such *A*;

nor to any *A* of the third class, since every such *A* is an ineunt of the sole *B* that is ϱ to it, while it is not an ineunt of *B*; nor to any *A* of the fourth class, since every such *A* is an exeunt of the sole *B* that is ϱ to it, while it is not an exeunt of *B*.

Hence *B* is not the sole ϱ to any *A* whatever. Q.E.D.

It follows that

$$\aleph_0 < 2^{\aleph_0} < 2^{2^{\aleph_0}} < 2^{2^{2^{\aleph_0}}} < etc.\ ad\ inf.$$

But suppose that we take together all the integers, all the collections of integers, all the collections of collections of integers etc. without end, as ineunts of a new collection, then, if there is any such collection, its multitude is the limit of the above series. For if the ineunts of any collection are taken together with all the possible collections of those ineunts, the resulting collection is of the same multitude as the collection of those collections of ineunts.

For the same reason, if $\beth_0$ (Beth-null) is a collection whose multitude is $\aleph_0$, $\beth_1$ a collection whose multitude is $2^{\aleph_0}$, $\beth_2$ a collection whose multitude is $2^{2^{\aleph_0}}$, etc., then the collection whose ineunts are all the ineunts of beths up to $\beth_n$ inclusive has the multitude of $\beth_n$; and therefore $\beth_{n+1}$ has the multitude of all possible collections formed of ineunts of all the beths up to $\beth_n$ inclusive. Hence, a collection which should have a multitude equal to the limit of the above series of interative functions, $2^{2^{2^{\cdot^{\cdot^{\aleph_0}}}}}$ etc, would be equal to the collection of all possible collections of its own ineunts. But this has just been shown is not the case with any collection. Hence, the taking together of the ineunts of all the beths, *is not a collection*, although all these ineunts form a distinctly definable class; and although there is a conceivable quantity which is the limit of the above iterative functions, it is not of the nature of a multitude.

I can even define a general class of forms each one of which can be written down in full, without any general description, and can be numbered by a finite number; and yet all these forms taken together do not constitute a collection. For that purpose, I use the notation $\frac{AB}{CD}$ to denote the collection whose ineunts are A and B and whose exeunts are C and D. I also write 0 for nothing. I start then to form a series of forms which, for convenience, I separate into orders.

The zero order consists of 0
The 1st order consists of $\frac{0}{0}$
The 2nd order consists of

$$\frac{0}{\frac{0}{0}} \qquad \frac{\frac{0}{0}}{0}$$

The 3rd order consists of

$$\frac{0}{0/0\ \frac{0}{0/0}\ \frac{0/0}{0}} \qquad \frac{\frac{0}{0}}{\frac{0}{0/0}\ \frac{0/0}{0}} \qquad \frac{\frac{0}{0/0}}{0\ \frac{0/0}{0}} \qquad \frac{\frac{0/0}{0}}{0\ 0/0}$$

$$\frac{0/0\ \frac{0}{0/0}}{\frac{0/0}{0}} \qquad \frac{0/0\ \frac{0/0}{0}}{\frac{0}{0/0}} \qquad \frac{\frac{0}{0/0}\ \frac{0/0}{0}}{\frac{0}{0}} \qquad \frac{\frac{0}{0}\ \frac{0}{0/0}\ \frac{0/0}{0}}{0}$$

The fourth order will consist of $2^{11} = 2048$ collections, the 5th of 2^{2059} collections. Call all the collections of all orders which could be so written down the αs. Now taking any one of the αs up to the nth order inclusive, there is one α of the $(n+1)$th order, and one only, which has the same ineunts as it. Moreover, any collection whatever of αs has for its ineunts the ineunts of some α. Hence calling the relation of 'having the same ineunts as', ϱ, every collection of αs is ϱ to some α to which no other collection of αs is ϱ. Hence if the αs formed a collection, the collection of collections of the αs would be no greater than the collection of αs, which is absurd. The *elenchus* consists in this, that that collection of αs which is ϱ to no α becomes absurd, when the αs are defined as including an α ϱ'd by each collection of αs. For it then becomes impossible to distinguish between the null-collection of the αs and an α.

B. PAUL CARUS (L 77)

Milford Pa 1899 Aug 17

My dear Dr. Carus:

I find I did an injustice to Kant in one of my *Monist* papers in which I discussed continuity. I was so much dominated by Cantor's point of view, that I failed to see the true nature of continuity, which is now quite clear to me, and also for the same reason mistook Kant. This was somewhat excusable since Kant himself made the same confusion between his real doctrine and the one I took for his, as can be seen at the bottom of p. 524 of 1st Ed. of the Critik. Namely he does not quite say that it is the same to say that Space is infinitely divisible, but he does not draw attention to the important distinction as he would have done, if he had seen it. His true doctine is not that space is divisible without end, but that it *cannot be so* divided as to reach an ultimate part as clearly stated in the last paragraph of p. 169. This definition is really no nearer right than the other, but it is better as looking at the matter from the right point of view and not trying to build up a continuum from points as Cantor does. To the obvious objection that points are ultimate parts of lines, Kant begins to make the right answer, that they are not parts but limits. But that he does not understand this rightly can be seen by p. 209 where he speaks of a change as passing through all the instantaneous intermediate states. He thus looks on the point as existing in the line, while the truth is they do not exist in the continuous line, and if a point is placed on a line it constitutes a discontinuity. As he conceives the matter, it becomes a mere quibble to say that a point is not part of a line. It is not a homogeneous part, but it is in the line. The case is an interesting example of Kant's sagacity about points he did not distinctly apprehend.

Very truly
C. S. Peirce

[P.S.] See Jour. of Spec. Phil. II. 200.

C. HENRY B. FINE (L 145)

Milford Pa 1903 July 17

Dear Sir:

A good way of putting all positive rational quantities into one-to-one correspondence with all ordinal numbers, so that the ordinal number of any given quantity or the quantity in any ordinal place can readily be calculated, is this:

Work the usual algorithm for finding the greatest common measure of the quantity and 1 so that the first quotient is zero if the quantity exceeds 1. Make the number of quotients even by (if necessary) diminishing the last by 1, when there will be an additional quotient, 1. Thus, any even series of quotients is possible, and of course any two series belong to different quantities.

Now, using the *secundal* system of numerical notation, put a fractional point and then a number of zeros equal to the first quotient, followed by a number of 1s equal to the second quotient, followed by a number of zeros equal to the third quotient, and so on. We thus get a fraction in the secundal notation.

Of any two rational quantities the larger will make this fraction the larger.

Remove the last unit and put it in the unit's place. Remove the fractional point and you will have the ordinal number required.

For example, what will be the millionth fraction?

x	2^x	x	2^x	x	2^x	x	2^x
0	1.	5	32.	10	1024.	15	32768
1	2.	6	64	11	2048.	16	65536
2	4.	7	128.	12	4096.	17	131072
3	8.	8	256.	13	8192.	18	262144
4	16.	9	512	14	16384	19	524288
5	32.	10	1024	15	32768.	20	1048576

$$\begin{aligned} 524288 &= 2^{19} \\ 262144 &= 2^{18} \\ 131072 &= 2^{17} \\ 65536 &= 2^{16} \\ 16384 &= 2^{14} \\ 512 &= 2^{9} \\ 64 &= 2^{6} \end{aligned}$$

$$1 0_2 0_2 0_2 0_4 0_3 0 = [11110100001001000000]$$

Our fraction is .11101000010010000001

Row of Quotients			0	3	1	1	4	1	2	1	6	1
Calc. of Numerator	1	0	1	3	4	7	32	39	110	149	1004	1153
Calc. of Denom.	0	1	0	1	1	2	9	11	31	42	283	325

Answer the millionth value is $\frac{1153}{325}$.

I write this, because it is a little point worth perhaps a letter, but not worth printing. I have much that *is** (* About Mengenlehre, with proofs of unproved propositions.); but it is voluminous.[1]

Very truly
C. S. Peirce

Prof. H. B. Fine
Princeton, N.J.

[1] A section of S-1 reads as follows:

"To find the value of the fraction in any ordinal place,

Write this place in the *secundal* system of numerical notation. Call this N.

Form a row of numbers to be called A_n where n is the place of a number in the row

A_1 is the number of zeros following the first 1 in N
A_2 is the number of ones following next
A_3 is the number of zeros and so on.
Only the value of the last A is greater by 1.
Then the value of the fraction is

$$\cfrac{1}{A_1 + \cfrac{1}{A_2 + \cfrac{1}{A_3 + \text{etc.}}}}$$

Thus what is the value of the 551st fraction?

551 = 1 0 0 0 1 0 0 1 1 1
3 1 2 4
1 0 1 1 3 13 13
0 1 3 4 11 48 48

What is the 365th fraction? 365 = 1 0 1 1 0 1 1 0 1

1 2 1 2 1 2

1 0 1 2 3 8 11 30 $\frac{30}{41}$

0 1 1 3 4 11 15 41

What is the 2nd fraction? 2 = 10

2

101 $\frac{1}{2}$

012

What is the 1st fraction? 1 = 1

1

101 $\frac{1}{1}$

011

What is the 13th? 13 = 1 1 0 1

0 1 1 2

1 0 1 1 2 5 $\frac{5}{3}$

0 1 0 1 1 3

1000	1001	1010	1011	1100	1101
.0001	.0011	.0101	.0111	.1001	.1011
3 1	2 2	1 1 1 1	1 3	0 1 2 1	0 1 1 2
1 0 1 1	1 0 1 2	1 0 1 1 2 3	1 0 1 3	1 0 1 1 3 4	101 1 2 5
0 1 3 4	0 1 2 5	0 1 1 2 3 5	0 1 1 4	0 1 0 1 2 3	010 1 1 3
$\frac{1}{4}$	$\frac{2}{5}$	$\frac{3}{5}$	$\frac{3}{4}$	$\frac{4}{3}$	$\frac{5}{3}$

What is the millionth value?

A million = 11110100001001000000
.11101000010010000001

0 3 1 1 4 1 2 1 6 1

1 0 1 3 4 7 32 39 110 149 1004 1153

0 1 0 1 1 2 9 11 31 42 283 325

The millionth value is $\frac{1153}{325}$

What is the 10th value? 1010
.0101

What is the zero? 0

101 $\frac{1}{0}$

010

What is the 2nd? 10
.01

11

1011 Ans. $\frac{1}{2}$

0112

What is the 1st? 1
.1
01
1011 Ans. $\frac{1}{1}$
0101

What is the value of the fraction whose ordinal number is 10001011? Striking off the last 1 and what follows, it is greater than the fraction whose number is 1000101. Striking off last 0 and what follows, it is less than 10001.

$$100010 < 1000101 < 10001$$

$$\begin{array}{ccccc} 1000 & < & 10001 & < & 100 \\ \| & & \| & & \| \\ \frac{1}{4} & & \frac{2}{7} & & \frac{1}{3} \end{array}$$

$$\begin{array}{ccccc} 1000 & < & 100010 & < & 10001 \\ \| & & \| & & \| \\ \frac{1}{4} & & \frac{3}{4} & & \frac{2}{7} \end{array}$$

$$1000101 = \frac{5}{18}$$

$$10001 = \frac{2}{7}$$

$$10001011 = \frac{7}{25}$$

A better rule for calculating the value is this:

Write down in a row the numbers of zero to the right of each 1 until you come to the next 1.

Increase the first by 1 and the others each by 2.

Let A_n be the number in the nth place in this row.

Make a row N_n taking $N_0 = 0 \quad N_{-1} = -1$

$$N_n = A_n N_{n-1} - N_{n-2}$$

Make another row taking $D_{-1} = 0 \quad D_0 = 1$

$$D_n = A_n D_{n-1} - D_{n-2}$$

Then the last $\frac{N_n}{D_n}$ will be the value of the fraction.

Thus to find value of	1	0	0	0	1	0	0	1	1	1	
				3			2	0	0	0	
				4			4	2	2	2	
		−1	0	1			4	7	10	13	13
		0	1	4			15	26	37	48	48 is the value.

Find value of	1	0	0	0	1	0	1	1	0	0
				3	1	0	2			
				4	3	2	4			
		−1	0	1	3	5	17	17		
		0	1	4	11	18	61	61 is the value."		

D. F. W. FRANKLAND (L 148)

P. O. Milford Pa 1906 May 8

My dear Frankland:

Your speech is admirable, and as far as I have been able to make out is quite true. I wish I were a New Zealander. I see the print talks of the banquet "tendered" to Mr. Seddon, which shows New Zealanders are not innocent of the sin of newspaper-English. The banquet was not only *tendered*, but accepted; not only accepted but given; not only given but celebrated.

On the leaf of Nature you enclose you have written a line which is not correct. Nobody ever doubted (by nobody I mean nobody worth considering) the possibility of the aleph multitudes and I am the author of the first proof of the general proposition that there is a multitude greater than any given multitude. But what has been doubted and what I *think* Russell is right in what he *means* when he says that the question is one of logic not of mathematics (and I have done a great deal of work upon it), is those [ω] multitudes (I mean the multitudes corresponding to [ω] numbers) of Cantor which surpass all the alephs, and of which the peculiarity is that two collections of such multitude can neither of them be related to the other so that if the one be called the collection of *A*s and the other the collection of *B*s, there is any relation in which every *A* stands to some *B* to which no other *A* is in that relation, nor the converse. (In my opinion Russell and Whitehead are blunderers continually confusing different questions. The best account of Cantor's results and best criticism of them is by Schoenfliess in the 2nd Heft of Vol. VIII of the Jahresbericht der Deutschen Mathematiker-Vereinigung, published in 1900.)

I translate Cantor's *Menge* by "collection"; Cantor's *Mächtigkeit* and *Cardinalzahl* by "Multitude." I call finite multitudes *enumerable*. The multitude of the simple endless series '*denumeral*' which according to Schoenfliess is the German "abzählbar"; but If I am not mistaken Cantor applies this word to any multitude which is either *enumerable* or is *denumeral*, which I call *numerable*. I call that series of multitudes of which the first is the multitude of collections that might each of them consist of the individuals of a denumeral collection, the 1st *abnumerable* multitude; and by the *n*th abnumerable multitude I mean the multitude of different possible collections each consisting of individuals belonging to one $(n-1)$thly abnumerable collection. These are really the alephs although differently defined.

Very faithfully
C. S. Peirce

P.S. I remember you used to say everything but *time* was finite. If this be meant for a proposition in *metaphysics*, I don't believe it, but anything may pass for tenable in metaphysics. In logic it is not tenable. W. James and F. C. S. Schiller maintain that God and everything else is finite, — a doctrine some people call *pragmatism*. To me it is as abhorrent as it is incredible.

But I should think that in any case you would wish to examine Cantor's work; and the report of Schoenfliess is the clearest and in every respect the best. I have convictions about the ethics of terminology in accordance with which I use the following terms.

By an *ascending fundamental series* I mean a collection of which it is true that there is a certain dyadic relation, which we may call being 'later than —' such that of any two different individuals of the collection one is later than the other; but if any is later than any one then the latter is not later than the former, and if any one, say θ is later than any one say A, then there is one and only one said to be *next later than A* which is later than A but is not later than any that is later than A, and there is one and only one said to be *next earlier than B*, than which B is later but is not later than any that is later than it. And if any one is later than another it is later than every one there may be that the latter is later than. And there is one that is not later than any. But there is none than which none is later. Thus the finite whole numbers form an ascending fundamental series calling a number later than another if and only if it is greater than that other.

A *descending fundamental series* is defined in the same way, only that the converse relation to being later than is everywhere substituted for being later than in the definition of the ascending fundamental series. Thus the whole numbers form a descending fundamental series if we call every smaller number *later* than every number greater than it.

A *limiting element* of a fundamental $\left\{\begin{matrix}\text{ascending}\\ \text{descending}\end{matrix}\right\}$ series is an individual object which does not belong to that series but which is $\left\{\begin{matrix}\text{later}\\ \text{earlier}\end{matrix}\right\}$ than every member of the series, and is a member of series of which that fundamental series is a part, this larger series being such that of any two different individuals of it one is later than the other while the other is not later than the former, and the limiting element is not later than any element of the larger series that is later than every element of the fundamental series.

A *Cantorian* (wohlgeordnet) series is a collection such that if every individual of a collection is a member of the Cantorian series, then there is an individual of this second collection than which every other individual of this same collection is *later*.

E. WILLIAM JAMES (L 224)

[April 1897]

My dear William:

The Renouvier is at hand and reading commenced. All such minute and exact analysis, if at all able, is valuable, even if not quite correct; and this author's great ability is manifest. To me, a writer who like Renouvier speaks the dialect (I refer to the conceptions not the phraseology) of Kant, addresses me in my vernacular. For the Critic of the Pure Reason, as you know, was my mother's milk in philosophy.

But here is a note you can file away touching Renouvier's razor, his "principle of number," — that an inenumerable collection is absurd. He says it is self-contradictory. He also hints (p. 48) that even were that refuted, he would stick to it that it was absurd.

I say it is not absurd. And, first, it is not self-contradictory. Let us use this diction, that *P* is *r* to *Q*, meaning that *PQ* is one of a class of ordered pairs implied by the general term *r*. Then, the proper definition of an enumerable collection is this: The *A*s form an enumerable collection if and only if, no matter what class of ordered pairs *r* refers to, either some *A* is *r* of two different *A*s, or there is some *A* to which no *A* is *r*, or every *A* is *r* of an *A*. For instance, either some *A* is an *offshoot* of two different *A*s, or there is some *A* which has no offshoot, or every *A* is an offshoot of an *A*. For if no *A* is offshoot of two different *A*s, and every *A* has an *A* as an offshoot, and yet there is a trunk *A* not an offshoot of any *A*, the ramifications or offshoots are endless.

I can easily show there is no contradiction involved. That no *A* is *r* of two different *A*s is, in the general algebra of logic, written thus (in a universe of *A*s):

$$\Pi_1 \, \Pi_2 \, \Pi_3 \, \bar{r}_{12} \, \psi \, \bar{r}_{13} \, \psi \, 1_{13}$$

That every *A* has an *A* as its *r* is written thus:

$$\Pi_5 \, \Sigma_4 \, r_{45}$$

That there is an A not r of any A is written thus:

$$\Sigma_6\, \Pi_7\, \bar{r}_{67}.$$

Now, if there is any contradiction involved, when we multiply these propositions into one another a sufficient number of times and suitably identify the indices, then the principle of contradiction, $x \cdot \bar{x} \prec 0$, must reduce the product to zero. That is,

$$\Pi_5 \Sigma_4\, \Pi_{5'} \Sigma_{4'}\, \Pi_{5''} \Sigma_{4''} \text{ etc. } \Sigma_6\, \Pi_7\, \Sigma_{6'}\, \Pi_{7'} \text{ etc.}$$
$$\Pi_{5'''} \Sigma_{4'''}\, \Pi_{5''''} \Sigma_{4''''} \text{ etc. } \Pi_1\, \Pi_2\, \Pi_3\, \Pi_{1'}\, \Pi_{2'}\, \Pi_{3'} \text{ etc.}$$
$$r_{45} \cdot r_{4'5'} \cdot r_{4''5''} \cdot r_{4'''5'''} \cdot r_{4''''5''''} \cdot \text{etc.} \cdot \bar{r}_{67} \cdot \bar{r}_{6'7'} \cdot \text{etc.}$$
$$(\bar{r}_{12} \,\psi\, \bar{r}_{13} \,\psi\, 1_{13}) \cdot (\bar{r}_{1'2'} \,\psi\, \bar{r}_{1'3'} \,\psi\, 1_{1'3'}) \text{ etc.}^{1}$$

must vanish on suitable identification of the numbers. But no 4 can be identified with a 6, because the quantification of both is particular (that is both are attached to Σ in the Peircian). Neither can one 4 be identified with another 4 (because the Σ follows the Π, the only other quantifier), and therefore both cannot at once be identified with a 1. Hence, there is no way at all in which the expression can vanish. In other words, we have proved that the hypothesis of an inenumerable collection involves no contradiction.

Though this proof is irrefragable, yet in order to throw further light upon it, let us see how it follows that it is absurd to suppose a collection inenumerable when it has some property which makes it enumerable. Observe, then, that by the above definition, a collection is only enumerable provided one or other of the three branches of the disjunction in the definition holds *no matter what* class of ordered pairs r refers to. If then we suppose that there is *a particular* class of ordered pairs of which something is true, we do not explicitly make it conform to the definition. Yet it may be such a supposition that it becomes absurd to suppose the collection inenumerable. Let us see how that happens. Suppose, for instance, there is a class of ordered pairs, such that no A is s to two different As, and another class of ordered pairs such that no A is t to two different As. Suppose however there is an A that is either s or t to every other A. Then, there are (as we see without any logical rules) only 3 As at most, and therefore it must now be absurd to say that the As are inenumerable. But how does this impossibility arise? The new premises are

$$\Pi_u\, \Pi_v\, \Pi_w\, \bar{s}_{uv} \,\psi\, \bar{s}_{uw} \,\psi\, 1_{vw}$$
$$\Pi_x\, \Pi_y\, \Pi_z\, \bar{t}_{xy} \,\psi\, \bar{t}_{xz} \,\psi\, 1_{yz}$$
$$\Sigma_8\, \Pi_9\, 1_{89} \,\psi\, s_{89} \,\psi\, t_{89}$$

[1] The subscripts are primed, i.e. 2′, 2″, 2‴, 2⁗. Recall that 1 is the identity element. Similar notation is found in *Collected Papers*, 3.395-3.463.

This last proposition *must be raised to the third power*, giving (the Σ preceding the Π)

$$\Sigma_8\, \Pi_9\, \Pi_{9'}\, \Pi_{9''}\, (1_{89} \psi s_{89} \psi t_{89}) \cdot (1_{89'} \psi s_{89'} \psi t_{89'}) \cdot (1_{89''} \psi s_{89''} \psi t_{89''}) \prec 1_{89} \psi 1_{89'} \psi 1_{89''} \psi s_{89} \cdot s_{89'} \psi s_{89} \cdot s_{89''} \psi t_{89'} \cdot t_{89''} \psi s_{89'} \cdot s_{89''} \psi t_{89} \cdot t_{89''} \psi t_{89} \cdot t_{89'}$$

We now multiply in the first two of the new premises three times each, and identifying u with 8, v with 9, w with 9′, u' with 8, v' with 9, w' with 9″, x with 8, y with 9′, z with 9″ etc. we get

$$\Sigma_8\, \Pi_9\, \Pi_{9'}\, \Pi_{9''}\, 1_{89} \psi 1_{89'} \psi 1_{89''} \psi 1_{99'} \psi 1_{99''} \psi 1_{9'9''}.$$

Now multiplying in the proposition $\Pi_5 \Sigma_4 r_{45}$, and identifying 9″ with 4 and 5 with 8, we get (transposing the third and fourth terms)

$$\Sigma_4 \Sigma_8\, \Pi_9\, \Pi_{9'}\, (1_{89} \psi 1_{89'} \psi 1_{99'}) \cdot r_{48} \psi 1_{84} \cdot r_{48} \psi 1_{94} \cdot r_{48} \psi 1_{9'4} \cdot r_{48} = \Sigma_4 \Sigma_8\, \Pi_9\, \Pi_{9'}\, (1_{89} \psi 1_{89'} \psi 1_{99'}) \cdot r_{48} \psi r_{88} \psi r_{98} \psi r_{9'8}$$

But if we identify 5 with 9 we get

$$\Sigma_4 \Sigma_8\, \Pi_9\, \Pi_{9'}\, (1_{89} \psi 1_{89'} \psi 1_{99'}) \cdot r_{49} \psi r_{89} \psi r_{99} \psi r_{9'9}$$

If we identify 5 with 9′, we get

$$\Sigma_4 \Sigma_8\, \Pi_9\, \Pi_{9'}\, (1_{89} \psi 1_{89'} \psi 1_{99'}) \cdot r_{49'} \psi r_{89'} \psi r_{99'} \psi r_{9'9'}$$

In these last three propositions we identify 9′ with 4′, and they become

$$\Sigma_4 \Sigma_8\, \Pi_9\, \Pi_{9'}\, 1_{89} \cdot r_{48} \psi r_{88} \psi r_{98} \psi r_{9'8}$$
$$\Sigma_4 \Sigma_8\, \Pi_9\, \Pi_{9'}\, 1_{89} \cdot r_{49} \psi r_{89} \psi r_{99} \psi r_{9'9}$$
$$\Sigma_4 \Sigma_8\, \Pi_9\, \Pi_{9'}\, 1_{89} \cdot r_{49'} \psi r_{89'} \psi r_{99'} \psi r_{9'9'}$$

In these we identify 4 with 9 and they give

$$\Sigma_8\, \Pi_9\, \Pi_{9'}\, r_{88} \psi r_{98} \psi r_{9'8}$$
$$\Sigma_8\, \Pi_9\, \Pi_{9'}\, r_{89} \psi r_{99} \psi r_{9'9}$$
$$\Sigma_8\, \Pi_9\, \Pi_{9'}\, r_{89'} \psi r_{99'} \psi r_{9'9'}$$

Multiplying these by the third power of $\Pi_1\, \Pi_2\, \Pi_3\, \bar{r}_{12} \psi \bar{r}_{13} \psi 1_{23}$ we get

$$\Sigma_8\, \Pi_9\, \Pi_{9'}\, (r_{88} \cdot r_{9'9} \cdot r_{99'} \psi r_{88} \cdot r_{99} \cdot r_{9'9'} \psi r_{98} \cdot r_{89} \cdot r_{9'9'} \psi r_{98} \cdot r_{9'9} \cdot r_{89'}$$
$$\psi r_{9'8} \cdot r_{89} \cdot r_{99'} \psi r_{9'8} \cdot r_{99} \cdot r_{89'}) \cdot T_{89} \cdot T_{89'} \cdot T_{99'}$$
$$\psi (r_{99} \cdot r_{9'9'} \psi r_{99'} \cdot r_{9'9}) \cdot T_{99'} \cdot (1_{89} \psi 1_{89'})$$
$$\psi (r_{88} \cdot r_{99'} \psi r_{9'8} \cdot r_{89'}) \cdot T_{89'} \cdot (1_{89} \psi 1_{99'})$$
$$\psi r_{99} \cdot 1_{89} \cdot 1_{89'}$$

This gives

$$\Pi_9 \Sigma_8 r_{98}$$

Multiplying by $\Sigma_6 \Pi_7 \bar{r}_{67}$ identifying 9 with 6 and 7 with 8, the whole vanishes.

It was only necessary to follow this out in order to show that the premises had to be multiplied into themselves at least three times. Had there been 4 objects, it would have been necessary to multiply them 4 times etc. If therefore the number of objects were endless, there would be no contradiction unless an endless collection of factors had been formed. But were an endless collection of factors formed, *that*, instead of proving an endless collection to be absurd, would on the contrary *disprove* this.

Thus, it is perfectly clear that an inenumerable collection involves no contradiction. But, says Renouvier, it is absurd anyway. I say it is not. Who is to decide. Shall we leave it to the mathematicians, the whole business of whose lives is to deal with relations of the kind considered? They now side with me; and though they were long beguiled, they always implicitly held the multitude of whole numbers to be inenumerable. If Renouvier still sticks to his opinion, he must admit that others fully as competent see no absurdity in it. He can hardly maintain then that the hypothesis is *necessarily* inconceivable. It is so only for his idiosyncracy. But that is hardly what we mean by absurd.

Then the example of the whole numbers makes it perfectly obvious that an inenumerable collection there *is*. To the reply that this is only a collection of *possibles*, and that *all* that is possible is *not* possible, the rejoinder is, that the question is whether *logical principles* exclude such a collection, not whether it is *physically* possible. Now logical principles can exclude nothing merely because of the predication of actual existence, which is mere brute fact with no essence or idea from which anything is deducible. Hence though all that is possible may not *coinhere in one subject*, yet it may as far as logical laws go, coexist. For a logic which could infer from what was true of one subject what was true of an independent subject would be a material, not a formal, logic.

From every point of view of logic, then, Renouvier's razor will cut no ice.

In fact, the only thing that gives it much semblance of truth, that there is *no last* to the series, is a peculiarity of the lowest grade of inenumerable collections in their natural order. In the higher grades, this feature is not found.

It is easy to sketch out the sort of nominalism that must result from Renouvier's maxim. Accordingly, looking ahead further than I have yet read I see a chapter about philosophical "fetishism." Yes, we know that talk. But the only thing that makes a fetish absurd is that it is fabricated by a *private party*. The terror of "fetishism" is a sort of philosophical "*bogueyism*." When you prove to me that a conception is an inevitable and fundamental "fetish," you have quite convinced one of its validity and truth, or approximate truth. That whatever is an element of thought has a corresponding element in being is a too absolute statement, but it is true to such an extent that no rule can say where or when it ceases to be true.

C. S. Peirce

P.S. Although the arguments used by Renouvier are wretched, I will take the trouble to refute them all.

The argument on p. 46 involves three errors either of two of which are sufficient to render it worthless. The first is where he says "Ou ces choses composent actuellement, toutes ensemble, un tout, ou elles ne composent pas un tout." This is not of much consequence. But it involves an ambiguity. The whole of a class, as when we say "Any A is B" is a whole in one sense, but it is not necessarily a whole in the sense of something whose synthesis is complete. The second error, which is fatal is where he says "si ces choses ne composent pas un tout, il est donc des choses qui sont et qu'on ne saurait considérer, sous le simple rapport de l'existence, conjointement avec d'autres choses qui sont." This is monstrous confounding $\Sigma x \Pi y$ with $\Pi y \Sigma x$. The third error, also fatal, is where he says: "Des choses qui sont, ... formeront toujours des nombres ... Sans cela, point de représentation ... d'un tout." This is not true in any sense of the *whole*.

p. 48. The idea that anything is logically impossible without involving contradiction is absurd.

"Un nombre sans nombre de parties" is a quibble about two senses of the word number. A collection is not necessarily such that the syllogism of transposed quantity is valid of it, as shown above.

What is the sense of dragging in space. That can have nothing to do with the question. There is no absurdity in things having any properties when real and actual that do not present not just the same absurdity when only regarded as arbitrary hypotheses or possibilities. As to the dilemma of p. 53 there is no absurdity in an actual infinity of things in themselves, nor is there any absurdity in an infinite collection of possible (or actual) representations.

An infinite number does not signify a number greater than any number "assignable" in fact or idea. An infinite collection is simply a collection of which the syllogism of transposed quantity does not hold good. An "assignable" number (which is a sophistical tool) has nothing to do with a collection anyway. Its use is to make a sophistical argument in the doctrine of limits. It is simply weak to say that an infinite collection means "un nombre qui n'est déterminable en aucune façon selon la pensée" etc. These are the arguments of geese.

p. 54. Mathematicians who are up to date hold there is no difficulty in reasoning mathematically about infinity; and they always did so in reality. But there is no need of examining what obsolete writers say.

p. 56. The same stupid ambiguity shows itself when he says "La réalisation du nombre infini suppose l'épuisement des nombres finis, et cet épuisement est incompatible avec le concept de nombre."

"L'assignable est la même chose que le possible." I never saw this manifest falsity stated so openly before.

His weak logic is amusingly illustrated when he says "Il n'y a pas de nombre plus grand que tout nombre possible, *puisque* tout nombre peut être augmenté." If instead of "puisque" he had put "non-obstant que," there would have been more sense in it.

At the bottom of this page he seems to think it absurd that infinite collections should be some greater than others. There is nothing surprising about that.

p. 57. His argument about square numbers only proves that an infinite part may be equal to its whole. If the whole is of the lowest grade of infinity it *must* be so. What of it? Can that be proved to be absurd? No; but it can be proved *not* to be so; and in fact his own argument is proof of that. To anybody having any distinct notions of *equality* and *infinity*, it is obvious at a glance. He says "il y a autant de nombres carrés qu'il y a de nombres en tout, *ce qui est contradictoire*." This marks Renouvier as one of those heads who call anything that muddles them *contradictory*. It is only necessary to consider that *equality* means that two collections can by some arrangement be brought into one-to-one correspondence and one sees at once that there is *no* contradiction. All mathematicians talk of the fundamental *theorem* of arithmetic. That theorem is no more than that this thing is not true of finite numbers. If it were an *axiom* it were not true of *any* numbers, it could hardly be a *theorem* it is not true of *finite* numbers.

What kind of a logician is it who says "Cette série contient plus de nombres qu'elle n'en a, puisqu'elle les contient premièrement tous avec leurs carrés, qui les égalent en nombre, et en outre les nombres non carrés." If he don't see that this is merely juggling with two meanings of the phrase *more than*, he is no logician. "More than" in the sense of "others besides a collection equal to" and in the sense of "being a collection greater than." What puerility! A man who had any sense of what sound reasoning *feels like* would reject such stuff.

Definition of Equality. Let $q_{a\beta c}$ mean a is in the relation β to c. Then, to say that the As form a collection equal to the collection of Bs, is the same as

$$\Sigma_\alpha \Sigma_\beta \Pi_1 \Sigma_2 \Pi_3 \Sigma_4 \Pi_5 \Pi_6 \Pi_7 \Pi_8 \Pi_9 \Pi_0$$
$$(\bar{q}_{5\alpha 6} \psi \bar{q}_{5\alpha 7} \psi 1_{67}) \cdot (\bar{q}_{8\beta 0} \psi \bar{q}_{9\beta 0} \psi 1_{89}) \cdot (\bar{A}_1 \psi q_{1\alpha 2} \cdot B_2) \cdot$$
$$(\bar{B}_3 \psi q_{4\beta 3} \cdot A_4)$$

That is, there is a relation α, such that to no two things is one thing α (or as we might say, to no two Bs is one A α), and there is a relation β such that no two things are β to the same thing (or no two As are β to the same B), and every A is α to a B, and to every B some A is β.

To say that the collection of Bs is greater than the collection of As, is

$$\Sigma_\alpha \Pi_\beta \Pi_1 \Sigma_2 \Sigma_3 \Pi_4 \Sigma_8 \Sigma_9 \Sigma_0 \Pi_5 \Pi_6 \Pi_7 (\bar{q}_{5\alpha 6} \psi \bar{q}_{5\alpha 7} \psi 1_{67}) \cdot$$
$$(\bar{A}_1 \psi q_{1\alpha 2} \cdot B_2) \cdot (q_{8\beta 0} \cdot q_{9\beta 0} \cdot T_{89} \psi B_3 \cdot (\bar{q}_{4\beta 3} \psi A_4))$$

That is, there is a relation α such that to no two things is one thing α and every A is α to a B; but whatever relation, β, be taken either two different things are β to something or there is some B to which no A is β.

Note to p. 62. Impenetrability is not essential to matter. Matter is that which 1st, occupies space in the sense of being connected with 3 dimensional continuous portions of space, 2nd, every part of it however small preserves a self-identity and otherness from the others. This is true not only of 3 dimensional parts but of parts of lower dimensions, 3rd, its place varies continuously with the time, 4th, it is absolutely, or nearly, permanent in all its parts, and they all preserve their dimensionalities. Impenetrability may be a property of matter, but if so, it amounts only to a repulsion. Attraction is the more general law; repulsion, if it exists at all, which is not absolutely certain, is a special phenomenon.

p. 64. Of course the Boscovichian hypothesis does away with impenetrability. But I suppose Renouvier understands that.

p. 65. There is no absurdity in supposing quantity formed out of *zeros*, two zeros making a unit, etc. This is only a dyadism.

I don't understand what he means by *elements*. Here he becomes obscure.

p. 66. I will answer the question which he says the partisans of the thing in itself (of whom I am not one, however) have never answered. The question is "comment un nombre sans nombre de parties d'étendue peuvent être parcourues en fait, et un nombre sans nombre de parties de durée s'écouler en fait." For this purpose we need not take the trouble to suppose space and time continuous. Had he asked the *right question* I should have been driven to define continuity; but as he puts the question, that is not needful.

It is only necessary to suppose that along any line of motion there is a point corresponding to every rational fraction positive and negative and also that in time there is an instant corresponding to every rational fraction positive and negative. Then, the velocity being a rational fraction, it is only necessary to multiply the lapse of time (a rational fraction) by the velocity (a rational fraction) and add to the rational fraction denoting the point of space, to get the rational fraction denoting the point of space that will be reached at the end of that lapse of time. Could anything be more elementary? To say there is any contradiction is not only an arbitrary assertion, but it is a *disprovable* assertion.

But he says the question relates separately to space and time; and what he wants to know is how an endless number can be completed. The answer to this is extremely simple, as soon as we define an endless number. Namely, an inenumerable collection is such a collection that the syllogism of transposed quantity does not *necessarily* hold. That can easily be shown to imply that *there is such a possible* arrangement of it in sequence that while there is a first and a unit next after each unit, there shall be no last. But it does *not* imply, as Renouvier evidently supposes, that in *every arrangement* there is no last. He commits the error that marks the loose reasoner always of confounding $\Sigma_x \Pi_y$ with $\Pi_y \Sigma_x$.

But it must be noted that it is not that I opine there is no contradiction while Renouvier opines there is; but that he says there is but fails to show how, while I *demonstrate* by rigid logical proof that there is not.

As to his statement of the argument of the flying arrow, p. 68, the definition of rest is (making it as near what he says as possible "si toujours une chose est en repos quand elle est dans un espace égal à elle-même.") *a thing which through a finite time occupies no space larger than itself is at rest during that time.* Second, in every now (that is, through every moment) the arrow is in a space no larger than itself. The conclusion is that the arrow performs no motion in a moment. A very sound conclusion. It takes time to perform a motion. In the same way a man and wife make a married couple, but neither the man nor the woman alone is a married couple. They must come *together*.

p. 71. "Vous renoncez à comprendre le mouvement relativement aux limites du temps; vous vous rejétez sur les intervalles de ces limites." What nonsense is this! There are no intervals between the instants of time. These instants are welded together into a continuum. The argument shows, what is evident to common sense without it, that the continuum is necessary to motion and that an instant broken away from all others, so far as this is possible without absurdity, admits no motion.

"Continus, c'est-à-dire divisibles sans fin." Probably no mathematician ever defended this definition of continuity. At any rate is is indefensible. But it is true that continuity implies infinite divisibility as one feature.

Then this argument having fallen to the ground, he tries another, the Achilles. This argument, which Renouvier justly says has not always been understood by those who undertake to answer it is, that an endless sum of finite times will not be sufficient for Achilles to overtake the tortoise. Very true. So the mathematician says. But what of it. After these times have elapsed, which are in sum finite, Achilles overtakes the tortoise. Zeno and Renouvier seem to think that because a series is endless therefore its sum is finite, because it is the sum of an infinite succession of finite times therefore it is infinite. This is the same old fallacy of reasoning from $\Pi_x\Sigma_y$ to $\Sigma_y\Pi_x$. It is reasoning as if because the collection of terms were infinite therefore it were greater than an infinite number of times the least of them. That is, it *thinks*, but never *says* (for that would be too manifestly absurd) because there is a finite term less than any given term therefore there is a finite term less than all the terms of an endless series.

The curious thing is that the Achilles never once touches the real difficulty of continuity. It is not an argument against continuity but against the succession of real numbers (or even of rational numbers) although such a succession only admits of further determination to become continuous, and the real difficulties concern this further determination. I say "real difficulties"; I mean questions difficult to answer, *not unanswerable*.

p. 78. It is difficult to say how far Aristotle had penetrated into the matter of continuity. It cannot be settled in any slight examination. But that is not necessary. I will now state what continuity really is. Whether time and space are really continuous or not is a question I will not touch here.

A relative *r* is said to be an *offshoot* relative, if and only if

$$\Pi_1\Pi_2\Pi_3\,\bar{r}_{12}\Psi\bar{r}_{13}\Psi 1_{13}$$

that is, nothing is *r* to two different things. Nothing is in an offshoot relation to two different things.

A *collection* is the same as a *predicate.* That is, it is an object which has a *being* as soon as a certain description is intelligible; but which has *existence* only when something of that description exists. If one description applies to everything of another description in existence, then the collection of the first description is said to *include* the collection of the other description; otherwise not. If two collections include one another, they are said to be identical (existentially). The existing subjects of a description are said to be the *units* of the collection of that description. It follows that every nonexistent collection is identical with every nonexistent collection. That is there is one and but one nonexistent collection (existentially).

Any two things not only *can be paired,* but *if they both exist,* they are paired. For existence means entrance into a given universe of discourse, or experience. (Experience is the irresistible and inevitable effect upon the mind of the course of life.) This entrance involves its opposition to everything in that universe. All possible combinations of ordered pairs have *being* as dyadic relations; for being is nothing but what is possible to be apprehended. Every relation logically possible has being, that is, every relation not involving contradiction.

From these principles it follows, I will not stop to show how, but by strict logic of relatives, that given any two collections the *A*s and *B*s, either there is an offshoot relation, *r*, such that every *A* is *r* to a *B*, or there is an offshoot relation, *r′*, such that every *B* is *r′* to an *A*.

If there is an offshoot relation, *r*, such that every *A* is *r* to a *B*, and there is an offshoot relation, *r′*, such that every *B* is *r′* to an *A*, the collection of *As* and the collection of *B*s are said to be *equal.*

If there is no offshoot relation, *r*, such that every *A* is *r* to a *B*, then the collection of *A*s is said to be *greater than* the collection of *B*s.

Every collection is equal to itself.

Given any collection the *A*s, there is an offshoot relation, *r*, such that to every *A* some *A* is *r*.

If no matter what offshoot relation *r* be taken, either there is some *A* to which no *A* is *r* or else every *A* is *r* to an *A*, then the collection of *A*s is enumerable, otherwise not.

If any collection, say the *A*s, be such that there is an offshoot relation, *r*, such that to every *A* some *A* is *r* and yet there is an *A* which is not *r* to any *A*, then the collection of *A*s is, as just said, inenumerable. If no *A* exists of any description involving no unanalyzed term except the general terms of logic and *r* and not referring to objects not *A*s except such descriptions of *A*s as are necessitated to exist by the hypothesis that the *A*s form an inenumerable collection, then the *A*s are said to form a *denumeral* collection; otherwise not. The relation *r* may be termed the *generating relation*. Thus, if the *A*s are a denumeral collection there must be for some relation *r* an *A* which is not *r* to any *A*, but there need not be an *A* not *r* to any *A* and other than some *A* not *r* to any *A*. Hence, there is no such *A*. That is, *r* being the generating relation of the denumeral collection of the *A*s, there is but one *A* which is not *r* to any *A*. In like manner there is but one *A* which is *r* to any given *A*. To define the denumeral collection, then, in convenient algebraical form, we write,

$$\Sigma_\alpha \Pi_a \Pi_b \Pi_c \Pi_d \Pi_e \Pi_f \Pi_g \Pi_h \Pi_i \Pi_j \Pi_k \Pi_l \, (\bar{q}_{a\alpha b} \psi \bar{q}_{a\alpha c} \psi 1_{bc}) \cdot (\bar{q}_{d\alpha f} \psi \bar{q}_{e\alpha f} \psi 1_{de}) \cdot q_{h\alpha g} \bar{q}_{i\alpha j} \cdot (1_{ik} \psi q_{i\alpha l})$$

This could be expressed with fewer indices; but this expression is most convenient.

The *combinations* of any description of things, as the *A*s, are the collections (including the non-existent collection) which are included in the *A*s. The collection of combinations of any description of things is greater than the collection of those things.

It follows that there is an endless sequence of collections each greater than those preceding it and all greater than a denumeral collection. I now call these multitudes the primipostnumeral, the secundopostnumeral, etc.

It can be proved that these are the only collections greater than denumeral collections.

Now I can define a continuous line. It is an object so connected with objects called its points that, in reference to either of two modes of description, given two different points of the line, there is a relation, *s*, such that, given any collection of objects whatever, a certain one of those points is *s* to the other and to every point of the line to which the other is *s* and also to a collection of *other* points of the line (to which the second point is not *s*) this collection being greater than the given collection.

Every unit of any collection, say the *A*s, is *discrete*, if there is a collection which differs from the collection of *A*s only in not including that unit.

If any collection, say the *A*s, has every one of its units discrete, then it is demonstrable that the collection of the combinations of *A*s has every one of its units discrete.

The points of a line if they form a collection form a collection greater than any collection all whose units are discrete. In fact, no point of a line is discrete, or removable by itself alone. A collection none of whose units can be removed by itself alone is said to be *continuous*.

The terminal point of a line cannot be removed by itself alone, leaving the line with all its other points, but without this one. If a line is broken into two parts, the point of breaking is doubled, or broken. It is impossible to say which because the point has no individual identity. A point is related to a line as the marriage of a man and woman is related to the family. As long as they are married, there is, *ipso facto*, a marriage. The marriage can exist alone without the man or woman as easily as a point can exist torn from a line. It becomes something else. It is a point, *in that sense*, no longer. If you divorce the pair, each has an imperfect marriage state. It is their old marriage, now broken.

In short the point is a logical abstraction connected with the line. Yet the line is wholly composed of points and nothing else.

In like manner, we may illustrate the thing otherwise. (Never mind whether the illustration represent *truths* or not.) I can imagine a *red* color and a thing can *be* red. I cannot be so precise in my imagination as to preclude the slightest vagueness whether of hue, of luminosity, or of chroma. Nor can the thing be for any length of time in precisely such a condition as to transmit precisely the same proportions of the different parts of the spectrum. Still, the *red* color is but the generalization of such absolutely determinate colors. As it exists and as I think it, there are certain limiting colors, it no more admits than includes. Everything is red or not red; but were a thing capable of absolute determination it might be just on the border of redness, and then would be neither red nor not red (or, if you please, both red and not red). Such an absolutely determinate object is a *nothing*. It therefore does not violate the dyadic laws of logic.

In like manner, in the sense in which a solid is real, a surface is a nothing; in the sense in which a surface is real, a line is a nothing; and in the sense in which a line is real, a point is nothing.

We must remember that the laws of logic are the laws of the cotruth of descriptions of any existing state of things. The idea of existence comes in, and existence is intrinsically dyadic. (That is, depends on the pairing of things.)

A logical abstraction like marriage may be quite real in its own universe, but it is *nothing*, — a *zero*, — for a universe of men and women. So the fortunateness or unfortunateness of a marriage is a real thing, but it has no place in the universe of marriages. As to them it is a *zero*.[1]

---

[1] A last page of another draft states:

"But Renouvier stands up and pronounces inenumerable collections absurd. As long, however, as they are not clearly seen to be absurd even by those whose chief business is to reason about such relations, it cannot properly be said they are *necessarily* absurd. And if not, they are not absurd at all.

Some say, they are not absurd as to future time, or as to what is merely possible, but they are so as to what is done and exists. The grain of truth in this, that the series has no *last* unit, applies only to one class of infinite collections. But that which is to be rejected as absurd must be rejected as inadmissible as a hypothesis concerning possibilities. Brute existence contributes no premise at all to deductive reasoning. So what is not absurd for possible collections, may be shown to be contrary to some principle of nature, but it cannot be shown contrary to deductive logic.

The series of possible whole numbers either has a last one or it has not. If it has, what number is it? If it has not, it is inenumerable.

It is easy to see what kind of nominalism must result from this razor of Renouvier's.

I look ahead and see he is going to terrify me with accusations of 'fetishism.' A fetish is something the *private individual* makes himself. Therein alone is its falsity. But this tiresome talk about 'fetishism' we hear so much of, is a sort of *bogueyism* in philosophy.

C.S.P."

[c. 18 April 1903]

My dear William:

In my Lowell lectures I want to deal with the supposed paradox of Achilles and the Tortoise. But the difficulty I meet with is that I cannot, for the life of me, see where the paradox, or difficulty, or contradiction, is supposed to lie. I think you could help me, if you would kindly read the following and answer my questions.

If we consider the decimal fraction written by a row of 7 ones immediately following the decimal point .1111111, this is equal to a fraction whose numerator is expressed by a row of 7 ones and whose denominator is expressed by a one followed by 7 zeros, or $\frac{1111111}{10000000}$. Is it not? The value of this will not be altered if we multiply numerator and denominator by 9; so that it is equal to a fraction whose numerator is expressed by a row of 7 nines and whose denominator is expressed by a nine followed by 7 zeros $\frac{9999999}{90000000}$. Is it not? Now $\frac{1}{9}$ is equal to a fraction whose numerator is expressed by a 1 followed by 7 zeros and whose denominator is expressed by a nine followed by 7 zeros or $\frac{10000000}{90000000}$. Is it not? And the number expressed by a one followed by 7 zeros, or 10000000 is one more than the number expressed by 7 nines or 9999999. The latter is the numerator of the fraction equal to the decimal fraction expressed by 7 ones immediately following the decimal point, while both fractions have the same denominator. That is $.1111111 = \frac{9999999}{90000000}$

$$\frac{1}{9} = \frac{9999999 + 1}{90000000}$$

So that $\frac{1}{9}$ is greater than the decimal fraction expressed by a row of 7 ones immediately following the point by $\frac{1}{90000000}$ or a fraction whose numerator is 1 and whose denominator is a 9 followed by 7 zeros. Is it not?

The same thing would be true if we considered any other decimal consisting of a row of *N* ones immediately following the decimal point, *N* being, instead of 7, any other *finite* number. Would it not?

Then $\frac{1}{9}$ is greater than any or all decimal fractions consisting of a *finite* row of ones following the decimal point. Is it not?

Is there any logical difficulty about that?

Is there any rational consideration to the contrary?

Suppose, then, that the Tortoise is $\frac{1}{10}$ of a kilometre ahead of Achilles and is moving away from Achilles at the rate of $\frac{1}{10}$ kilometre per hour while Achilles is moving after the tortoise at the rate of 1 kilometre per hour.

Then the tortoise being ahead .1 kil, it will take Achilles .1 hour to reach that point.

The tortoise will in that .1 hour have gone forward .01 kil. and it will take Achilles .01 hour more or .11 hour in all to reach that point. But when he does reach it in that .01 hour the tortoise will have advanced .001 kil making .011 in all and to get over this .001 kil will take Achilles .001 hour making .111 hour in all. But during this .001 hour the tortoise will have advanced .0001 kil and thus Achilles will have to pass successively over

.1 kil
.01 kil
.001 kil
.0001 kil
.00001 kil
ad infinitum

(by *ad infinitum* we mean of course as long as the row of zeros is finite, there being an endless series of *finite* numbers)

which will take him more than
.1 hour from the start
or than .11 hour from the start
or than .111 hour from the start
or than .1111 hour from the start
or than .11111 hour from the start

and so on ad infinitum, that is to any *finite* number of ones in the row whatsoever.

But each of those fractions of an hour is less than $\frac{1}{9}$ of an hour since the start and therefore there seems to be no reason why at the end of $\frac{1}{9}$ of an hour he should not have overtaken the tortoise, having then run $\frac{1}{9}$ kilometre.

Now is not this conclusive?

At the end of a fraction of an hour expressed by a decimal fraction of a million ones he will be behind the tortoise by a fraction of a kilometre expressed by a fraction having 1 for numerator and for denominator a nine followed by a million zeros.

At the end of a ninth of an hour he will have passed through all the endless series of points at distances of

.1 kil
.11 kil
.111 kil
.1111 kil
et ad infinitum (as before the number of ones being any *finite* number of which there is an endless series)

Now what I want to know is whether the demonstration about the decimal fractions (which has absolutely nothing to do with continuity) is not clear and evident.

Whether when we make those fractions express fractions of kilometres and hours, it becomes less evident.

Whether it does not evidently follow that Achilles will come indefinitely near to catching up with the tortoise in less than $\frac{1}{9}$ of an hour.

Whether there is any counter consideration and if so how that counter consideration affects the argument that $\frac{1}{9}$ is greater than any decimal fraction consisting of a finite row of ones immediately following the decimal point.

In short, what, from this point of view, is the difficulty? There is supposed to be a contradiction of logic. *Whose* logic? If it were mine that was in so flagrant violation of fact I would change it for one that died not lead from a true premiss to a false conclusion. This remark is in my opinion the very hinge of the whole matter.

Very faithfully
C. S. Peirce

P.S. I have contented myself with showing that the fact that Achilles will not have overtaken the tortoise in any one of the infinite series of times and places only shows he will not overtake him in less than $\frac{1}{9}$ of an hour having run less than $\frac{1}{9}$ of a kilometre. In $\frac{1}{9}$ of an hour the tortoise will have gone $\frac{1}{90}$ kilometre which added to his start of $\frac{1}{10}$ or $\frac{9}{90}$ of a kilometre will make $\frac{10}{90}$ or $\frac{1}{9}$ of a kilometre and Achilles in this $\frac{1}{9}$ of an hour will have passed over this $\frac{1}{9}$ kilometre.

34 Felton Hall. 1903 Apr. 18

Dear William:

The Achilles sophism virtually pretends to show that the sum of an endless series of finite intervals of time must exceed any finite interval of time. For suppose the time it takes the tortoise to run any distance to be T times as great as the time Achilles requires to run the same distance, and suppose that Achilles is s seconds in reaching the point where the tortoise starts. Then if that distance is m metres, the tortoise while Achilles runs m metres will run $\frac{m}{T}$ metres; and since Achilles runs m metres in s seconds, he will run $\frac{m}{T}$ metres in $\frac{s}{T}$ seconds, and the tortoise in these $\frac{s}{T}$ seconds while Achilles runs $\frac{m}{T}$ metres will run $\frac{m}{T^2}$ metres which it will take Achilles $\frac{s}{T^2}$ seconds to cover. Therefore before Achilles overtakes the tortoise a number of seconds must elapse equal to

$$s+\frac{s}{T}+\frac{s}{T^2}+\frac{s}{T^3}+\text{etc. } \textit{ad infinitum}$$

If the sum of this endless series of finite times is finite, the *Achilles* argument proves nothing. It is therefore equivalent to an attempt to prove that the sum of an endless series of finite intervals must exceed every finite interval. But there is no peculiar property of time which is made a premise to the argument. Therefore, if it is good for anything at all, the sum of any endless series of finite quantities must exceed any finite quantity.

But now let $\frac{a}{b}$ be any fraction and $\frac{c}{d}$ any fraction greater than $\frac{a}{b}$. Then $\frac{a+c}{b+d}$ will be a fraction of intermediate value; and by the same token $\frac{a+(a+c)}{b+(b+d)}$ will be intermediate in value between $\frac{a}{b}$ and $\frac{a+c}{b+d}$ and so $\frac{3a+c}{3b+d}$ will be intermediate in value between $\frac{a}{b}$ and $\frac{2a+c}{2b+d}$ and so on *ad infinitum*. Therefore, the interval between the values of two fractions one greater than the other is the sum of an endless series of intervals. And if the first is finite, all are finite.

Therefore, if "Achilles and the tortoise" proves anything, it proves that one fraction cannot be greater than another by a finite amount. But $\frac{c}{d}$ exceeds $\frac{a}{b}$ by $\frac{bc-ad}{bd}$ which is plainly finite; and therefore the Achilles argument does not prove anything.

Now I want to know whether it is too much to ask that the Argument should be so stated that there should be an *appearance* of necessity in the proposition that the sum of an endless series of finite quantities must exceed every finite quantity?

The real difficulty of dealing with it is that it presents nothing in particular to refute, because it has no appearance of being sound reasoning.

Very faithfully
C. S. Peirce

P.S. As you stated the matter today, before an incomplete process is completed something has to be done which leaves the process incomplete. Well, *what of it*? The stages of the process are merely earlier and later times, and earlier and later times are merely smaller and greater numerical *dates*. Therefore to say that before an incomplete process is completed, something has to be done which leaves the process incomplete is merely as much as to say that the amount by which a lesser quantity falls short of a greater one can be separated in all cases into the sum of two parts. Very true. But you take this as showing that how a process can be completed transcends human reason, that is that because every difference between two quantities can be subdivided, therefore how one quantity can exceed another transcends human reason. It can be readily understood even in the case of rational fractions, *where there is no continuity*. (All admit *that*.)

34 Felton St. 1903 April 18

Dear James:

This is my second epistle of even date on the subject of *Achilles*. The argument is substantially as follows:

Let A_0 be the initial time of which *Achilles* starts and Ω the time of catching up with the tortoise [Fig. 1].

But one can specify an instant, or point of time between A_0 and Ω. Call it A_1 [Fig. 2]. And one can specify a point, A_2, between A_1 and Ω [Fig. 3], and in general between A_n and Ω there is room for a point A_{n+1}, because every line is divisible. *Ergo*, the length of time between A_0 and Ω must be infinite.

Fig. 1

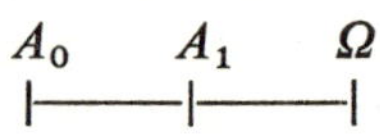

Fig. 2

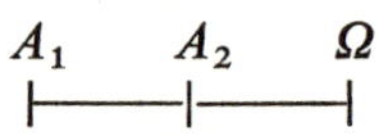

Fig. 3

Just so if A_0 is the value $\frac{a}{b}$ and Ω is the value $\frac{x}{y}$, there is a value $A_1 = \frac{a+x}{b+y}$, a value $A_2 = \frac{a+2x}{b+2y}$, a value $A_3 = \frac{a+3x}{b+3y}$ and so on endlessly, and *therefore* the difference $\frac{x}{y} - \frac{a}{b}$ must be infinite.

Where the necessity of the connection between premiss and conclusion lies is not explained. It is a jingle of words.

It is obvious that the series of fractions $\frac{a+x}{b+y}, \frac{a+2x}{b+2y}, \frac{a+3x}{b+3y}$, etc. is just as multitudinous as the series of whole numbers, and no more so.

But it is easy to prove, so far from this series making up a continuum, it is easy to describe a series of quantities between the same limits that is more multitudinous.

For all those fractions can be distinguished by assigning an ordinal number to each. Namely, we may regard $\frac{a+Nx}{b+Ny}$ as the Nth fraction. If then I prove that there is a series of quantities between the same limits which the ordinal numbers are *inadequate to distinguish*, this will prove this series to be more multitudinous than those fractions.

Namely, if we assume $\frac{a}{b}=\frac{0}{1}$ and $\frac{x}{y}=\frac{1}{1}$ so that $\frac{a+x}{b+y}=\frac{1}{2}$, $\frac{a+2x}{b+2y}=\frac{2}{3}$, $\frac{a+3x}{b+3y}=\frac{3}{4}$, etc., I say that all that series of quantities which can be indefinitely approximated to by decimal fractions in the expression of which only the digits 0 and 1 appear, cannot be distinguished by ordinal numbers. For if they could be suppose this were done, and I will show that the hypothesis involves a contradiction. For supposing it done, I will separate all those integer numbers which are used for numbering the decimal fractions into two classes, the *unarians* and the *nullarians.* Namely, taking any integer say N, then on examination of the decimal fraction which is alone numbered the Nth (supposing any is so numbered) if there is in the Nth decimal place of this fraction a 1, I call the number N a *unarian*; but if in that Nth decimal place of the *Nth* decimal fraction there is a 0, I call N a *nullarian.*

Perhaps not all the integer numbers have been used but the hypothesis is that every decimal fraction whose value can be indefinitely approximated to by putting in each decimal place either 0 or 1, receives an ordinal number that is given to no other such decimal fraction. Such is the hypothesis; but I will prove it absurd by mentioning a decimal fraction that has received no ordinal number.

Namely, the decimal fraction I mean is that one which in the first place of decimals has 1 provided 1 is a nullarian but 0 if 1 is a unarian, in the 2nd decimal place has 1 if 2 is a nullarian but 0 if 2 is a unarian, and in general in the Nth decimal place has 1 if N is a nullarian but 0 if N is a unarian.

Then I say that this fraction has no ordinal number. For if it has, let it be the Qth. But Q cannot be a unarian, since if it were in the Qth decimal place of the Qth fraction there would be 1, whereas by the definition of this fraction, Q being a unarian, in its Qth decimal place is 0. Nor can Q be a nullarian; since if it were, in the Qth decimal place of the Qth fraction, would be 0, whereas by the definition of this fraction, Q being a nullarian, in its Qth decimal place, there is 1. Q is therefore neither unarian nor nullarian and therefore is not one of the integers which gives an ordinal number to a fraction, contrary to the hypothesis.

Thus the fractions intermediate between $\frac{a}{b}$ and $\frac{x}{y}$, far from constituting a continuum are (though infinitely multitudinous) comparatively small in multitude.

So that the *Achilles* argument, if it were sound, would not only overthrow continuity but also an arrangement of a comparatively small multitude of points if this arrangement were such as to give a point between any two points.

But whatever integer N may be the fraction $\frac{a+Nx}{b+Ny}$ is a finite fraction. Unless, therefore, you are prepared to say that *there is not an endless series of possible finite whole numbers*, you must acknowledge that it is possible, without contradiction, to insert a point (or fraction) between any two.

Now if an endless series of possible finite whole numbers be impossible there must be some definite finite multitude to which the addition of a unit involves contradiction. But certainly there is no such largest possible whole number.

Therefore, you are compelled to say that no contradiction [exists] in Achilles overtaking the tortoise *even without continuity*, but merely supposing that points of time and of space can be specified corresponding to all *rational fractions*.

All this time those who are puzzled by Achilles and the tortoise, though they say some contradiction is involved (for that is the argument) have *never* been able to point out precisely where the contradiction lies. A mere jingle of words deceives them.

They take the ground that mathematicians reason badly; and this bad reasoning of the mathematicians has the curious property of leading them from truth to truth, while the "better" reasoning of the philosophers leads them from truth to falsity and from reason to absurdity.

[August 1905]

Dear William:

I liked your suggestion to express my phaneroscopy in French. Years ago, when it was more at my pen's point perhaps, — at any rate I was better up in it, — I often used the language in my private notebook of logical studies, where there are many pages of it. I spent a whole day in beginning to carry out your suggestion, and then saw that I must not, — least of all just now, — spend time in what does not bring money. Then came Royce's paper about 0 collections. I remember you said that you carefully read all my letters. Therefore I will take pains to express myself so that your anti-mathematic mind may catch my drift. I will first give you Royce's definition of an 0-collection. A *collection* he makes to consist of *elements*. A single element is also a collection. He denotes collections by Greek minuscules, α, β, γ, δ etc., and elements (whether they enter into collections or not), by Italic lowercase letters. He writes (abc) for the collection whose sole elements are a, b, and c; and he writes ($\alpha\beta$) for the collection whose elements are the elements of the collections α and β, whether these have any common elements or not. To be sure he talks about an element's occurring repeatedly in a collection, but without elucidating this unclear conception, he dismisses it and so far as I have read, does not resuscitate it. He writes $0(\alpha)$ or $0(\beta\gamma)$ or $0(\alpha\beta)$ etc. to express an *assertion*, namely, the assertion that α, or ($\beta\gamma$) or ($\alpha\beta$) is a collection of the 0-form. His definition of the 0-form is expressed in 6 propositions; but the first two by themselves express all that is essential to the 0-form and the other 4 describe the form of the universe, or "system," constituted by the totality of the "elements" that he imagines. I will first give his first two clauses defining the 0-form in general, just as he gives them, and will then analyze their meaning and make it lucid. His first proposition is that if $0(\alpha)$, that is, if any collection, α, is an 0-collection, then no matter what collection γ may be, $0(\alpha\gamma)$, that is, the collection composed of all the elements of α and γ, is also an 0-collection.

In order to show what this means, I will remark that he writes $E(\alpha)$; $E(\alpha\gamma)$ etc. to assert that α, or ($\alpha\gamma$), or whatever collection follows the E, is *not* an 0-collection. Therefore his first clause may otherwise be stated, thus:

No matter what collections α and γ may be, if $E(\alpha\gamma)$ is true then $E(\alpha)$ is true. Or, if we write κ for $\alpha\gamma$, it will read thus:

No matter what collection κ may be, so long as α is a part of it, if κ is an *E-collection* so also is α.

That is, to say that κ is an E collection implies (besides what may be implied in Clause II) that the members of this collection have a relationship which if it subsists between all, subsists also between any some of them.

This is precisely Aristotle's definition of a universal predication, which has been adopted by all logicians. I gave it on p. 179 of the April *Monist*. Therefore to say that any collection is an E-collection, or is of the E-form, implies (besides what clause 2 may imply) only that some predicate is true of *all* the members of that collection.

This predicate may however not be an entirely definite one. There are two different kinds of indefiniteness with which it may be affected. It can be demonstrated that if a universal predicate is indefinite (or vague), its indefiniteness is equivalent to its being a *definite relative predicate*, with an indefinite correlate; and the two kinds of indefinacy are owing to the difference in the order of the selectives. In the one case, the universal predication is of the form

Every member of α is P to some Q *or other*.

In the other case, the form is

Relatively to *some one* Q, every member of α is P.

In either case, Clause I holds good. (Both kinds of indefinacy might be present together.) There is no need of further notice of the *former* kind, but the latter is important, and owing to what follows, it will be well to *duplicate* it, and to say that,

To assert that a collection α is an E-collection implies (and implies *only*, except for Clause 2) that relatively to some one of a class of correlates called the Rs and relatively to some one of a class of correlates called the Qs, every member of the collection α is in the relation P.

It will be easier to "catch on" to clause 2 if we call the Rs the "essential respects," or for short, "the respects," and call the Qs the "determinations" of P in those several respects.

Royce expresses Clause 2, as follows:

Be β and δ any two collections, then, if $0(\beta)$ is true, and if (β consisting of the elements b_1, b_2, b_3, etc.) no matter which b we take $0(\delta b)$ is true, then $0(\delta)$ is true.

Expressing this in terms of the *E*-form, it is as follows:

If δ is of the *E* form, while no collection composed of δ and of a single member of β is of the *E* form, then β is of the *E* form.

In order to see what this implies in addition to what Clause 1 implies, let us express the proposition in terms of the implication just made out. Namely, Clause 2 expresses that

If in an essential respect all elements of δ have a single determination, while in no essential respect have all the elements of the same determination as *any* element of β (no matter which element it be), then in some essential respect all the elements of β have the same determination.

Let us see how this can be. Let us write 0, 1, 2, 3, etc. for the different determinations, and let us write the determinations in different respects in different horizontal rows and determinations of different elements in different columns. Then δ may be represented by this block:

	δ Elements
	0 0 0 0 0
Essential	0 0 1 2 2
respects	1 1 1 2 2
	0 0 1 2 3
	2 3 3 0 1

In an essential respect (top row) all the elements of δ have a single determination so that $E(\delta)$ is true.

	β Element
	1 2 3 4
Essential	0 1 2 0
respects	1 1 1 1
	2 3 4 0
	0 0 1 1

The other condition holds also, since the top line of δ is the only one having all the determinations the same, namely 0, and there is no 0 on the top line of β. It *accidentally* happens that the consequent also is true because of the 3rd line of β. But it need not have been so. How then could it become logically necessary. Evidently, to say that if all the determinations of the first line of β are other than zero they must be alike, is to say that there are only two alternative possible determinations, and this is precisely the meaning of clause 2. So then the whole definition of $0(\alpha)$ is that to say that is true is to say that in every respect some two elements of β have the *two sole possible* determinations.

The notation of Royce is simple but wholly devoid of mathematical power. Every proposition is proved by an operose and abstruse operation. My representation by a block of 0s and 1s need only to be carried in the head, and one sees in a moment the necessary truth of the theorem. It is difficult indeed to suppose that Royce was not aware of the meaning of his 0-form as I have explained it. Perhaps he thought all this pomp of abstruse demonstrations was "scientific." If so, my opinion is quite the other way.

But it is desirable that a notation should connect the theory of this relationship with the general logic of relations. This could be done in various ways. I will mention only one way. The matter is interesting, because it calls for a logical representation of an indefinite, or vague, conception, and the examination of the logic of such conceptions.

A chemist in studying a delicate organic compound takes care not to use too strong reagents. It is true that with the weak reagents the reactions are so readily reversible that pure products are not obtained; but then, even with the most powerful reagents, it is impossible that mass-action should not enter at all, and the strong reagents so tear the compound to pieces that its structure is not brought to light, while with the weak reagents, the proximate products are all present and can be observed. Just so, the rough distinctions which we are accustomed to make in logic, although they yield very impure results, are far more informative than more powerful analysis would be. Roughly we divide terms into the universal or general, the singular, and the indefinite, or vague, or "particular." A proper name is set down as singular, although in absolute strictness, of course it is not so. The plural of a common noun is usually universal or general. For every object of thought is One. The plural is one object, one collective object distributively taken. "Men are mortal," "pigs have great pointed snouts." The singular of a common noun is usually indefinite. "A man married a wife," i.e. a man not specified. The singular term determinately denotes the very object. The general term leaves the object partly indeterminate, and leaves the person addressed (and there always is a person addressed, for solitary thought is dialogue) to make the further determination at his pleasure. "Men are mortal" = "Any man *you please* is mortal." The indefinite term is partly indeterminate, but the speaker reserves the determination. "A man marries" = "A man *who might be instanced* marries." The general term, in so far as it is, and in those respects in which it is, general, is not subject to the principle of excluded middle. That "Men are wise" and that "men are foolish" ("men" being taken generally) may be alike false. The indefinite term, in so far as, and where, it is indefinite, is not subject to the principle of contradiction. I have already noticed that there is no such thing as an absolutely singular term. *Nor* is there any such thing as an absolutely indefinite term, which might denote anything but undistributively, i.e. in the singular number. "Something," which comes as near to it as any, obviously would be of no use if it were absolutely indefinite. To such a term the principle of contradiction would have *no* application whatever. It could truly be said to be anything. As such it would be *absurd*. For the absurd is that of which anything whatever is necessarily true. We do not say that any object is in itself "absurd," but only that a term, or apparent term, is absurd. Nor is anything absolutely universal, since of such a term nothing whatever could be truly asserted, so that it would be quite *meaningless*. The universe, in the sense of our environment, is not absolutely universal; yet taking it as including every kind of environment, it is not the object of any term. For by "universe" we mean usually optical environment. Notwithstanding the impossibility of strict universality, indeterminacy, and singularity, the distinctions are perfectly clear. My remark about nouns was of course merely a remark about human languages. Now grammar is governed by logic modified to practical requirements. Practical considerations prevent plural nouns from being truly general, even in a limited way. For a truly general proposition does not assert existence. For it is what is required as the major premiss of a direct syllogism of the first figure. That is what entitles it to a place in logic. Now such a premiss need not assert existence. But plural nouns generally do so. We speak not of phenixes and dragons but of *the* phenix and of this or that species of dragon. (?) A genuinely general proposition is a necessary proposition. Any phenix *would* rise from its ashes. Consequently, a particular proposition, which is the precise denial of a general proposition, having the same subject, does not assert *existence*, but as the denial of a necessity expresses merely possibility. "A phenix need not necessarily rise from its ashes." All our knowledge of the future is general, for we have no experience of it. My confidence that I know the sun will rise tomorrow consists in my thinking it *must* and under the circumstances (vaguely apprehended) always would. It is true that the proposition derives a sort of singularity from our notion of Time as a *single* continuum, so that tomorrow morning is a sort of proper name (which daily changes its denotation). How fundamental Kant made this circumstance in his philosophy without the slightest attempt to analyze it! The *continuity* is an exceptionless generality of relations, especially involving the perfect non-metrical homogeneity of all its parts; while the *unity* obviously is owing to continuity forbidding any splitting of the stream of time, combined with the perception that there is a single stream in the present. Thus, the singleness consists in the inferential extension *by law* or necessity of the singleness of the present to other times. It is specifically different but generically similar, to a sort of singularity found in all generality which is inferred from observation. What more did Kant mean by calling Time *Anschauung*? First, that it was a matter of feeling; and secondly, that it is given before the formation of the percept. All conceptions, and especially all natural conceptions are accompanied with peculiar feelings. As for time being given in the first sensation, before the apprehension though this is asserted by Kant, he never that I remember offers the least proof of it; and I should like to know how he supposed himself to know this. Anyway, our knowledge of the future is prevailingly general, and that of the present is prevailingly singular. As to our knowledge of the past, it is prevailingly indefinite, as all reproductive imagination is. The distinction between the Indefinite, the Singular, and the General is obviously only another application of the distinction between the Possible, the Actual, and the Necessary, for which the Germans have invented the convenient name *Modality*. Time is clearly a special kind of real modality. The distinction between a *state*, an *occurrence*, and a *law* is another kind of real Modality which is more fundamental. An Occurrence is a determination in the singular Mode of being. Time is a Modality of Occurrences. Exactly how the one Modality is related to the other I do not undertake to say. But certainly any Occurrence now determined to be about to take place is Necessary, and any Necessary Occurrence involves some Futurity. A *Potentiality* is a positive possibility, not a mere negation of necessity. A Potential Occurrence, or occurrence which is germinal, is a relatively Past occurrence. But I will not follow further these indigested ideas. I have made the indefinite sufficiently clear to serve our purpose.

It is also to be remembered that I showed in 1870 that any proposition may be analytically expressed in a form in which the only terms are the verb "—is in the relation—to—" and monadic rhemes "—is such and such."

We now have all that is requisite. Let x be an indefinite term. Then I will express $E\alpha$ in the form "In some respect either every element of α is x or no element of α is x" and will express 0α thus: "In every respect some element of α is x and some element of α is not x." These are written as existential graphs by writing $\genfrac{}{}{0pt}{}{A—}{B—}$is α and $\genfrac{}{}{0pt}{}{D—}{C—}$is x to mean "in the respect A, B is an element of α" and "in the respect C, D is x." Observe that it is the upper tail of the "=is α" and the lower tail of the "=is x" that denotes the $E(\alpha)$ [Fig. 1]. The graph for $0(\alpha)$ [Fig. 2] is merely that for $E(\alpha)$ in an enclosure. I have varied the arrangement, only. The graph of $0(\alpha)$ is permissively transformable into the graph of $0(\alpha\gamma)$ [Fig. 3], thus satisfying Clause I. For Permissions 5 and 6 of my "Syllabus" permit us to surround "—is α" with a double cut in both instances. Then Permissions 2 and 6 authorize us to insert —(is γ) within 3 enclosures in both instances. Then the same two permissions supplemented by Nos. 8 and 9 authorize our joining the line of —(is γ) to the line of —(is α). For since whatever transformation may be made on the sheet may be reversed in an enclosure, it necessarily follows that it can equally be reversed in 3 enclosures. That, whatever element of β b may be, $0(\delta b)$ is true, is scribed as follows [Fig. 4]. For this denies that there is any respect in which either some element of β is x while every element of δ is x or in which no element of δ is x while some element of β is not x.

In order to assert that if $0(\beta)$ is true then $0(\delta)$ is true, i.e. to deny that $0(\beta)$ and $E(\delta)$ are both true, we scribe this graph [Fig. 5]. I will now show that the last graph but one is permissively transformable into the last, as it should be according to Clause II. For the sake of abbreviation, I will write P for 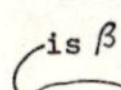 Q for 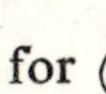 R— for 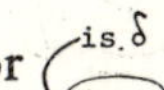 and S— for 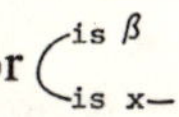

Then the last graph but one appears thus [Fig. 6]. Now by the principle

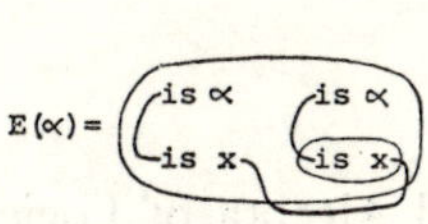

Fig. 1

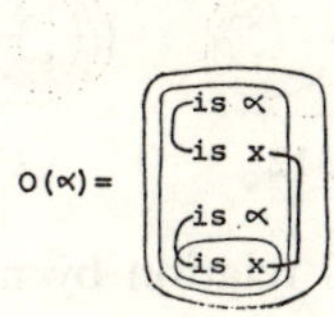

Fig. 2

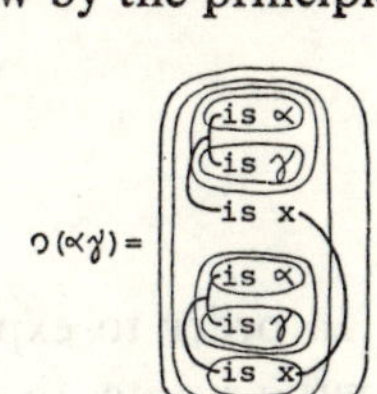

Fig. 3

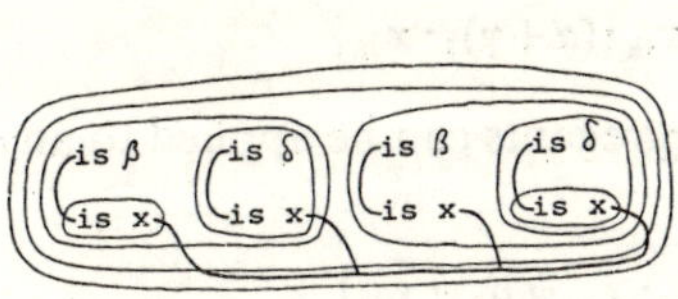

Fig. 4

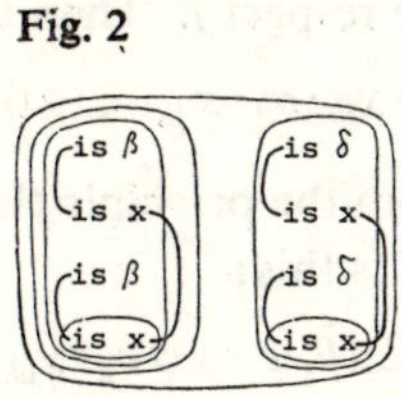

Fig. 5

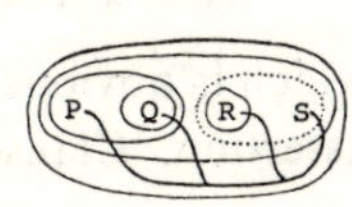

Fig. 6

of iteration which is deducible from Permissions 3 and 6, supplemented by 8 and 9 we can scribe this 2nd graph [Fig. 7]. I have made the cut round S wavy and have dotted the one round the second Q for the sake of perspicuity. Using the same principle again, we get the 3rd graph [Fig. 8]. Now using twice over the principle of erasure within even numbers of cuts we get the 4th graph [Fig. 9].

Now Permissions 5 and 6 (supplemented by 8 and 9) permit us to erase *two* oddly enclosed double cuts composed of the cuts 2nd and 3rd for one, and 4th and 5th for the other, counting from the outside. (The double cuts could have been erased if they had been evenly enclosed just the same, but by Permission 4 instead of No. 5 we get Graph 5 [Fig. 10].) Next by Permissions Nos. 4 and 6, we make a double cut round P and S so as to enclose a line of identity joining them, thus getting graphs 6 or 7 [Figs. 11 and 12]. Next by Permissions 2 and 6 (supplemented by 8) we erase a portion of the line of identity between the double cut round PS giving graph 8 [Fig. 13]. This is the desired conclusion.

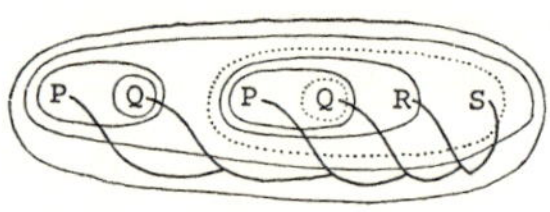

Fig. 7

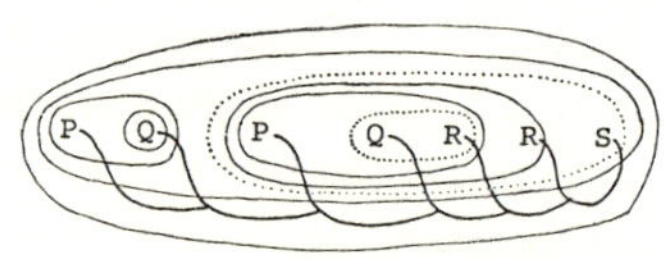

Fig. 8

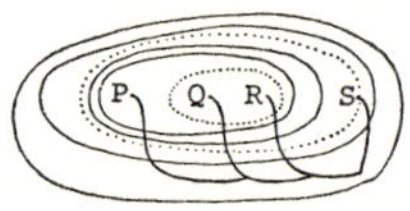

Fig. 9

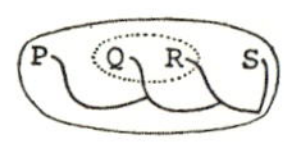

Fig. 10

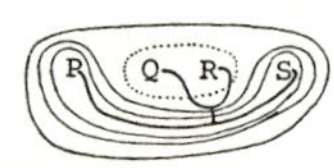

Fig. 11

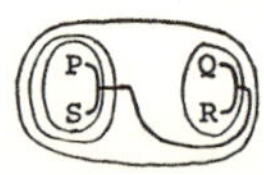

Fig. 12

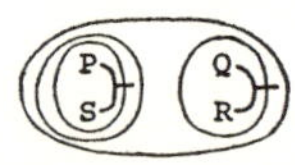

Fig. 13

In order to express the 0 relation by my General Algebra of Logic, I write α_i, etc. to mean that i is an element of α etc. and I write x_{ij} etc. to mean i is x in the respect j. Then Clause I is

$$\Pi_h\Sigma_i\Sigma_j\{\alpha_i\cdot x_{ih}\cdot\alpha_j\cdot\bar{x}_{jh} \prec (\alpha\psi\gamma)_i\cdot x_{ih}\cdot(\alpha+\gamma)_j\cdot x_{jh}\}$$

which is obvious from the principle that aggregants can be applied to any assertion. Clause II is this:

$$\Pi_h\Pi_i\Pi_j\Sigma_u\Sigma_v(\bar{\beta}_i\psi x_{ih}\psi\delta_u\cdot x_{uh})\cdot(\delta_v\cdot\bar{x}_{vh}\psi\bar{\beta}_j\psi\bar{x}_{jh}) \prec \Pi_h\Sigma_k\Pi_i\Pi_j\Sigma_u\Sigma_v\bar{\beta}_i\psi x_{ik}\psi\delta_u\cdot x_{uh}\cdot\delta_v\cdot\bar{x}_{vh}\psi\bar{\beta}_j\psi\bar{x}_{jk}$$

The first step toward deducing the consequent from the antecedent proceeds on the commonest of all principles that $(a \psi b) \cdot c \prec a \psi b \cdot c$. This gives the consequent except that k is not introduced, h remaining everywhere as in the antecedent. The principle by which we pass from this to the conclusion is $\Pi_x l_{xx} \prec \Pi_x \Sigma_y l_{xy}$. I do not know whether I ever stated this in print or not. It is a part of a principle thoroughly developed by me in a memoir which Newcomb practically refused to print in 1885 or 1886 which is the reason why I have never since printed anything on logic which could not be put in popular form. I there called the principle (of which this is a very small part) the principle of identification and diversification. It holds good strictly even if there is no x. From "every phenix would burn itself" it follows that "every phenix would burn something." It is somewhat remarkable since $\Pi_x l_x$ does *not* warrant $\Sigma_x l_x$.

In order to use my algebra of dyadic relations, as given in my Note B, we may write

$$O(\alpha) = \breve{\alpha} x \cdot \breve{\alpha} \breve{\bar{x}} \,\mathrm{f}\, 0 \qquad E(\alpha) = (\breve{\bar{a}} \,\mathrm{f}\, \bar{x} \psi \breve{\bar{\alpha}} \,\mathrm{f}\, x) \infty$$

Clause I is self-evident. Clause II is

$$(\breve{\bar{\beta}} \,\mathrm{f}\, x \,\mathrm{f}\, 0 \psi \delta x \,\mathrm{f}\, 0) \cdot (\delta \bar{x} \,\mathrm{f}\, 0 \psi \breve{\bar{\beta}} \,\mathrm{f}\, \bar{x} \,\mathrm{f}\, 0) \prec$$
$$(\breve{\bar{\beta}} \,\mathrm{f}\, x) \infty \psi (\bar{\beta} \,\mathrm{f}\, \bar{x}) \infty \psi (\delta x \,\mathrm{f}\, 0) \cdot (\breve{\delta} \bar{x} \,\mathrm{f}\, 0)$$

My Rules for Parentheses,

I. An added product needs no parenthesis

$$\begin{aligned} a \cdot b \psi c &= (a \cdot b) \psi c \\ a \cdot b \,\mathrm{f}\, c &= (a \cdot b) \,\mathrm{f}\, c \\ ab \psi c &= (ab) \psi c \\ ab \,\mathrm{f}\, c &= (ab) \,\mathrm{f}\, c \end{aligned}$$

II. A relative compound component of a similar non-relative compound needs no parenthesis

$$\begin{aligned} a \,\mathrm{f}\, b \psi c &= (a \,\mathrm{f}\, b) \psi c \\ ab \cdot c &= (ab) \cdot c \end{aligned}$$

I might have written the consequent of Clause II in the slightly neater form

$$(\breve{\bar{\beta}} \,\mathrm{f}\, x \psi \breve{\bar{\beta}} \,\mathrm{f}\, \bar{x}) \infty \psi \breve{\bar{\delta}} x \cdot \breve{\delta} \bar{x} \,\mathrm{f}\, 0$$

From the antecedent the usual formula gives

$$\breve{\bar{\bar{\beta}}}\, f x f\, 0\, \psi\, \breve{\bar{\bar{\beta}}}\, f \bar{x}\, f 0\, \psi\, \breve{\delta} x \cdot \breve{\delta} \bar{x} f\, 0$$

In Note B, I did not give the formula

$$z f 0 \prec z \infty$$

doubting whether it was correct, without adding some further premiss. But I do not now see any room to doubt it. With this, we reach the consequent.

I will now consider the four propositions which define the form of the universe of elements.

The first, his [Royce's] No. III, asserts that there is an element.

The second, his No. IV, that every element is unequal to some element; (an awkward way of saying that there are unequal elements).

The graph of the first is $-\infty$ or $-$, where ∞ means "is an element of the system".

The graph of the second is [Fig. 14a] or, more simply, [Fig. 14b].

The third, his No. V, is that if *i* and *j* are any two unequal elements, there is an element *k* such that *E*(*ik*), *E*(*jk*), and *O*(*ijk*).

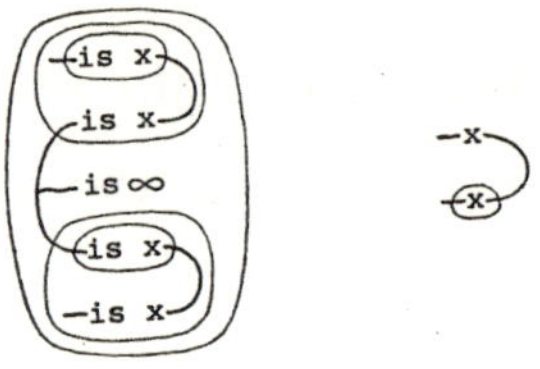

Fig. 14a, b

Whatever elements *i* and *j* may be [in Fig. 15], either *i* and *j* are alike in every respect or else there is an element *k* such that

in some respect *k* is like *i*	and in some respect *k* is like *j*

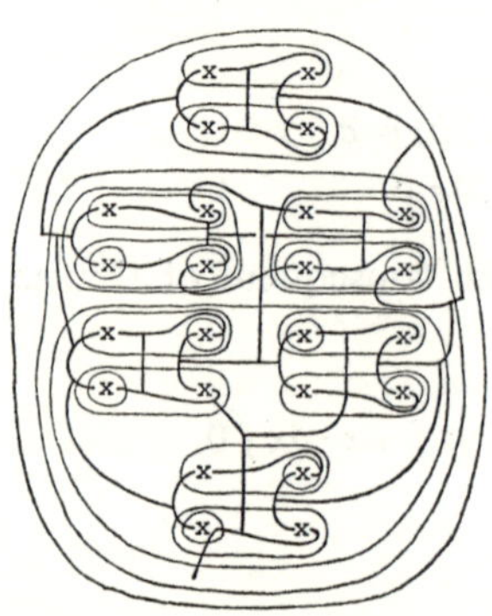

Fig. 15

and in every respect either k is unlike i or k is unlike j or i and j are unlike each other.

It follows that two unlike elements are unlike in at least two respects. And consequently there cannot be less than two respects.

This clause might be expressed follows:

Along with any two elements that are at all unlike, is a third which differs from them in every essential respect in which they agree, but is like each of them in some respect or other.

The fourth and last clause of the definition of the form of the system, Royce's No. VI, is stated by him substantially as follows:

Whatever collection and element of the system β and i may be, if $0(\beta i)$, then there is an element j of the system such that $0(\beta j)$ and whatever element of β b may be $0(bij)$.

This is the same as to say that unless there is some respect in which all the elements of β are like i, there is always in the system an element j which is like i in every respect in which all the elements of β are like one another and is unlike i in every other respect. Here is the graph [Fig. 16]:

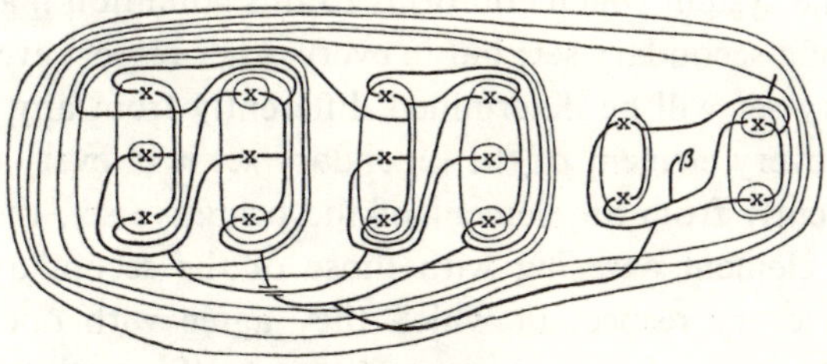

Fig. 16

Owing to carelessness in arranging the parts, I have had to make the ligatures of i and j intersect. I have made a bridge.

I will now restate his postulates. These are concerned with 5 categories of objects; namely, *systems*, *collections*, *elements*, *respects*, and *determinations*.

There is a non-symmetrical, transitive relation called belonging to, such that no two different objects belong reciprocally to each other. Collections belong to systems (whether one collection can belong to different systems or not he does not say, as far as I have read). A collection may belong to different collections not belonging one to another. For any set of elements and collections there is one sole collection to which they and no other element all belong. An element may belong to different collections not belonging one to another. An element cannot belong to another element. Respects and determinations do not, in the sense considered, belong to anything. For each pair composed of an element and a respect there is a single determination. But this is indefinite, and the only definiteness there is about it is that in any one respect any two elements have either like or unlike determinations. Two elements whose determinations in the same respect are both like or both unlike that of a third in the same respect have like determinations the one to the other in this same respect.

(To every system belongs some element. But this, which is his III, is an impertinent remark that does not affect any theorem.)

His principles IV, V and VI can be expressed as follows:

Of any collection of characters, *a*, *b*, *c*, etc. let every one which denotes anything at all denote an element of the system, and call the aggregate of what they denote "a primitive set." Then besides each primitive set there will be a set or sets in the system that we may call a "secondary" set or sets of elements in the system; and in whatever respect there is one determination affecting every element of the primitive set, every element of every secondary set will be determined otherwise and any element of the system which conforms to this condition may be taken as an element of a secondary set; but in every other respect every element of the secondary set will be determined differently from any other, except that in case every element of the secondary set is in every respect determined differently from an element of the primitive set, there will be in addition an element agreeing with those of the secondary set already described in every respect in which they agree with one another but differing from each of them in some respect. (Since there are but two "determinations," a secondary set will consist of two elements except in the exceptional case when it will contain 3.)

This may be called two principles tacked together, the extra member of the secondary set covering V in these cases in which it is not otherwise satisfied.

Any primitive set joined to one of its secondary sets forms a new primitive set of which any element of the system may be taken as an element of a secondary set and the other element will be its reverse (or "obverse"). Therefore every element has its reverse in the system. Now joining to any element another element to form a primitive set, the reverse of either element of the secondary set agrees with the first in all the respects in which the second does and in more besides; and this reverse may be joined to the first to make a new primitive and therefore the multitude of respects is innumerable, and so is the multitude of elements.

I blundered about this at first; for I have great ability as a blunderer.

There are, however, two points concerning which Royce is wrong in thinking that I am wrong. One is that he professes to show that "the principles of logic can be developed solely in terms of a polyadic symmetrical relation." I have not read so far; but I know it is wrong and can readily prove it. In my own system of existential graphs, the only formal signs are the sheet, the cut or enclosing line, and the line of identity, each of which expresses a dyadic symmetrical relation. But that does not amount to developing the principles of logic from symmetrical relations alone. The first relation of logic that of antecedent and consequent is unsymmetrical. Now an unsymmetrical relation cannot result from any combination of symmetrical relations alone, although a friend of a cousin is not necessarily a cousin of a friend. But when you put together unlike things the relation is unsymmetrical. As long ago as 1870 in my first paper developing the logic of relations, I showed that $A:B$ is the original type of unsymmetric dyadic relations. Here there is no dissymmetry (pertinent to the question) in the colon. I chose it precisely for that reason. It showed that dissymmetry has its origin in the matter, and that *unlike correlates can be in no other than a dissymmetric relation.* Such a relation as "not" is symmetrical only because for every element $A:B$ it contains another $B:A$. It is thus what you might call symmetrical by composition but not *thoroughly* symmetrical. If two such relations are joined, one being say $A:B \Psi B:A$, and the other $A:C \Psi C:A$, the product is $B:C$, which is unsymmetrical. But this unsymmetrical character has not resulted from compounding symmetry, but because symmetry can be compounded from dissymmetry and in this case is so; and half being excluded the rest is unsymmetrical. Everybody knows that symmetrical forms can be built up out of unsymmetrical forms [Fig. 17]. But everybody can see that symmetrical forms, *put*

Fig. 17

together symmetrically, will never make an unsymmetrical form. Why not? Because *symmetry* is a special kind of *equality*. Now equality can be built up out of inequalities; but is is evident that inequality can never result from a chain of equalities, for if one thing is equal to a second, and this second to a third, the first is equal to the third and you are precisely where you were at the outset. So far as I have read I may not have got all of Royce's reasoning, but I have come across a palpable fallacy. He writes $(y\beta|\gamma x)$. He puts an *F* first, but we shall do better to drop it, as it has nothing to do with the argument. His point is that the relation expressed by the vertical line is a symmetrical one; and so it is a *compound* symmetrical relation. The same is true of the relation expressed by writing the letters together. But the relation between *x* and *y* is unsymmetrical. Really, I should think anybody could see that that dissymetry is due to the inherent dissymmetry of the relation of β in itself to γ in itself. Yet he says "the unsymmetrical dyadic relation in question has been entirely derived from the wholly symmetrical relations." Plainly it has not been so derived. This argument has been borrowed from Kempe who brought it forward in his great memoir of 1886, though Kempe was much more cautious in drawing his conclusion. Kempe drew this diagram [Fig. 18). "Now," says Kempe, "the relation

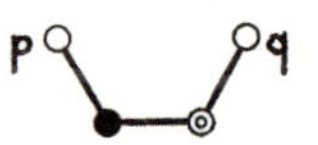

Fig. 18

p to *q* is an unsymmetrical one; and the figure shows that, *by means of the introduction of auxiliary units*, such relations can be represented by spots united by simple links alone." Observe his putting in the auxiliary units as essential, though Royce in copying him overlooks this. Still Kempe, had he stopped an instant to think would not have said that there was but one kind of links in the diagram; for he himself remarks in another place that in every diagram of the sort the spots that are not linked by lines drawn *are* linked by the absence of such lines and he redraws a diagram uniting in the second form by black lines every pair of spots not so joined in the other and conversely and expressly says that they are the same diagram essentially. Let us do this in the present case and [Fig. 19a] becomes [Fig. 19b]. He is quite right in saying that the relation

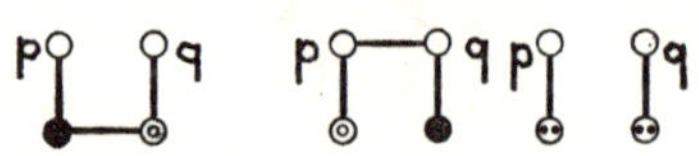

Figs. 19a, b, c

between *p* and *q* is an unsymmetrical one. It is equally so represented in both forms of the diagram. But the second form brings more forcibly to our attention the fact that dissymmetry is due to a difference in the matter of correlates. What is the distinction of matter and form in logic? Why, it consists in this: the *matter* is all the circumstances of real cases which do not concern the logician and which he puts out of view, while the *form* is what he takes into account. If we are to attend to the circumstance that in Kempe's diagram the red and brown spots are differently colored, that becomes a matter of form and we then cannot say that the relation between p and q is built up of symmetrical relations; since there will be the *formally unsymmetrical* relation of brown to red. If however this is to be relegated to matter, then we must not and cannot attend to the difference and they might as well be of the same color. [Fig. 19c]. Here, however, the dissymmetry disappears. This fallacy is certainly far less deceptive than the one that the relation of being a *cousin's spouse* is different from that of being a *spouse's cousin*. For here it seems that the relation is compounded of nothing but the two symmetrical relations of being *spouse* and of being *cousin*. It requires something more than ordinary intelligence, it requires the proper subtlety of the logician, to perceive that something more is involved, namely the relation expressed by relative multiplication, which is asymmetric. That is to say, it is asymmetric unless the relations multiplied are converses of each other. To be sure, relative multiplication involves no asymmetric relation between the two aspects of the individual at its centre or at any rate not between any two different correlates, the second of the first relative term, and the first of the second. But it is a factor of the resulting relation, and it is the asymmetry at this point to which the asymmetry of the product is due. This is a refutation against which the [geists] of Hegel shall not prevail. If Royce, instead of arguing the question upon a popular platform, had resorted to the only scientific method, that of employing my matrix of relatives of 1870, he could not have fallen into his fallacy. He would have seen that elements of the irreducibly symmetric form $A{:}A$, $B{:}B$, etc. can never be of any aid in producing asymmetry. That $A{:}B \;\Psi\; B{:}A$ etc. will produce asymmetry when multiplied into $A{:}C \;\Psi\; C{:}A$ or when multiplied into $A{:}C \;\Psi\; C{:}A \;\Psi\; C{:}B \;\Psi\; B{:}C$. That $(A{:}C \;\Psi\; C{:}A \;\Psi\; D{:}B \;\Psi\; B{:}D)\; (C{:}B \;\Psi\; B{:}C \;\Psi\; D{:}A \;\Psi\; A{:}D)$ is symmetrical, etc. etc.

The other doctrine of mine which Royce attacks, as remarkably shows how unscientific his training has been. He attacks my one-two-three doctrine in the very field where it is most obviously defensible, that of formal logic. You must remember that in my very first paper on the subject, in 1867, which is almost perfect, I stated distinctly that thirdness is inconceivable without secondness, just as secondness is without firstness.

But I never heard of anybody's denying that relation is something entirely distinct from quality, or of anybody's maintaining that relation could be conceived apart from quality. The distinction is possible *because quality can be conceived without relation.* In like manner, symmetry cannot be conceived without plurality of parts. Yet nobody maintains that symmetry is nothing more than plurality of parts. Nobody can conceive of viscosity without fluidity but viscosity is distinct from fluidity. Nobody can conceive of three things without two things, and it is essentially Royce's position to say that consequently there is no particular difference (for he is hopelessly vague in his position) between three things and two. I also, if not in my original memoir on relations of 1870, at least in my Note B of 1882, expressly mentioned and extensively applied the fact that a dyadic relation could be regarded as a triadic relation by formally adding a correlate common to all dyadic relations, just as I regarded a non-relative term such as *man* as a relative by conceiving it as a 'man for everything' or as a 'man for something' and so spoke of the converse of such a term meaning the relation of "coexisting with a man that is—" But this is a mere device of symbolization. It is not pretended, and is not true that a class-name for objects really is a relative term or that a dyadic relation is a triadic relation. Of course, as objects exist they are in various relations dyadic and triadic. Therefore if you choose to take classes of single objects, such as 'man' and to consider them in connection with some circumstance which varies yet is in large classes constant, such as their dwelling place, or their first distinguished ancestor, you get such relative terms as 'inhabitant of—' and 'descendant of—,' while if you take one of these terms and turning back insist on considering *of what*, you get Italian, French, etc. on the one hand, Romanoff, Bourbon, Stuart, etc. on the other. You convert the relative into a non-relative by considering the correlate as fixed, while if you choose to take into consideration a variable circumstance you convert non-relative terms into relative terms. This is the *A*, *B*, *C* of the subject. But a man who should infer from this that the distinction between relative and non-relative terms was of little importance in logic would simply show that he did not understand what logical importance consists in. The whole logic of non-relative and of relative terms is so different that ordinary syllogistic can be worked with a machine while relative syllogistic cannot be so worked until you can get a machine which will register an infinity of different conclusions or will select what line of reasoning to pursue. For from every premiss in relative logic an infinite multitude of different conclusions can be drawn. Yet it is from just such considerations that

Royce argues that the difference between dyadic and triadic relations is "superficial." I shall not retort with that word, which he leaves as a mere expression of emotion. Again he argues always from haphazard examples never considering the question in the only scientific way. The points are (1) that no dyadic relation can be composed of non-relative factors alone, and (2) no triadic relation can be composed of dyadic and monadic factors alone, although (3) every tetradic and higher relation can be composed of monadic, dyadic, and triadic factors. I proceed to prove these three propositions. The first amounts to this, that from any number of single objects you can never produce a pair without either splitting one of the monads, which would prove it to be a dyad, or else putting two monads together. It is true you may put three together and then remove one of them. But the very fact that you can remove one and thus produce two shows that the twoness is inseparable from threeness, and that therefore in putting three together you have put two together. It is true also that degenerate dyadic relations, that is, irreducibly symmetrical relations, such as 'man that is—' are in fact mere monads looked upon as dyads. Nothing is really in a relation to itself, although it is in a relation to any *part* of itself. But a part and its whole are different objects. A collection and its members are different objects, although the mere thinking the members together creates the collection, which is thus an *ens rationis*. If a man were merely a collection of cells, he would be an *ens rationis*. A bushel of wheat would be an *ens rationis* if it weren't that the grains are more or less alike and, what is more to the purpose, were not very strikingly near together. A bushel of wheat of which the different grains are in different parts of the world, have no common ownership, or any circumstance that connects them except our vague thought is quite an *ens rationis*. This is to show that a part and its whole are different objects and so removes an objection to my statement that nothing is in a real relation to itself. Therefore, if any kind of dyadic relation could consist of monadic element or elements alone, it would be one of these degenerate relations which are mere limitations of identity. Yet the very fact that they are converted from monads into dyads by merely thinking them so, makes it particularly obvious that unless this element of *Thought* be introduced they remain dyads. For genuine dyadic relations of course the fact is more real. Take a physical relation, say that of striking. Remove the objects from one another by billions of miles and each remains all that it is in itself. But in that situation none can strike another. It is proved then that secondness is something essentially different from firstness; and the fact that secondness cannot be conceived without firstness is no argument to the contrary.

You will see for yourself, if you consider the matter, that I could prove my second proposition in the same way; but I will vary the matter by taking a somewhat different method. A function of one variable stands to that variable in a dyadic relation. Of course, if we fix the value of the variable, the function becomes a constant, and is a function no longer. But that does not show that functionality is nothing different from value, which would be contrary to the proposition just proved. A function of two variables, say $x + y\sqrt{-1}$. If we fix the value of x or y and making $x = a$ or $y = b$ reduces the function to $a + y\sqrt{-1}$ or $x + b\sqrt{-1}$ we reduce it to a function of one variable; but that does not prove that the *combining* power of a function of two variables can be made out of functions of single variables. So, for $a + b\sqrt{-1}$, one may substitute first $x + b\sqrt{-}$ or $a + y\sqrt{-}$, and afterward $x + y\sqrt{-1}$ setting first one variable, and then the other, free. I can see nothing but this in some of Royce's argument to show the "superficiality" of the difference between functions of one and of two variables, or between dyadic and triadic relations, which is the same thing.

It is self-evident that a function of two variables or a triadic relation *combines* the effects of different variables (*what one throws over into the matter* of the *correlates* becomes *ipso facto* variable, and what one retains in the matter of the *relation* becomes *ipso facto* fixed), but no single variable function or function of a function *combines* the effects of different variables, nor does (what is the same thing) a dyadic relation.

I will propose to Royce the most favorable possible case to show, if he can, that there is no essential difference between a dyadic and a triadic relation. Namely, a positive rational fraction is a function of two positive integers. Can he show that the rational values are functions of a single variable? They have the same multitude as the integers, and therefore each rational fraction may be made to correspond to a distinct positive integer. But can the *relationship* between the rational fractions be stated as resulting from the relationship of the integers that correspond to them, in general terms? The sole definite relation between integers is their order of succession. I can prove, and will therefore grant, that the sole definite relation between the values of positive rational fractions is their order of succession. For, from the fact that the value of $\frac{a+b}{c+d}$ is intermediate between those of $\frac{a}{c}$ and $\frac{b}{d}$, whatever positive integers a, b, c, d, may be, the entire arithmetic of rational fractions can be deduced (or assuming only besides that $\frac{1}{0} > \frac{0}{1}$ and that $\frac{2}{1}\ \frac{3}{1}\ \frac{4}{1}$ etc. are 2, 3, 4 etc.; but these last assumptions are not strictly indispensable *except* to connect the arithmetic of fractions with that of integers). This is readily made out. Of course, definitions of addition and multiplication have to be extended or created, always observing the one equation furnished by the fundamental hypothesis; namely that intermediate in value between any fraction and itself, that is, equal to that fraction, is the same with numerator and denominator doubled, and therefore that such fractions as $\frac{7a}{7b}$ and $\frac{5a}{5b}$ are equal. The reasoning by which you are forced to the definitions is very pretty. I haven't space for it here. Now, then, let us start with $\frac{0}{1}$ and $\frac{1}{0}$ and put in successive lines all the intermediates and number them in the order in which they are inserted. I express these ordinal numbers in the secundal system of numeration, because there is thus a simple rule connecting the ordinal numbers of two fractions and their values.[1] Thus, if Royce is right, he should have every facility here for expressing the value of a rational fraction as a function of the ordinal number of it. That is why I developed this idea. It is easy to prove that every positive rational quantity will appear once and only once in an indefinite extension of this table. In fact, I may as well point out a way to think about the matter. Let the points on a plane sheet represent *fractions* (not fractional *values*). Let the rule be that taking any two points representing two fractions, say $\frac{5}{7}$ and $\frac{8}{11}$, every other point on this line shall represent a fraction whose numerator differs from either of those by some multiple of 3 (since $8-5$ is 3), and whose denominator differs from the same one by the *same* multiple of 4 (since $11-7=4$) so that the fractions on that ray (by which I mean an unlimited straight line) will be $\frac{2}{3}\ \frac{5}{7}\ \frac{8}{11}\ \frac{11}{15}\ \frac{14}{19}\ \frac{17}{23}\ \frac{20}{27}$ etc. Let us begin the diagram by assuming a point to represent $\frac{0}{0}$ (which has no particular value) and let us assume three other points (so that no three of the four shall be on a straight line) to represent $\frac{0}{1}\ \frac{1}{0}$ and $\frac{1}{1}$. These can be placed anywhere but most conveniently about as I have placed them in the little figure [Fig. 20]. Then rule a line through $\frac{0}{0}$ and $\frac{0}{1}$. On that will lie every fraction whose numerator is 0 and whose value is 0. Rule a line through $\frac{0}{0}$ and $\frac{1}{0}$. On that will lie every fraction whose denominator is 0, and whose value is therefore ∞. Rule a line through $\frac{0}{0}$ and $\frac{1}{1}$. On that will lie every fraction whose numerator is equal to its denominator, and whose value is therefore 1. Rule a line through $\frac{0}{1}$ and $\frac{1}{1}$. On that will lie every fraction whose

[1] Peirce here refers to a table not now in the manuscript. It may have resembled that in Section 6D of this volume.

denominator is 1, and, of course, these will all have integer values. This ray will cut the ray through $\frac{0}{0}$ and $\frac{1}{0}$ at a point which will equally represent $\frac{\infty}{1}$ and $\frac{\infty}{0}$. Rule a line through $\frac{1}{0}$ and $\frac{1}{1}$. On that will lie every fraction whose numerator is 1. It will cut the ray through $\frac{0}{0}$ and $\frac{0}{1}$ in a point which will equally represent $\frac{0}{\infty}$ and $\frac{1}{\infty}$. Rule a line through those intersections. On it will lie every fraction of which numerator or denominator is infinite. We will call it the infinity-ray. It will cut the ray through $\frac{0}{0}$ and $\frac{1}{1}$ in a point $\frac{\infty}{\infty}$ (or as we say through $\frac{n}{n}$ where $n = \infty$). Let us call the ray through any fraction-point and the $\frac{0}{0}$ point the "value-ray" of that fraction since all fractions represented by points of it have the same value. Let us call the point at which any ray cuts the infinite line the "infinity-point" of that line. Then, given any two fractions, the point of intersection of the rays from each to the infinity-point of the value-ray of the other will represent the fraction whose numerator and denominator are respectively the sums of the numerators and the denominators of the two fractions. Each line of the table inserts a new fraction between every pair of adjacent value-lines already obtained. Consequently, no two fractions of the same value can be inserted. But now look at the succession of numerators. In successive lines they are [found in Fig. 21]. Each begins by repeating all the numerators previously obtained. Consequently, since we have seen that no two fractions can have the same value, and since in every line each numerator occurs a number of times equal to its totient (which is the number of numbers less than the number and prime to it) it follows that *all fractional values occur*. (There are 2 leaks in the proof, but I cannot stop to plug them.) It is pretty to see how they are picked up. Thus on any line with numerator 5 you will find denominators $n5-1$ and $n5+2$ and $n5+3$. Where is $n5+1$?

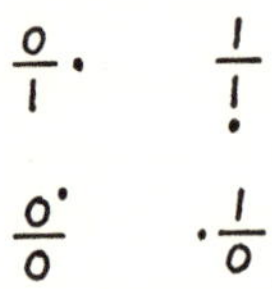

Fig. 20

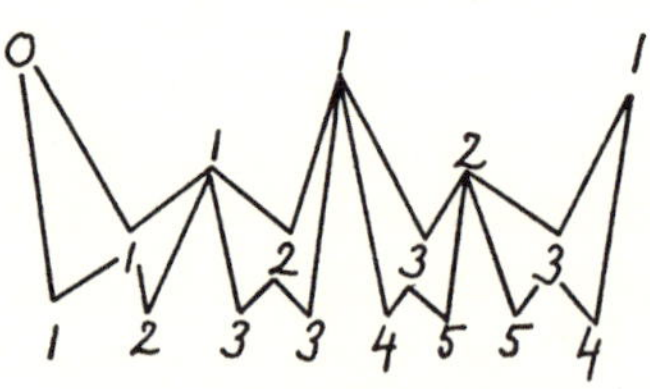

Fig. 21

The *next* line picks that up. Thus in the last line of the enclosed table you find $\frac{5}{9}$ $\frac{5}{12}$ $\frac{5}{13}$ but you don't find $\frac{5}{4}$. No, but the *next* line will have it. In the enclosed table I have used the secundal system of arithmetical notation (1 = our 1, 10 = our 2, 11 = our 3, 100 = our 4, 101 = our 5, etc.) to denote the order in which the fractions were written. If the first 1 in each symbol be transposed to the end and then the full number of figures be made out by affixing zeros, the numbers so changed will show the order of values. Thus,

100000 will become 000001 or 1		
10000 will become 000010 or 10	i.e. 2	
100001 will become 000011 or 11	i.e. 3	
1000 will become 000100 or 100	i.e. 4	
100010 will become 000101 or 101	i.e. 5	
10001 will become 000110 or 110	i.e. 6	
100011 will become 000111 or 111	i.e. 7	
100 will become 001000 or 1000	i.e. 8 etc.	

I may also mention that if to the numerator of any line we add the denominators (the row which is the row on numerators reversed) and prefix that row of numerators, you have the numerators of the next line. Thus

Num. of 1st line	1
Denom. of 1st line	1
Sum	2
Sum with previous line prefixed / Numerators of 2nd line	12
Denom. 2nd line	21
Sum	33
Num. 3rd line	1233
Denom. 3rd line	3321
Sum	4554
Num. 4th line	12334554
Denom. 4th line	45543321
Sum	57877875
Num. 5th line	1233455457877875
Denom 5th line	5787787545543321

The last half of the last row of fractions

Sum / Last row of denoms. $\frac{6}{5}\ \frac{9}{7}\ \frac{11}{8}\ \frac{10}{7}\ \frac{11}{7}\ \frac{13}{8}\ \frac{12}{7}\ \frac{9}{5}\ \frac{9}{4}\ \frac{12}{5}\ \frac{13}{5}\ \frac{11}{4}\ \frac{10}{3}\ \frac{11}{3}\ \frac{9}{2}\ \frac{6}{1}$

etc.

With all these extraordinary facilities and I could mention sundry others if Royce cannot reduce the values of fractions to a function of their ordinal place in this table, he had better admit that a function of two variables cannot always be reduced to a function of one variable; i.e. that a triadic relation cannot always be represented as a dyadic relation with the same correlates (whatever that may mean). I have a difficulty in expressing his idea, if he has any. I cannot speak of a "combination of dyadic elements alone" since the word "combination" means precisely something involving a triadic relation.

But I am tired of arguing a truism. Who but Royce, *hegelisé* as he must be, can fail to see that "—gives—to—" *unites* for each correlate the effect of two others, while "—benefits—" does not? Nor does "—by the gift of a clock in a glass case with an alarum attachement and a calendar which only requires correction once in four hundred years, enriches—." It is not what is thrown into the *matter* as a *constant* but what remains in the *form* as a *variable*, that is the question. If I said that *form* was constant, *matter* variable, as I may have done, that was from an opposite point of view. Perhaps I didn't in fact say so; but it is true.

My third proposition is not so raw a truism. It is a truism, too, when rightly understood. But one is tempted to extend it beyond its strict meaning, because more than its strict meaning happens to be true and (as far as I know) unsaid. This proposition is that every tetradic, pentadic, and higher form of relation can be resolved into a combination of triadic relations. We can now speak of "combination" without self-contradiction. Now to begin with, I will point out some things that I do *not* mean. I do not mean to deny that a carbon atom with its four bonds can have such a relation as it has in nicotine, whose graph I will write [in Fig. 22],

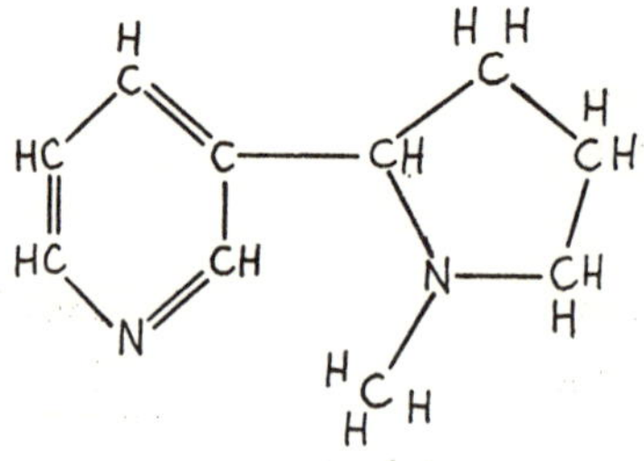

Fig. 22

and put a red dot upon the *C*-atom meant: (It is not a *condensed* aromatic body like other alcaloids). This marked *C* is united by its four bonds to four different sorts of things, and so, being a tetrahedron in a space of neither less nor more than 3 dimensions reverses its character when *any two* of these are interchanged, like this determinant when either two letters or two numbers are interchanged.

$$\begin{vmatrix} a_1 & b_1 & c_1 & d_1 \\ a_2 & b_2 & c_2 & d_2 \\ a_3 & b_3 & c_3 & d_3 \\ a_4 & b_4 & c_4 & d_4 \end{vmatrix}$$

Perhaps you will say that the tetrahedron differs from the determinant in only having a single set of four correlates concerned, while the determinant has *two* sets of four *umbrae* (as Sylvester called them) each. (Umbrae are strictly analogous to logical correlates. Quantities are not so strictly so. They are rather relative terms. The method of existential graphs renders this clear.) But I reply that there *are* two sets of four correlates in the case of the tetrahedron, since its reversed character depends on its being in tridimensional space, which is quaternionic. That is, *four* is the lowest number of points which in such a space are not in a space of fewer dimensions. Hence, just as what is true, as to interchanges, of the determinant of the fourth order, is equally true of determinants of orders 3 and 2, so what is true of the tetrahedron in 3 dimensional space is equally true of a triangle on a perissid surface or like an arrow on a bounded line. Of course, I do not dream of disputing the possibility of such relations, although their peculiarity cannot appear for any set of interchangeable correlates less in multitude than the set of fixed correlates. Far from disputing their existence, I should be disposed to regard this as the fundamental form of relationship, — as what we call in mathematics the "general" form, just as we say that

$$u_{x+3} - (\alpha + \beta + \gamma)\, u_{x+2} + (\alpha\beta + \beta\gamma + \gamma\alpha)\, u_{x+1} - \alpha\beta\gamma u_x = 0$$

is the "general" form of a linear difference equation (with constant coefficients) of the third order, whose solution is

$$u_x = C_1\alpha^x + C_2\beta^x + C_3\gamma^x$$

in contradistinction to the form

$$u_{x+3} - (\alpha + 2\beta)\, u_{x+2} + (2\alpha\beta + \beta^2)\, u_{x+1} + \alpha\beta^2 u_x = 0$$

whose solution is

$$u_x = C_1\alpha^x + (C_2 + C_3 x)\, \beta^x.$$

There are a good many propositions which might easily get erroneously supposed to be implied in the proposition which I undertake to prove; and some of these are in fact *true*, but are not what I mean. For instance, it is true that any calculable function of more than two variables can be reduced to a function of functions where all the functions (both the uniting and the united) are functions of two variables. The reason is that since whole numbers form a simple succession, addition is the only elementary mode of combining them. Extremely important as this great fact is, it is not what I here mean. Nor is it implied in my proposition that the truth of an algebraic equation consists in the truth of one or another of a collection of quadratic equations, which is only true of equations which express dyadic relations, i.e. equations of one unknown.

For what I mean is, like my other two propositions, *not* a theorem, but a great fundamental self-evident truth, or principle. Such was the second, that *combination* cannot be resolved into *dependence*. The present amounts to saying that in all cases a proposition concerning a set of objects is equivalent to a special proposition that something combines a truth concerning an object described in terms of any part of that set and an object described in terms of the rest of the set.

Suppose, for example, I say that the 32nd proposition of the first book of Euclid's *Elements* follows from the 16th proposition and the 5th postulate. Of course I mean to take for granted various matters that I leave indefinite which constitute a fourth correlate. Then I can begin by combining any two of my correlates, and shall then have three, which I can take as correlates of a triadic relation. Thus:

In the universe of geometry if the 5th postulate be admitted, the 32nd proposition follows from the 16th.

In the universe of geometry if the 16th Proposition be assumed the 32nd follows from the 5th Postulate.

If the 5th Postulate and the 16th Proposition be assumed then in the universe of geometry the 32nd holds good.

If the 5th Postulate is true while the 32nd is false then in the universe of geometry the 16th Proposition must be false.

If in the universe of geometry the 32nd proposition is false then from the truth of either the 5th Postulate or the 16th Proposition follows the falsity of the other.

Take any tetradic relation R whatever, and let A, B, C, D, be four objects which, taken in this order, are in that relation. Then, to begin with, we will suppose the relation to be such that this proposition does not imply that those four objects are in that same relation in any other order. In that case, there are four distinct places in that relation which we may call α, β, γ, δ. So that the statement is that while A is in the α place, B in the β place, C in the γ place, D is in a particular case in the δ place of the relation R. Now I will call A's being in the a place the αA situation, so that the αA situation is in a triadic relation to α and A. I will call B's being in the B place the βB situation, and so with the others. I will call the fact that in this particular case the αA situation holds, fact 1. This is quite indefinite, meaning that when suitable objects not specified are assigned to the other places, I will call the fact that the βB situation subsists along with fact 1, fact 2. This is a little more definite leaving only 2 places unspoken of. I will call the fact that situation γC subsists along with fact 2, fact 3. Finally then I say that situation δD subsists along with fact 3 and this precisely expresses that in the particular case A is α, to B being β, C being γ and D being δ to one another in the relation R.

Now if the relation between A, B, C, D in this order is also a relation between them in another order, this merely adds another relation to be combined with the one given. Thus, in any case, it is self evident that all tetradic relations can be reduced to triadic relations, and upon the same principle, so also can all higher forms of relation.

But perhaps Royce may say (and I do not think it unlikely) that every dyadic relation is, according to this, a triadic relation. Certainly if he insists upon taking all that can be *thought* of a dyadic relation into account, this may be made out. But this only carries me to the higher aspects of my categories. I intended only to show that in formal logic they were all three required, and this I have done. But the Hegelians are not satisfied with the position of formal logic, although it is the only scientific attitude. It is simply absurd to attempt, as they do, to attain absolute perfect truth. Of course, no proposition of theoretical science is true in practice. In other words it is only true of an ideal world that differs from the actual world. What of that? It is the only way to attain any kind of mastery over the real world. But when he says (if he does say it) that every thought is triadic, I reply, Yes but a relation may be apprehended *without* thought, not indeed under a general category, but as something positive in the actual case. A man making an effort or otherwise reacting with the outer world does not necessarily think, but he knows that relation, that purely brute dyadic action in the particular case before him, and he has no need to look further.

There is mighty little in the C. S. Peirce of 1905 of identity with the C. S. Peirce of 1867. I feel entitled to speak of him as quite another person. But my opinion is that the paper on A New List of Categories is one of the most perfect gems of all philosophy. I have not been able to find any positive error in it. There is a good deal that was not then worked out; but the leading features were made out correctly. However far we carry thought there remains a dyadic element not transmuted or taken up into triadicity, into tertianity. The *phaneron*, as I now call it, the sum total of the contents of human consciousness, which I believe is about what you (borrowing the term of Avenarius) call *pure experience*, — but I do not admit the point of view of Avenarius to be correct or to be consonant to any pragmatism, nor to yours, in particular, and therefore I do not like that phrase. For me *experience* is what life has forced upon us, — a vague idea no doubt. But my *phaneron* is not limited to what is *forced* upon us; it also embraces all that we most capriciously conjure up, not *objects* only but all modes of contents of cognitional consciousness.

Now I ask what are the kinds of elements, undecomposable parts of the phaneron? Surely no mode of division of them can be more important than that according to the degrees of complexity of their combinations, — like the division of the chemical elements according to their valency. I daresay it may have been Kekulé's *Kohlensloffverbindungen* that set me thinking in that direction. *A priori* they can only be of those three kinds, the Priman (uni-valent), the Secundan (or bivalent) and the Tertian (or trivalent) (one might state them as non-valent, uni-valent, and bivalent, thinking of the relation of one correlate to others, — but that is only a difference of the point of view. Thinking of the fact, the univalent, bivalent, and trivalent is better. Just so the function of 2 variables is the *trivalent* relation etc.) So much *must be, a priori*. Now look at the actual contents of the phaneron and see what ones of these possibilities are realized. You find they *all* are realized. How could it be otherwise? One may well ponder that. However, to come down to fact, when we look we find it is so. Take that Kantian division of the soul which has found so much employment, in spite of being abused incessantly by those who have employed it, that it must approach a great truth — Pleasure-Pain, Conation-Cognition. Let us correct its misapprehensions and leave its kernel of truth. Pleasure and pain are not qualities of feeling *per se* as this supposes them to be. They are those feelings which attract and repel. They therefore belong to the domain of conation. But the feelings, the qualities of feeling, are eminently priman, positive somethings regardless of aught else, like nothing else, indescribable as nothing else is indescribable. Conation, action, reaction, is in *surprise* in a more passive form. For action cannot take place without reaction, effort without resistance. If therefore we are acted on in sensation, we must react. But it is but the two sides of Secundity, — as two opposite sides there necessarily are, — that distinguish centrifugal and centripetal innervation. Finally it is not the mere sensation which is any kind of cognition. It is the entrance of concepts, which makes the third branch. If we say *thought*, since all thought is dialogue and since symbols alone fully develop the idea of triadic relation, then Feeling, Consciousness of Reaction, and Thought will be the Priman, Secundan, and Tertian elements of consciousness. If I were really the pop-gun that people think and always will think I am, and were capable of pluming myself on having given birth to the valency idea, — the cenopythagorean idea as I sometimes call it, — I might, seeking the fortune of my idea, wish to account for the striking contrast between sensation and volition as being due to physiological causes. But evidently the physiological difference is on the contrary due to what was required to make an animal, to the final causation that it is the fashion to blind oneself to. But the difference is striking and emphatic because duality is the natural habitat of contrast. There is also a threefold distinction in thought for which the theory calls. But it is not so crying a contrast as that between sensation and volition. (The existence of final causes, which every voluntary act conclusively evidences, is far more certain than any theory of science whatever; but the idea is that they must be frowned upon, because the business of physiology is to account for vital phenomena by physical causes. No doubt. But if the neglect of the study of logic had not reduced men to the level of children, they would see that the recognition of final causes in nowise interferes with the business of physiology. I doubt if there is any other among many silly notions that are current which is quite so silly as this is.) Thought is 1, creative thought; 2, performing thought, which develops an idea internally, and insistently carries it out into consequences, theoretical and practical; 3, docile thought, the thought that listens, that restrains itself, that seeks to be brought to judgment. To think of one's spending one's whole life in trying to make "thinkers" see anything so simple and self-evident, and failing! It makes one want to sing Crambambuli.

1909 Feb 26.[1]

My dear William:

Your Appendix C seems to give a fair idea, as far as it professes to go, of opinions of mine which I adhere to today more than ever. I do not remember to have read anything of Bergson's; but I cannot imagine anything more contrary to my notions than what I understand to be meant, what you represent to be meant, and what ought according to the usages of philosophical lingo to be meant by saying that "Reality changes" or "Truth changes." Of course, I admit that changes take place. But I do not admit for an instant that logic fails. I can imagine nothing more absurd. But it is necessary to be exact in logic. You begin by talking of the "dictum de omni et nullo." The substance of a sentence in the first chapter of Aristotle's Analytics frequently went by this name during the nineteenth Century especially after 1827. Throughout the middle ages it was more accurately called the "Dici de omni." It is susceptible of complete proof that it was so called because the Latin translation of Aristotle's sentence begins with the words "Dici de omni"; and it is equally certain, as the great Thomistic treatise on Logic declares (I have it here at hand), that this sentence was meant as a definition of the Greek words corresponding to *Dici de omni.* Its sense is "To be predicated *of all* a subject is to be predicated of all of which that subject is predicable." Now it cannot truly be said of any science, not even of Logic, that its "foundation" is a definition. But it may be said of a system of signs, such as an algebra that its foundation is the definition of its chief signs and because Syllogistic was always looked upon as the doctrine of what sort of statements, as premises, did signify, that which the syllogistic conclusions from them signified, that it came to be said by the medieval logicians that the "Regula dici de omni," as it was most commonly called (i.e. the rule of universal predication), was the foundation, as some said, of the direct moods of the first figure, *Barbara*, *Celarent*, *Darii*, *Ferio*, and as a few others said (contrary to Aristotle) was the foundation of syllogism generally. I can hardly believe that any logician ever made that definition the foundation of all Logic. If he did, I don't regret

[1] This is the 40-page letter mentioned in *The Thought and Character of William James* by R.B. Perry (2, 439).

having forgotten him, — so long as I can't bear in mind everything I have read. But as for you, though you *talk* of the "Dictum de Omni et de nullo," what you seem to have in mind is what is called the *Nota notae* i.e. the rule *Nota notae est notra sei ipsius* given in Cap. III of the Categories, which is not given as a definition but as a property of predication. But this again goes no further than syllogistic. It is as much as to say that predication is a "transitive relation" of which you give several other examples. I first remarked that all relations have this property which are of the form, "is lover whatever is loved by," "outweighs whatever is outweighed by" etc. etc.; and it is a fact that *every* transitive relation *involves* such form of composition. Schröder, however, made the curious remark that not *all* transitive relations are of that form; and that to get a form that would hold good of *all* transitive relations, that form must be *restricted* (which sounds paradoxical). He shows that every transitive relation is of the form "— is both lover of and lover of everything loved by —." This is a relation which not everything is in to itself. For it is conceivable that a being should not love himself. Just as a thing is not in the transitive relation of "being larger than" to itself. But my original form is limited to such transitive relations in which everything stands to itself. I had, however, previously noticed in my Note B in "Studies in Logic by Members of the Johns' Hopkins University," that $(l \text{ ƒ } \breve{\bar{l}}) \cdot l\breve{\bar{l}}$ which means 'lover of everything loved by and of something else,' conformed to the syllogistic formula. In fact, that any transitive relation whether consistent with identity or not would give a form of conclusion similar to the syllogistic formula had been noticed by logicians long and long before. But I first analyzed the matter and showed that it depended on the following points.

First, writing *lb* for lover of a benefactor I called it (though I was not the first so to regard it) *relative multiplication* of the relative *l*, "lover of" into *b*, "benefactor of."

Second, I was the first to introduce what I called *relative addition*, writing *l* ƒ *b* for "lover of everything except benefactors of" and I proved that relative addition and relative multiplication were subject to this rule; namely, the *s* being any third relative, say "servant of" from the premiss *A* is (*l* ƒ *b*)*s* to *B*, i.e. *A* stands to some servant of *B* in the relation of loving everything except benefactors of him, there follows *A* is *l* ƒ *bs* to *B* that is *A* is a lover of everything except benefactors of servants of *B*. And in the same way from *s*(*l* ƒ *b*) we can infer *sl* ƒ *b*.

In the third place I noticed that from *A* $l\breve{\bar{l}}$ to *B*, "i.e. (the straight mark over the second *l* signifying *negation* and the curved mark signifying the *converse* relation) from A is a lover of something not loved by B, we can infer A is n to B, i.e. A is other than (or *not*) B; while in the 4th place, from A is $l\,f\,n$ to B, i.e. A stands to B in the relation of being lover of everything except what is other than him we can infer A is l to B or A is a lover of B.

Combining these four points we have, as I showed, in the first place from the premises A is $(l\,f\,\breve{l})$ to B or 'A stands to B in the relation of loving everything except what is not loved by him,' and B is $(l\,f\,\breve{l})$ to C, the conclusion

$$A \text{ is } (l\,f\,\breve{l})\,(l\,f\,\breve{l}) \text{ to } C$$

or A stands to C in the relation of loving everything loved by something that loves everything loved by him (C).

But then by my second point this gives

$$A \text{ is } l\,f\,\breve{l}(l\,f\,\breve{l}) \text{ to } C$$

or A stands to C in the relation of loving everything except what is not loved by something that stands to C in the relation of loving everything except what is not loved by him.

And by the same rule I can further infer

$$A \text{ is } l\,f\,\breve{l}l\,f\,\breve{l} \text{ to } C$$

or A stands to C in the relation of loving everything except what stands to everybody except perhaps not to what is not loved by C in the relation of not being loved by something that loves him (i.e. that *any* body).

And this by the 3rd point justifies the inference

$$A \text{ is } l\,f\,n\,f\,\breve{l} \text{ to } C$$

i.e. A stands to C in the relation of loving everybody except (perhaps) what is other than everything except what is not loved by C.

And then further by the 4th point we can further infer

$$A \text{ is } l\,f\,\breve{l} \text{ to } C$$

I further showed that using $\prec$ to mean "implies" and a dot as above explained, whatever relatives (simple or complex) x, y, and z may be

$$(x \cdot y)\, z \prec (xz) \cdot (yz)$$

and whatever relatives u, v and w may be

$$u(v \cdot w) \prec (uv) \cdot (uw)$$

whence

$$\begin{aligned}[l\breve{l} \cdot (l\,f\,\breve{l})]\,[l\breve{l} \cdot (l\,f\,\breve{l})] &\prec l\breve{l}l\breve{l} \cdot l\breve{l}(l\,f\,\breve{l}) \cdot (l\,f\,\breve{l})\,l\breve{l} \cdot (l\,f\,\breve{l})\,(l\,f\,\breve{l}) \\ &\prec l\breve{l}(l\,f\,\breve{l}) \cdot (l\,f\,\breve{l})\,l\breve{l} \cdot (l\,f\,\breve{l})\,(l\,f\,\breve{l}) \\ &\prec l(n\,f\,\breve{l}) \cdot (l\,f\,n)\,\breve{l} \cdot (l\,f\,n\,f\,\breve{l}) \\ &\prec l\breve{l} \cdot (l\,f\,\breve{l})\end{aligned}$$

So that this is also a transitive relation.

But it seems to me that in referring to the passage in your psychology, you should call attention to its reference to a kind of relations that have quite escaped the notice of the logicians, which may be called *relations of diminuent transitivity* of which there are three kinds: 1st, the *extensively diminuent*, such *lover of MOST of the things loved by*; 2nd those of *intensively diminuent transitivity* loving in a lower degree the things loved by. Then there are relations both extensively and intensively diminuent in transitivity. The extensively diminuent will include relations of *probable* transitivity. All are plainly hinted at in your paragraph. When you say the principle cannot be applied to concrete objects with any certainty, I cannot see anything peculiar to these relations in that. No *a priori* principle, — I may say no principle at all, — can be applied without a preliminary inquiry into the question of its applicability to the case in hand. There is no other failure of the principle as far as I see. It is nonsense to say that a mathematical axiom "fails" because the actual conditions are not those that the axiom specifies; and a person who undertakes to apply mathematics in such a way must be a downright lunatic, or at any rate a mind of the circle-squaring order.

Now as to Reality, I discussed my theory fully in a paper I wrote at your suggestion in Cambridge in which I explained in fresh language the doctrine of logical extension, denotation, or *Umfang* and comprehension, signification, or *Inhalt*. But although you admitted it was perfectly lucid, you professed yourself unable to comprehend a word of it. I shall however repeat a little of it in order to show how very far I am from what appears to be the doctrine of Bergson and Schiller. It is necessary first of all to keep sharply distinguished in one's mind what nevertheless are very often confounded. The only sure way of avoiding all confusion is to start with a definition of a sign. My definition is:

A Sign is a Cognizable that, on the one hand, is so determined (i.e. specialized, *bestimmt*), by something *other than itself*, called its Object (or, in some cases, as if the Sign be the sentence 'Cain killed Abel,' in which Cain and Abel are equally Partial Objects, it may be more convenient to say that that which determines the Sign is the Complexus, or Totality, of Partial Objects. And in every case the Object is accurately the Universe of which the Special Object is member, or part), while, on the other hand, it so determines some actual or potential Mind, the determination whereof I term the Interpretant created by the Sign, that that Interpreting Mind is therein determined mediately by the Object.

This involves regarding the matter in an unfamiliar way. It may be asked, for example, how a lying or erroneous Sign is determined by its Object, or how if, as not infrequently happens the Object is brought into existence by the Sign. To be puzzled by this is an indication of the word determine being taken in too narrow a sense. A person who says Napoleon was a lethargic creature has evidently his mind determined by Napoleon. For otherwise he could not attend to him at all. But here is a paradoxical circumstance. The person who interprets that sentence (or any other Sign whatsoever) must be determined by the Object of it through collateral observation quite independently of the action of the Sign. Otherwise he will not be determined to thought of that object. If he never heard of Napoléon before, the sentence will mean no more to him than that some person or thing to which the name "Napoléon" has been attached was a lethargic creature. For Napoleon cannot determine his mind unless the word in the sentence calls his attention to the right man and that can only be if, independently, habit has been established in him by which that word calls up a variety of attributes of Napoleon the man. Much the same thing is true in regard to any sign. In the sentence instanced Napoleon is not the only Object. Another Partial Object is Lethargy; and the sentence cannot convey its meaning unless collateral experience has taught its Interpreter what Lethargy is, or what that is that 'lethargy' means in this sentence. The Object of a Sign may be something to be created by the sign. For the Object of "Napoleon" is the Universe of Existence so far as it is determined by the fact of Napoleon being a Member of it. The Object of the sentence "Hamlet was insane" is the Universe of Shakespeare's Creation sofar as it is determined by Hamlet being a part of it. The Object of the Command "Ground arms!" is the immediately subsequent action of the soldiers sofar as it is affected by the molition expressed in the command. It cannot be understood unless collateral observation shows the speaker's relation to the rank of soldiers. You may say, if you like, that the Object is in the Universe of things desired by the Commanding Captain at that moment. Or since the obedience is fully expected. It is in the Universe of his expectation. At any rate, it determines the Sign although it is to be created by the Sign by the circumstance that its Universe is relative to the momentary state of mind of the Officer.

Now let us pass to the Interpretant. I am far from having fully explained what the Object of a Sign is; but I have reached the point where further explanation must suppose some understanding of what the Interpretant is. The Sign creates something in the Mind of the Interpreter, which something in that it has been so created by the sign has been, in a mediate and *relative* way, also created by the Object of the Sign, although the Object is essentially other than the Sign. And this creature of the sign is called the Interpretant. It is created by the Sign; but not by the Sign quâ member of whichever of the Universes it belongs to; but it has been created by the Sign in its capacity of bearing the determination by the Object. It is created in a Mind (how far this mind must be real we shall see). All that part of the understanding of the Sign which the Interpreting Mind has needed collateral observation for is outside the Interpretant. I do not mean by "collateral observation" acquaintance with the system of signs. What is so gathered is *not collateral.* It is on the contrary the prerequisite for getting any idea signified by the sign. But by collateral observation, I mean previous acquaintance with what the sign denotes. Thus if the Sign be the sentence "Hamlet was mad" to understand what this means one must know that men are sometimes in that strange state; one must have seen madmen or read about them; and it will be all the better if one specifically knows (and need not be driven to *presume*) what Shakespeare's notion of insanity was. All that is collateral observation and is no part of the Interpretant. But to put together the different subjects as the sign represents them as related — that is the main [result] of the Interpretant-forming. Take as an example of a Sign, a *genre* painting. There is usually a lot in such a picture which can only be understood by virtue of acquaintance with customs. The style of the dresses for example, is no part of the *significance*, i.e. the deliverance of the painting. It only tells what the *subject* of it is. *Subject* and *Object* are the same thing except for trifling distinctions (and the German fashion of making them mark a great cleavage in thought, is either a great blunder or it is a shocking instance of disregard of the morals of science, — which in fact they are apt not to "give a dam" for). But that which the painter aimed to point out to you, presuming you to have all the requisite collateral information, that is to say just the quality of the sympathetic element of the situation, generally a very familiar one — and something you probably never did so clearly realize before — *that* is the Interpretant of the Sign, — its "significance."

Now all this is, so far, very muddled for the lack of certain distinctions which I proceed to point out, though it will be hard to make them fully comprehended.

In the first place, it should be observed that so far as the Sign denotes its Object, it calls for no particular *intelligence* or *Reason* on the part of its Interpreter. To read the Sign at all, and distinguish one Sign from another, what is requisite is delicate perceptions and acquaintance with what the usual concomitants of such appearances are, and what the conventions of the system of signs are. To know the Object what is requisite is previous experience of that Individual Object. The Object of every sign is an Individual, usually an Individual Collection of Individuals. Its *Subjects*, i.e. the Parts of the Sign that denote the Partial Objects, are either *directions for finding the Objects* or are *Cyrioids*, i.e. signs of single Objects. [I form the word from Κύριον ὄνομα, a Proper Name, literally the *regular*, *legitimate*, *literal* name from Κῦρος, authority; which L. and S. says is with all its cognates entirely Post-Homeric, in spite of the fact that the root is the same as that of Latin *fortis*. But it seems very doubtful whether it has anything to do with Latin *fortis* = *forctis*.) Such for example are all *abstract* nouns, which are names of single characters, the personal pronouns, and the demonstrative and relative pronouns etc. By directions for finding the Objects, for which I have as yet invented no other word than "*Selectives*," I mean such as "Any" (i.e. any you please), "Some" (i.e., one properly selected), etc. To know the Interpretant, which is what the sign itself expresses, may require the highest power of reasoning.

In the second place, to get more distinct notions of what the Object of a Sign in general is, and what the Interpretant in general is, it is needful to distinguish two senses of "Object" and three of "Interpretant." It would be better to carry the division further; but these two divisions are enough to occupy my remaining years. Others must carry the study further when I am gone, which will be, I fear, all too soon for me to explain what work I have done. (Indeed, there are some studies of years of which I never expect to be able to say a word: I can only pick out what few I can. If the Carnegie or some such rich institution would put me in the position in which happier students are placed it would be different. It is most irksome to me to feel myself dependent on persons who need all their money themselves. And at the same time, I cannot deny that it is a deep happiness to feel that I have the warm friendship of such a soul as you; while I wish I could dispense with putting you to unknown amounts of trouble and expense and annoyance.)

As to the Object, that may mean the Object as cognized in the Sign and therefore an Idea, or it may be the Object as it is regardless of any particular aspect of it, the Object in such relations as unlimited and final study would show it to be. The former I call the *Immediate* Object, the latter the *Dynamical* Object. For the latter is the Object that Dynamical Science (or what at this day would be called "Objective science) can investigate. Take for example, the sentence "the Sun is blue." Its

Objects are "the Sun" and "blueness." If by "blueness" be meant the Immediate Object, which is the quality of the sensation, it can only be known by Feeling. But if it means that "Real," existential condition, which causes the emitted light to have short mean wave-length, Langley has already proved that the proposition is true. So the "Sun" may mean the occasion of sundry sensations, and so is Immediate Object, or it may mean our usual interpretation of such sensations in terms of place, of mass, etc. when it is the Dynamical Object. It is true of both Immediate and of Dynamical Object that acquaintance cannot be given by a Picture or a Description, nor by any other sign which has the Sun for its Object. If a person points to it and says, See there! *That* is what we call the "Sun," the Sun is *not* the Object of that sign. It is the *Sign* of the sun, the *word* "sun" that his declaration is about; and that *word* we must become acquainted with by collateral experience. Suppose a teacher of French says to an English speaking pupil, who asks "Comment appelle-t-on ça?" pointing to the Sun; and the teacher replies "C'est le soleil," he begins to furnish that collateral experience by speaking in French of the Sun itself. Suppose, on the other hand he says "Notre mot est 'soleil'" then instead of expressing himself in language and *describing* the word he offers a pure *Icon* of it. Now the Object of an Icon is entirely indefinite equivalent to "something." He virtually says "Our word is like this:" and makes the sound. He informs the pupil that the word (meaning, of course, a certain *habit*) has an effect which he *pictures* acoustically. But a pure picture without a legend only says "*something*" is like this: True he attaches what amounts to a legend. But that only makes his sentence analogous to a portrait we will say of Leopardi with Leopardi written below it. It conveys its information to a person who knows who Leopardi was, and to anybody else it only says "something called Leopardi looked like this." The pupil is in the state of a person who was pretty sure there was a man Leopardi; for he is pretty sure there must be a word in French for the sun and thus is already acquainted with it, only he does not know how it sounds when spoken nor how it looks when written. I think by this time you must understand what I mean when I say that no sign can be understood, — or at least that no *proposition* can be understood unless the interpreter has "collateral acquaintance" with every Object of it. As for a mere *substantive*, it must be borne in mind that it is not an indispensable part of speech. The Semitic languages seem to be descendants of a language that had no "common nouns." Such a word is really nothing but a *blank form* of proposition and the Subject is the blank, and a blank can only mean "something" or something even more indefinite. So now I believe I can leave you to consider carefully whether my doctrine is correct or not.

As to the Interpretant, i.e. the "signification," or "interpretation" rather, of a sign, we must distinguish an Immediate and a Dynamical, as we must the Immediate and Dynamical Objects. But we must also note that there is certainly a third kind of Interpretant, which I call the Final Interpretant, because it is that which *would* finally be decided to be the true interpretation if consideration of the matter were carried so far that an ultimate opinion were reached. My friend Lady Welby has, she tells me, devoted her whole life to the study of *significs*, which is what I should describe as the study of the relation of signs to their interpretants; but it seems to me that she chiefly occupies herself with the study of words. She also reaches the conclusion that there are three senses in which words may be interpreted. She calls them *Sense*, *Meaning*, and *Significance*. Significance is the deepest and most lofty of these, and thus agrees with my *Final Interpretant*; and Significance seems to be an excellent name for it. *Sense* seems to be the logical analysis or definition, for which I should prefer to stick to the old term *Acception* or *Acceptation*. By *Meaning* she means the *intention* of the utterer. But it appears to me that all symptoms of disease, signs of weather, etc. have no utterer. For I do not think we can properly say that God *utters* any sign when He is the Creator of all things. But when she says, as she does that this is connected with Volition; I at once note that the volitional element of Interpretation is the *Dynamical Interpretant*. In the Second Part of my Essay on Pragmatism, in the Pop. Sci. of 1877 Nov. and 1878 Jan. I made three grades of Clearness of Interpretation. The first was such Familiarity as gave a person familiarity with a sign and readiness in using it or interpreting it. In his consciousness he seemed to himself to be quite *at home* with the Sign. In short, it is Interpretation *in Feeling*. The second was Logical Analysis = Lady Welby's *Sense*. The third was Pragmatisistic Analysis [which] would seem to be a Dynamical Analysis but identified with the Final Interpretant.

At present, I am decidedly inclined to think that when we speak of Logical Analysis of a Concept and a Real Definition of it what we generally mean is analogous to Color-analysis; and in Color Analysis we take a sensation of color, and throw two lights together upon a surface so that physically they certainly do not affect one another at all, and then the *color* so produced which is the effect of matching under a given intensity of illumination, that is, when the photometrically measured luminosity measure (which will hardly differ in the final result from any other measure of it that makes the luminosity of the aggregate of two lights equal to the sum of their luminosities when separate) this color which will be of widely different apparent hues according as the luminosity varies is said to be "composed" of the two lights that make it up. But this gives the same result as if, always working at the same apparent photometric luminosity we start with any 4 colors such that no one can result from mixing any two of the others in any proportion whatsoever, and calling them A, B, C, and X, we lay down on paper 4 points to represent them. There are then two cases. Either there is one of the 4 which can be produced by mixing the other 3 in that case we call that one X and represent it by a point in the interior of the triangle ABC; or else there is no such color among the four. In that case the colors can be separated into two pairs such that there is a color which can be produced by mixing the two colors of either pair. In that case we represent the colors by the vertices of a quadrilateral each pair of colors being represented by two opposite vertices of the quadrilateral. In the former case, we may arbitrarily write $X = \dfrac{A+B+C}{3}$. In the latter case, the two pairs being AC and BX we may write $X = A + C - B$. We may then go on in either case on the principle that every color is represented by a trinomial in A, B, C which has the sum of its three coefficients equal to 1 and of any three colors, of which one can be produced by a mixture of the other 2, that one shall be represented by the middle point of three points in one straight line, the two colors mixed being the other two of the three points. And moreover the algebraic representation of the three shall be related as $\dfrac{pA + qB + rC}{p+q+r}$, $\dfrac{uA + vB + wC}{u+v+w}$, and $\dfrac{(mp+nu)A + (mq+nv)B + (m+nw)C}{m(p+q+r) + n(u+v+w)}$. We can then go on and represent *all* colors in that way. Thus in the first case [Fig. 1].

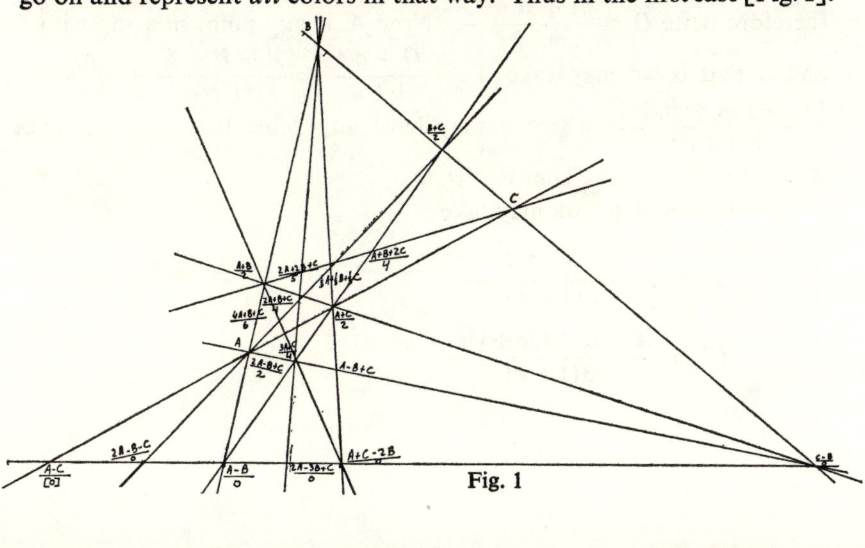

Fig. 1

By this beautiful *barycentric* notation, as it is called, several fundamental theorems can readily be proved, that geometers had a great deal of trouble in proving otherwise except by dragging in considerations about the lengths of lines which considerations are utterly irrelevant to the theorems. Here are two of these called the nine-point and the ten-point theorems.

Through one point, 0, draw three rays (i.e. unlimited straight lines) and mark two other points on each; say A and A' on one, B and B' on another, and C and C' on the third [Fig. 2]. Complete the triangles ABC

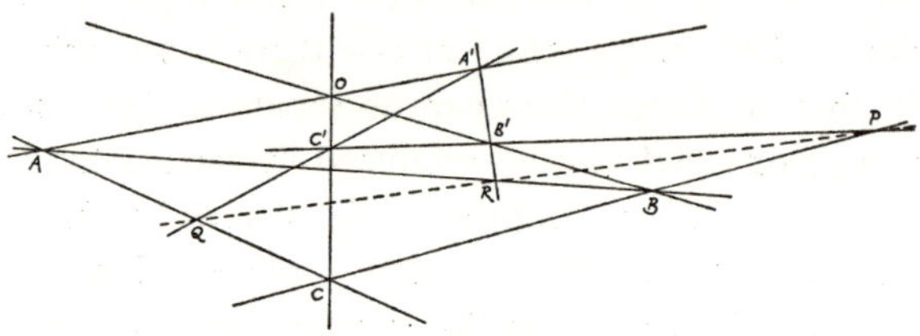

Fig. 2

and $A'B'C'$ by drawing (I write two points between parentheses to denote the ray through them, and I write the signs of two rays between square brackets to denote their point of intersection) by drawing, I say, the six rays $(AB)(BC)(CA)$ $(A'B')(B'C')(C'A')$ then the three points $P = [(BC)(B'C')]$, $Q = [(CA)(C'A')]$, and $R = [(AB)(A'B')]$ lie on one ray. That is, $(PQ) = (QR) = (RP)$.

The proof is as follows: We may put $0 = \dfrac{uA + vB + wC}{u + v + w}$ and give u, v, w any values we please so long as none of them are zero. For we can always do this for any one fourth point besides those that we take as standard wherewith to express all other points in the plane. We will therefore write $O = \dfrac{A + B + C}{3}$. Now A' is any point in a ray with O and A that is we may take $A' = \dfrac{O + aA}{1 + a} = \dfrac{A + B + C}{3(1 + a)} + \dfrac{aA}{1 + a} = \dfrac{(3a + 1)A + B + C}{3(1 + a)}$ where a may have any value that will *not* make $A' = O$ nor $A' = A$. That is $a \neq O$.

Then by symmetry we may take

$$B' = \frac{A + (3b + 1)B + C}{3(1 + b)}$$

$$C' = \frac{A + B + (3c + 1)C}{3(1 + c)}$$

where $b \neq O$ and $c \neq O$. Now P is on a ray through B and C. It must therefore be of the form $\xi B + \bar{\xi} C$ where I make a line over a letter to denote 1 *minus* that letter. But P is also on the ray through B' and C'. There must therefore be some value of x such that $P = \xi B + \bar{\xi} C = x\dfrac{A + (3b+1)B + C}{3(1+b)} + \bar{x}\dfrac{A + B + (3c+1)C}{3(1+c)}$. The sum of the coefficients of A must disappear because there is no A in the first expression for P and there is only one trinomial in A, B, C to denote any one point. Consequently $\dfrac{x}{3(1+b)} + \dfrac{\bar{x}}{3(1+c)} = 0$ or $\dfrac{x(1+c) + (1-x)(1+b)}{3(1+b)(1+c)} = 0$. That is $x(1+c) - x(1+b) + (1+b) = 0$ or $x = \dfrac{b+1}{b-c}, \bar{x} = \dfrac{c+1}{c-b}$.

$$\text{Hence } P = \left[\frac{b+1}{b-c}\cdot\frac{3b+1}{3(b+1)} + \frac{c+1}{c-b}\cdot\frac{1}{3(c+1)}\right]B + \left[\frac{b+1}{b-c}\cdot\frac{1}{3(b+1)} + \frac{c+1}{c-b}\cdot\frac{3c+1}{3(c+1)}\right]C = \frac{3b}{3(b-c)}B - \frac{3c}{3(b-c)}C.$$

And by symmetry

$$Q = \frac{3c}{3(c-a)}C - \frac{3a}{3(c-a)}A$$

$$R = \frac{3a}{3(a-b)}A - \frac{3b}{3(a-b)}B$$

Now then if there is any quantity t such that $tP + \bar{t}Q = R$, [then] P, Q and R lie on one ray. Take $t = \dfrac{b-c}{b-a}, \bar{t} = \dfrac{c-a}{b-a}$

$$tP + \bar{t}Q = \frac{3b}{3(b-a)}B - \frac{3c}{3(b-a)}C + \frac{3c}{3(b-a)}C - \frac{3a}{3(b-a)}A = \frac{3a}{3(a-b)}A - \frac{3b}{3(a-b)}B = R$$

Q.E.D.

That is the ten-point theorem.[2] The nine-point theorem is as follows. On any two rays in the plane, take any three points, A, B, C on one and any three A', B', C', on the other. Draw the rays (AB') $(B'C)$ (CA')

[2] Also known as Desargues' Theorem —

Triangles ABC and $A'B'C'$ are perspective from 0

$\therefore$ They are perspective from a line (PQR).

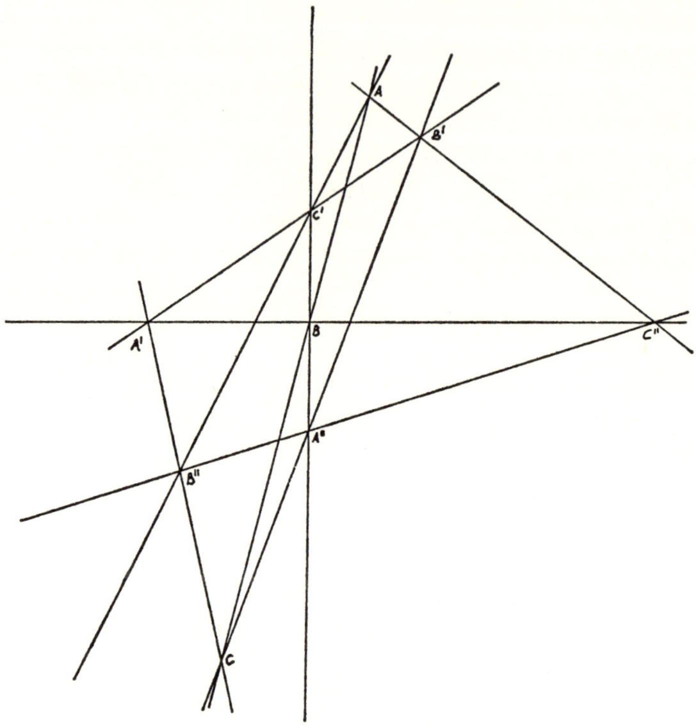

Fig. 3

$(A'B)\ (BC')\ (C'A)$. Then $[(AB')\ (A'B)]$, $[(BC')\ (B'C)]$, and $[(CA')\ (C'A)]$ lie on one ray. ... [Fig. 3].[3]

If you have ever had your attention called to this famous nine-point theorem, you have doubtless seen a much better and easier demonstration of it than the one I am about to give, since it is only a corollary from Pascal's *hexagram*. But my purpose being to illustrate the elementary handling of the barycentric notation I prefer the one I give because it does not soar into conic sections.

And by the way, as algebra I believe is one of your *bêtes noires*, let me say that it wouldn't pay at all to dig through the algebraic work. I give it, because this letter may come to be read by somebody who would be better satisfied to see it; but if it is not *easily* grasped in a moment, it would only distract the attention. Suffice it, for most readers, to be assured that it is here.

I now restate the theorem. *In a plane* (this is not immediately extensible

[3] In the manuscript Peirce draws a substitute diagram and colors the lines as indicated in the presentation of the Pascal hexagram which follows.

to 3 dimensions as the 10-point theorem is. The latter is easiest proved by first showing that it holds when the three rays first drawn are *not* in a plane) let there be two black rays $(A'B'C')$ and $(A''B''C'')$ and let them meet in $[P]$ [use Fig. 3]. On the former, mark any three different points $[A']$, $[B']$, $[C']$; and on the latter any three all different $[A'']$, $[B'']$, $[C'']$. Draw the six rays $(A'B'')$, $(B''C')$, $(C'A'')$, $(A''B')$, $(B'C'')$, $(C''A')$, and mark the three points $[(A'B'')(A''B')]$, $[(B'C'')(C'B'')]$, $[(C'A'')(A'C'')]$ [Points C, A, B in Fig. 3]. Then these three points will lie on one ray. Let us express the conditions of the proposition barycentrally in terms of $[P]$, $[B']$, $[B'']$ taking a', a'', c', c'' so that

$$[A'] = \bar{a}'P + a'B', [A''] = \bar{a}''P + a''B''$$
$$[C'] = \bar{c}'P + c'B', [C''] = \bar{c}''P + c''B''$$

Taking ϱ_1, ϱ_2, ϱ_3 for coefficients varying along the red (ῥόδεος) rays $(A'B'')$, $(C'A'')$, $(B'C'')$ and κ_1, κ_2, κ_3 for coefficients varying along the blue (κυανοῦς) rays $(B''C')$, $(A''B')$, $(C''A')$

Any point on $(A'B'')$ is $\bar{\varrho}_1\bar{a}'P + \bar{\varrho}_1 a'B' + \varrho_1 B''$
Any point on $(B''C')$ is $\bar{\kappa}_1\bar{c}'P + \bar{\kappa}_1 c'B' + \kappa_1 B''$
Any point on $(C'A'')$ is $(\bar{\varrho}_2\bar{c}' + \varrho_2\bar{a}'')P + \bar{\varrho}_2 c'B' + \varrho_2 a''B''$
Any point on $(A''B')$ is $\bar{\kappa}_2\bar{a}''P + \kappa_2 B' + \bar{\kappa}_2 a''B''$
Any point on $(B'C'')$ is $\varrho_3\bar{c}''P + \bar{\varrho}_3 B' + \varrho_3 c''B''$ and
Any point on $(C''A')$ is $(\bar{\kappa}_3\bar{c}'' + \kappa_3\bar{a}')P + \kappa_3 a'B' + \bar{\kappa}_3 c''B''$

Consequently, for the point $[(A'B'')(A''B')]$ we have the equation

$$\bar{\varrho}_1\bar{a}'P + \bar{\varrho}_1 a'B' + \varrho_1 B'' = \bar{\kappa}_2\bar{a}''P + \kappa_2 B' + \bar{\kappa}_2 a''B''$$

that is to say $\bar{\varrho}_1\bar{a}' = \bar{\kappa}_2\bar{a}''$, $\bar{\varrho}_1 a' = \kappa_2$, and $\varrho_1 = \bar{\kappa}_2 a''$, so that, for this point

$$\kappa_2 = \frac{a'(1-a'')}{1-a'a''} \quad \text{and} \quad \bar{\kappa}_2 = \frac{1-a'}{1-a'a''}$$

and $$[(A'B'')(A''B')] = \frac{(1-a')(1-a'')}{1-a'a''}P + \frac{a'(1-a'')}{1-a'a''}B' \div \frac{(1-a')a''}{1-a'a''}B''$$

In the same way, for the point $[(B'C'')(B''C')]$ we have the equation

$$\varrho_3\bar{c}''P + \bar{\varrho}_3 B' + \varrho_3 c''B'' = \bar{\kappa}_1\bar{c}'P + \bar{\kappa}_1 c'B' + \kappa_1 B''$$

that is to say,

$$\varrho_3\bar{c}'' = \bar{\kappa}_1\bar{c}', \ \bar{\varrho}_3 = \bar{\kappa}_1 c', \ \varrho_3 c'' = \kappa_1$$

whence $$\kappa_1 = \frac{(1-c')c''}{1-c'c''} \quad \bar{\kappa}_1 = \frac{1-c''}{1-c'c''}$$

and $[(B'C'')(B''C')] = \frac{(1-c')(1-c'')}{1-c'c''}P + \frac{(1-c'')c'}{1-c'c''}B' + \frac{(1-c')c''}{1-c'c''}B''$

In the same way again, for the point $[(C'A'')(A'C'')]$ we have

$$(\bar{\varrho}_2\bar{c}' + \varrho_2\bar{a}'')P + \bar{\varrho}_2c'B' + \varrho_2a''B'' = (\bar{\kappa}_3\bar{c}'' + \kappa_3\bar{a}')P + \kappa_3a'B' + \bar{\kappa}_3c_3''B''$$

that is to say,

$$\bar{\varrho}_2\bar{c}' + \varrho_2\bar{a}'' = \bar{\kappa}_3\bar{c}'' + \kappa_3\bar{a}', \ \bar{\varrho}_2c' = \kappa_3a', \ \varrho_2a'' = \bar{\kappa}_3c''$$

whence $\kappa_3 = \frac{(a''-c'')c'}{a'a''-c'c''}, \quad \bar{\kappa}_3 = \frac{(a'-c')a''}{a'a''-c'c''}.$

and

$$[(C'A'')(A'C'')] = \frac{(a'a''\bar{c}'\bar{c}'' - \bar{a}'\bar{a}''c'c'')P + a'c'(a''-c'')B' + (a'-c')a''c''B''}{a'a''-c'c''}$$

It will now be convenient

to call the point $[(A'B'')(A''B')]$ by the single letter C
to call the point $[(B'C'')(B''C')]$ by the single letter A
and to call the point $[(C'A'')(C''A')]$ by the single letter B

They are so marked on the diagram. They there appear to lie on one ray. But that may be owing to an inexactitude of the drawing. To find out whether or no they be so, then if we take any one of the three terms in each of their expressions in terms of P, B', B'' and find out in what proportions the terms, say, in B'' have to be multiplied in order that the sum of the products may vanish. Then if the same multiplier will cause the sum of the products of the terms in B' to vanish, it must necessarily be that the same will be true of the terms in P, since the sum of the coefficients is always 1. Calling the multipliers c, a, b respectively then, we shall have $cC + aA + bB = 0$ where $c + a + b = 0$ which shows that A, B, C, lie on one ray, for then $c = -(a+b)$ and dividing by $a+b$ we get $-C + \frac{a}{a+b}A + \frac{b}{a+b}B = 0$ or $C = \frac{a}{a+b}A + \frac{b}{a+b}B$, which is the expression of C as a point on (AB).

I have now sufficiently exhibited the characteristics of the *barycentral* notation. For a considerable time I was much occupied by the question whether or not a notation similar to this would not represent the modes in which concepts are, or should be, represented as compounded in definitions, with a leaning to the affirmative. Among the arguments which supported that leaning were the following:

Argument 1st. That the working of this notation has a remarkable similarity with that of a logical calculus, in that there is no measurement in it. Thus, $\frac{9}{10} A + \frac{1}{10} B$ represents any point we choose on the ray (*AB*) provided the rules of algebra are observed. Nor is there any need of working out the algebraic transformations in detail. This *feeling* of the affinity between the two weighed much with me.

Argument 2nd. I have long felt that it is a serious defect in existing logic that it takes no heed of the *limit* between two realms. I do not say that the Principle of Excluded Middle is downright *false*; but I *do* say that in every field of thought whatsoever there is an intermediate ground between *positive assertion* and *positive negation* which is just as Real as they. Mathematicians always recognize this, and seek for that limit as the presumable lair of powerful concepts; while metaphysicans and oldfashioned logicians, — the sheep and goat separators, — never recognize this. The recognition does not involve any denial of existing logic but it involves a great edition to it to recognize such a mode of logic that it takes no heed of the *limit* between two realms. I do not say composition of concepts as that of the barycentric calculus would be one way of recognizing the idea of the limit of lower dimensionality between any two mutually exclusive fields.

Argument 3rd. I have long urged that, in itself considered, any one concept is just as simple as any other. Now if we recognize that concepts are compounded to make other concepts in the same way, or in any analogous way, with that in which in the barycentric calculus two points are compounded to make a point, is to recognize that.

But notwithstanding these and other arguments, I find myself bound, in a way which I discovered in the sixties, to recognize that there are concepts which, however we may attempt to analyze them, will always be found to enter intact into one or the other or both of the components into which we may fancy that we have analyzed them. Such concepts then are simple in the sense in which the chemical elements are simple, namely, in the sense that we are powerless to analyze them. Moreover, the system of Existential Graphs is the only system that does to perfection that which all logical algebras have aimed to do. It has just the number of fundamental signs that are requisite. It recognizes no transformations as primarily permissible except such as must in any case be recognized; namely *erasures* under certain simple conditions and *insertions* in equally simple conditions. Its sole apparent violation of its Idea is that if the innermost ligature binding together two graphs is not wholly within the innermost area containing both those graphs, it has no effect thus 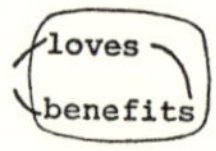 means more than 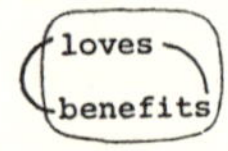

But 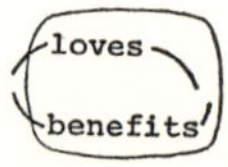 means *no* more than 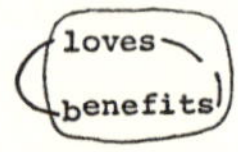

I expect, however, that I shall ultimately discover that this rule only holds under certain special conditions; and that in short, it only appears to hold because a requisite premiss has not been noticed. While that "Algebra of Dyadic Relations" with which Schröder fell so in love was obliged to provide four fundamental symbols of operation to account for the composition of concepts by non-relative aggregation, by non-relative multiplication, by relative aggregation, and by relative multiplication, the System of Existential Graphs includes all of these under the sole mode of composition it recognizes, — and without *any* special symbol; with but the ligature. The following are examples

In Aryan Syntax	In the algebra of Dyadic Relatives	In Existential Graphs
A loves, but does not benefit, *B*.	$\breve{A}(l \cdot \bar{b})B$	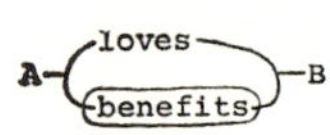
A loves a non-benefactor of *B*.	$\breve{A}l\bar{b}B$	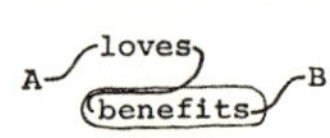
A loves something unbenefitted by *B*.	$\breve{A}l\breve{\bar{b}}B$	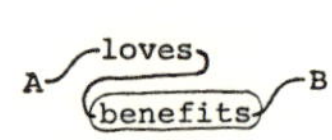
A loves whomsoever benefits *B*.	$\breve{A}(l \mathbin{\text{ʃ}} \bar{b})B$	
A loves nobody but benefactors of *B*.	$\breve{A}(\bar{l} \mathbin{\text{ʃ}} b)B$	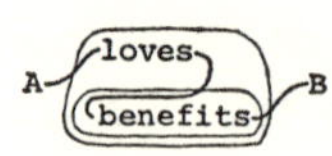

The location of the spots is immaterial.

Existential Graphs, besides, expresses much that I should be puzzled to express in the Algebra of Dyadic Relatives; such as the following. "*A* maligns nobody but lovers of those to whom *B* does not prefer the person to whom *A*'s malignment of them is addressed." [Fig. 4].

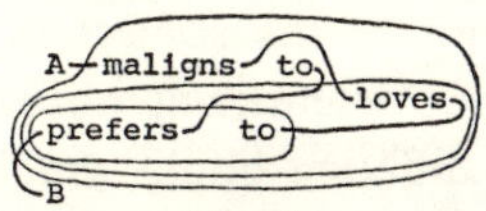

Fig. 4

Yet the System of Existential Graphs escapes the opposite fault into which my "General Algebra of Logic" falls; namely that of representing as equally simple assertions that, in fact, are far from being so. The comparison of the three following will illustrate this, which ramifies in various directions, on account of that Algebra being determined by mere external or formal relations, and forcing everything into a Procrustean bed [Fig. 5].

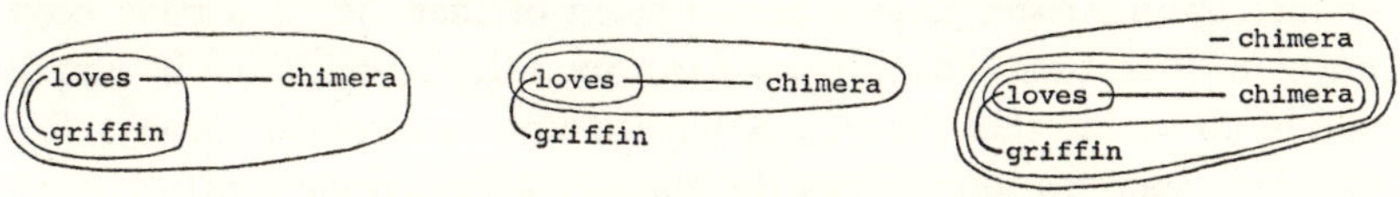

Fig. 5

These diagrams in "general algebra of logic" represent

$$\pi_j \Sigma_i (g_i \cdot l_{ji} \Psi \bar{c}_j), \Sigma_i \pi_j g_i \cdot (l_{ji} \Psi \bar{c}_j), \Sigma_i \pi_j g_i \cdot l_{ji} \Psi \bar{c}$$

In "Aryan syntax" they represent each in turn:

Whatever chimera there may be is loved by some griffin or other.

One griffin loves whatever chimera there may be.

All the chimera there may be are loved by some griffin.

I hope and believe that it is not seriously indispensable that I should resort to padding in order to give this letter a decent length. Otherwise, I might here make an Excursus about the relation of Genius to Insanity. I could say that on the one hand I do not think that definition of genius which, in my college days, I used to hear attributed to Dr. Johnson, although I do not find it in the (pre-Todd-ian) edition of his Dictionary, "Genius consists in great general powers accidentally determined in a particular direction," needs much emendation to render it satisfactory. I think *good* general powers would be sufficient, but that *activity* of mind is essential, and that the "accidental determination" ought to be explained as either congenital or acquired. On the other hand, I think Lombroso's thesis, — though I don't know why it should be called his, — that, "as Dryden says, "great wits are sure to madness near allied," is usually true, too. Lombroso's Book is, with the exception of some circle-squaring productions quite the most illogical I ever read, — in spite of all the metaphysics, cosmology, and logic that I have dug into; and his first chapter is a fair sample of the whole. I have, to be sure, only the English translation, but in that translation he appears as pretending to quote from the works of Aristotle in one of which he seems to say that the Stagyrite said that Plato was disposed to Insanity. It is true that Aristotle makes some slight reference to the subject in three of his works, the Rhetoric, the Poetic and the Problems; but the two passages that Lombroso represents as drawn from different books are from the same chapter of the problems, one on p. 953, the other on p. 954 of the Berlin edition. Aristotle quotes lines 200-202 of the 6th Book of the Iliad about Bellerophon, translated by Cowper thus:

> "But when Bellerophon, at last, himself
> Had anger'd all the gods, feeding on grief
> He roam'd alone the Aleian field, exiled
> By choice, from every cheerful haunt of man;"

and thereupon Aristotle remarks that men of that diathesis have been known in "our time," as Empedocles, Socrates, and Plato. It simply amounts to saying that Plato shut himself up and avoided society. We know that Socrates didn't even do that. It would argue Aristotle mad himself if he had called Plato mad; but he has never been so understood.

But Lombroso does not quote the most oft-quoted ancient support of his doctrine, the sentence of Seneca's in his De Tranquillitate Animi: "Nullum magnum ingenium sino mixtura dementiae fuit," nor yet the still more striking, true line of Juvenal: "Nemo mathematicus genium indemnatus habebit." However, the mathematical cuss who is cursed with genius *does* have some gratifications in the delusion that his genius is a part of himself and that if *it* is admirable, *he* must be something very superior. (He who carries the burden of a genius for logical analysis sees perfectly, sees too clearly, that it is nothing but a mental parasite, no more implying any worth of character, nor any good mark for himself than being drawn for a juror; but, like that, demanding considerable inconvenience, self-restraint, and responsibility; while he is forbidden to feel all that to be other than a privilege.)

Another (apparently) strong argument in favor of the theory that concepts are combined in a form substantially like that of the combinations of points in the barycentric calculus, is one about which my recollections have suggested my remarks about genius. I may call such a mode of combination "barycentral," and I may remark that it is in barycentral composition that forces are combined according to the principle of the parallelogram of forces. I will now recount the history of my knowledge of the subject. About 1869 my studies of the composition of concepts had got so far that I very clearly saw that all *dyadic* relatives could be combined by ways capable of being represented by *addition* (and of course subtraction), by a sort of multiplication such that $(x + y)z = xz + yz$ and

$x(y+z) = xy + xz$ and $(xy)z = x(yz)$ (although xy was not generally the same as yx any more than they are the same in quaternions) and by two kinds of involution which I represented, the following year, by x^y and ${}^xy$, which called "forward" and "backward" involution, subject to the formulae $x^{y+z} = x^y \cdot x^z$ and ${}^{x+y}z = {}^xz \cdot {}^yz$, $(x^y)^z = x^{(yz)}$ and ${}^x({}^yz) = {}^{(xy)}z$ and $({}^xy)^z = {}^x(y^z)$. But I found my mathematical powers were not sufficient to carry me further, though I saw with tolerable distinctness, that the system was capable of extension to *triadic* relatives; only in that case there would have to be occasionally introduced a sort of multiplication in which $(xy)z$ was not equal to $x(yz)$, just as we write $\sin(yz)$ which is not equal to $(\sin y)z$. I therefore set to work talking incessantly to my father (who was greatly interested in quaternions) to try to stimulate him to the investigation of all the systems of algebra which, instead of the multiplication-table of quaternions which is [Fig. 6], had some other

	l	i	j	k
l	l	i	j	k
i	i	$-l$	k	$-j$
j	j	$-k$	$-l$	i
k	k	j	$-i$	$-l$

Fig. 6

more or less similar multiplication table. I had hard work at first. It evidently bored him. But I hammered away, and suddenly he became interested and soon worked out his great book on "Linear Associative Algebra." The next year I was in Europe and wrote several letters to my father on the subject. But one long and important epistle was lost in the mail. The rest he had printed. Though the lost one contained a demonstration of a theorem which demonstration I subsequently printed, the theorem itself was known, though not known to be certainly true, except to myself. It was that any such algebra *of a finite number* of fundamentally different kinds of quantities so that no one of them, say i, could be expressed as equal to a polynomial in the others with numerical coefficients; as $i = a \cdot j + b \cdot k + cl +$ etc. where a, b, c are numbers and j, k, l etc. are relatives could be brought to such a form that every quantity in it is expressed in the form

$$\begin{aligned} &a_{11}(A:A) + a_{12}(A:B) + a_{13}(A:C) + \text{etc.} \\ &+ a_{21}(B:A) + a_{22}(B:B) + a_{23}(B:C) + \text{etc.} \\ &+ a_{31}(C:A) + a_{32}(C:B) + a_{33}(C:C) + \text{etc.} \\ &+ \text{etc.} \end{aligned}$$

where the small letters with subjacent numbers are real numbers (or as my father understood me real or imaginary, but that is of no consequence, for imaginaries are easily so expressed as reals) and where $(A:A)$, $(A:B)$ etc. which I will here call capital-pairs, follow this rule of multiplication $(X:Y)(Y:Z) = (X:Z)$. That is the product of two capital-pairs, which are such that the second capital of the multiplier is the same as the first capital of the multiplicand, is the capital-pair whose first capital is the same as that of the multiplier and its second that of the multiplicand. But $(X:U)(V:Z) = 0$ i.e. the product of two capital pairs in which the second capital of the multiplier is different from the first capital of the multiplicand is zero.

The proof of this is the simplest conceivable. Yet Sylvester, who though a great mathematician and an amiable man to those he liked was petty to the last degree, and in particular, like some other mathematicians I have met, seemed to think that a man who had a divine genius for mathematics was *ipso facto* a superior man. Sylvester, I say, had such a colossal genius for *not* seeing the force of any demonstration that excited his jealousy that he continued to say that he could not see that I had proved my theorem, until he began to discern that he was becoming the laughing-stock of all who had studied his subject, and then he only held his peace, but did not explicitly acknowledge the proof. Here that proof is. It relates to those systems of algebra which have a number of *linearly independent units*; that is to say, if i_1, i_2, i_3 etc. are a set of such, no one of them as i_1 can be expressed as a polynomial in the others $i_1 = c_2 i_2 + c_3 i_3 +$ etc. where c_2, c_3, etc. are numbers. If there is one such set there are an infinite number each of such a nature that every other quantity of the system can be expressed in the form $c_1 i_1 + c_2 i_2 + c_3 i_3 +$ etc. The whole character of the system is expressed in its multiplication-table. As a general rule $i_1 \times i_2$ is *not* equal to $i_2 \times i_1$ but the multiplication is *associative*; that is to say if α, β, γ are any quantities of the system and if $\alpha\beta = \mu$, $\beta\gamma = \nu$ then $\alpha\nu = \mu\gamma$ or $(\alpha\beta)\gamma = \alpha(\beta\gamma)$. The multiplication-table may be expressed in a general way by writing $i_m \cdot i_n$ where m and n may be any two numbers and i_m is the mth of the set of independent units while i_n is the nth unit of the set.

$$i_m \cdot i_n = [m, n, 1]i_1 + [m, n, 2]i_2 + [m, n, 3]i_3 + \text{etc.} = \Sigma_p[m, n, p]i_p$$

where the quantities in brackets are the numerical coefficients so expressed that one can see by their expression what they are coefficients of, the first of the 3 numbers denoting the multiplying unit, the second the multiplied, and the third that of which the number expressed by the 3 letters is the coefficient. To express that $i_l(i_m i_n) = (i_l i_m) i_n$ we see first that $i_m i_n = \Sigma_p[m, n, p] i_p$ where Σ_p means that all the different units are successively to be put in place of i_p with the proper coefficients and added, as just shown. Then $i_l(i_m i_n) = i_l \Sigma_p[m, n, p] i_p = \Sigma_p[m, n, p] i_l i_p = \Sigma_p[m, n, p] \Sigma_q[l, p, q] i_q$. But $i_l i_m = \Sigma_r[l, m, r] i_r$ and $(i_l i_m) i_n = \Sigma_r[l, m, r] i_r i_n = \Sigma_r[l, m, r] \Sigma_s[r, n, s] i_s$. By the association principle, $i_l(i_m i_n) = (i_l i_m) i_n$. Therefore

$$\Sigma_p[m, n, p] \Sigma_q[l, p, q] i_q = \Sigma_r[l, m, r] \Sigma_s[r, n, s] i_s$$

But owing to the independence of the units, this cannot be unless the two sums the one according to q and the other according to s are equal term by term; so that, putting $s = q$,

$$\Sigma_p[m, n, p]\,[l, p, q] = \Sigma_r[l, m, r]\,[r, n, q]$$

which is another expression of the associative principle.

I will now create a set of symbols A, I_1, I_2, I_3 etc. one more in multitude than the set of independent units, i_1, i_2, i_3 etc. But I do not define them at all, further than to say that they can each be multiplied by numbers, and the products, or any of them, can be added, and one such sum can be subtracted from another. I will further create a set of symbols

$(A:I_1), (A:I_2), (A:I_3)$, etc.,
$(I_1:A), (I_2:A), (I_3:A)$, etc.,
$(I_1:I_1), (I_1:I_2), (I_1:I_3)$, etc.,
$(I_2:I_1), (I_2:I_2), (I_2:I_3)$, etc.,
$(I_3:I_1), (I_3:I_2), (I_3:I_3)$, etc.,
etc.

and of these I will only say that they can be multiplied by numbers and can be multiplied *into* the symbols A, I_1, I_2, I_3 etc. and the products can be added, and such sums can be subtracted one from another. By multiplying *into*, I mean that when the commutative principle does not hold, so that pq is not generally equal to qp, then the first written of the two factors of a product, as p in pq, is called the *multiplier*, and the other is called the *multiplicand*; and I say the multiplicand (q in pq) is multiplied *by* the multiplier, while the multiplier is multiplied *into* the multiplicand. That is Hamilton's terminology. Now any of the last set of symbols, which I will call capital pairs, being multiplied into its second capital, so as to give for instance $(I_1:I_3)I_3$ gives for the product the *first* capital of the pair $(I_1:I_3)I_3 = I_1$ and in general $(I_m:I_n)I_n = I_m$ but multiplied into *any other* capital gives zero $(I_1:I_3)I_2 = 0_1\,(I_1:I_3)I_1 = 0$. I will further assume that a sort of associative principle holds so that

$$([(I_1:I_3)I_3]:I_5) = ((I_1:I_3)\,(I_3:I_5)) = I_1:I_5$$

But $$([(I_1:I_3)I_4]:I_5) = (0:I_5) = 0 = (I_1:I_3)\,(I_4:I_5)$$

So that that last set of [symbols given above] can be multiplied together; and if the last capital of the multiplier is the same as the first capital of the multiplicand, the product is a capital-pair whose first capital is the first capital of the multiplier, and its second is the last capital of the multiplicand. But if the last capital of the multiplier is *not* the same as the first of the multiplicand, the product is *zero*.

I will further create a set of symbols j_1, j_2, j_3 etc. equal in multitude to the original set of independent units i_1, i_2, i_3 etc. to be defined as, for example, follows:

$$\begin{aligned} j_7 = (I_7:A) &+ [7,1,1]\,(I_1:I_1) + [7,1,2]\,(I_2:I_1) + [7,1,3]\,(I_3:I_1) + \text{etc.} \\ &+ [7,2,1]\,(I_1:I_2) + [7,2,2]\,(I_2:I_2) + [7,2,3]\,(I_3:I_2) + \text{etc.} \\ &+ \text{etc.} \end{aligned}$$

or in general

$$j_u = (J_u:A) + \Sigma_v\Sigma_w\,[u, v, w]\,(I_w:I_v)$$

Any two *operations* whatsoever (using the word "operation" not for the individual deed but for whatever deed may be done under a fully explicit *rule* or *prescription*, covering a multitude of possible situations) are *equal*, i.e. the same *en effet*, if and only if, the *result* of either is, in *every conceivable case*, indistinguishable from that of the other. They are identical, that is, the same in essence, if and only if, whatever is prescribed in the defining fully explicit *rule* for doing the one is prescribed in that of the other. Now it is plain that the last set of symbols j_1, j_2, j_3 etc. are essentially symbols of operation, or, as we say, are *operators*, the ultimate *operands* being the units A, I_1, I_2, I_3 etc; and likewise the original set of independent units i_1, i_2, i_3, etc. are operators, since they are no otherwise defined than by their multiplication-table. I am now going to prove that these two sets of symbols are equal. For that purpose, I *propose* to prove that every combination of symbols of the set j_x are equal provided they have the same effect when multiplied *into* A. Let us suppose, for example, that $j_x j_y A = j_u j_v A$ then I have only to show that whatever number w may be, $j_x j_y I_w = j_u j_v I_w$. For if that be the case, since the only form of combination of the units denoted by single capitals with numerical subjacent indices is by multiplying them by numbers and adding the products, or subtracting one such product from another, and since the operators that are small letters with numerical subjacents are *distributive*, that is, such that $j_x(I_k + I_l) = j_x I_k + j_x I_l$ we shall have

$$j_x(j_y A) = j_x I_y = [x, y, 1] I_1 + [x, y, 2] I_2 + [x, y, 3] I_3 + \text{etc.} = \Sigma_g[x, y, g] I_g$$

$$j_u(j_v A) = j_u I_v = [u, v, 1] I_1 + [u, v, 2] I_2 + [u, v, 3] I_3 + \text{etc.} = \Sigma_h[u, v, h] I_h$$

so that since $j_x j_y A = j_u j_v A$ and all the units I_1, I_2, etc. are independent, it must be that $[x, y, 1] = [u, v, 1], [x, y, 2] = [u, v, 2]$ etc. or whatever number f may be, $[x, y, f] = [u, v, f]$. But in the original set of units i_1, i_2, etc.

$$i_x i_y = [x, y, 1]\, i_1 + [x, y, 2]\, i_2 + [x, y, 3]\, i_3 + \text{etc.} = \Sigma_g[x, y, g]\, i_g$$
$$i_u i_v = [u, v, 1]\, i_1 + [u, v, 2]\, i_2 + [u, v, 3]\, i_3 + \text{etc.} = \Sigma_h[u, v, h]\, i_h$$

It follows, then, from the equalities of all the coefficients of units in the development $i_x i_y$ with the corresponding coefficients of the development of $i_u i_v$ that from the assumed equality of $j_x j_y A$ and $j_u j_v A$ follows the equality of $i_x i_y$ to $i_u i_v$. But since $i_x i_y = i_u i_v$ we have also $i_x i_y i_m = i_u i_v i_m$, whatever unit i_m may be. Now $(i_x i_y)\, i_m = i_x (i_y i_m)$

$$= i_x([y, m, 1] i_1 + [y, m, 2] i_2 + \text{etc.}) = i_x \Sigma_\xi[y, m, \xi) i_\xi =$$
$$= \Sigma_\xi[y, m, \xi]\, (i_x i_\xi) = \Sigma_\xi[y, m, \xi] \Sigma_\xi[x, \xi, \zeta] i_\zeta$$

On the other hand,

$$j_x j_y j_m A = j_x[j_y(j_m A)] = j_x[j_y I_m] = j_x[[y, m, 1] I_1 + [y, m, 2] I_2 + \text{etc.}] = j_x \Sigma_t[y, m, t] I_t$$
$$= \Sigma_t[y, m, t] j_x I_t = \Sigma_t[y, m, t]\, ([x, t, 1] I_1 + [x, t, 2] I_2 + \text{etc.})$$
$$= \Sigma_t \Sigma_\pi[y, m, t][x, t, \pi] I_\pi$$

Now put this result beside the [preceding], putting t, and p as the summatory variable in the one and τ and π in the other; and they are

$$i_v i_y i_m = \Sigma_t \Sigma_p[y, m, t][x, t, p]\, i_p$$
$$j_x j_y j_m A = \Sigma_\tau \Sigma_\pi[y, m, \tau][x, \tau, \pi] I_\pi$$

and it is obvious that their multiplication-tables are exactly alike in form. Now the form of the multiplication-table is the only meaning or sense that these otherwise undefined letters have.

It is thus proved that any linear associative system of units may be put into matricular, or dyadic-relative form by simply regarding the coefficients of its multiplication-table as coefficients of capital-pairs in the precise manner in which the formula

$$j_u = (J_u\!:\!A) + \Sigma_v\Sigma_w[u, v, w]\,[I_w\!:\!I_v]$$

where the bracketed coefficients take the values read off from the multiplication table by the rule

$$i_p i_q = \Sigma_r[p, q, r]\,i_r$$

A concrete example will give the familiarity-kind of clearness to this. Required to put the algebra called by my father (bd_5) into matricular, or dyadic-relative, form. Its multiplication table is [in Fig. 7.]

	i_1	i_2	i_3	i_4	i_5
i_1	i_2	0	i_4	0	0
i_2	0	0	0	0	0
i_3	$i_2 + ri_4$	0	$i_1 + i_5$	ri_2	$-i_2 - ri_4$
i_4	0	0	i_2	0	0
i_5	$(r^2-1)i_2$	0	$-i_4$	0	$-r^2 i_2$

Fig. 7

Now the first term of the value of i_x in the new form, whatever x may be is always $(I_x\!:\!A)$. The other terms are found by looking along the horizontal rank of the table that has i_x set down at the left of it and if a unit say i_y is found in the vertical file headed i_z, with it[s] coefficient k, then the matricular value of i_x must have a term $k(I_y\!:\!I_z)$. This rule will give in the present case

$$i_1 = (I_1\!:\!A) + (I_2\!:\!I_1) + (I_4\!:\!I_3)$$
$$i_2 = (I_2\!:\!A)$$
$$i_3 = (I_3\!:\!A) + (I_2\!:\!I_1) + r(I_4\!:\!I_1) + (I_1\!:\!I_3) + (I_5\!:\!I_3) + r(I_2\!:\!I_4) - (I_2\!:\!I_5) - r(I_4\!:\!I_5)$$
$$i_4 = (I_4\!:\!A) + (I_2\!:\!I_3)$$
$$i_5 = (I_5\!:\!A) + (r^2-1)(I_2\!:\!I_1) - (I_4\!:\!I_3) - r^2(I_2\!:\!I_5)$$

This will evidently give the multiplication table. For to begin with i, as a multiplier, we find it contains capital-pairs ending only in A, I_1, and I_3. We ask, then, where among these values of the is are capital-pairs beginning with A, I_1 and I_3. A begins no capital pair. But I_1 begins $(I_1\!:\!I_3)$ in the value of i_3; so that $(I_2\!:\!I_3)$ is a term of i_1i_3 and I_1 begins no other capital-pair. I forget, or overlook $(I_1\!:\!A)$ as a term of i, so that i_1 containing $(I_2\!:\!I_1)$ and $(I_1\!:\!A)$, i_1^2 evidently contains the term $(I_2\!:\!A)$ which equals i_2. I set these products down in a blank multiplication-table in their right places, and so gradually form the result which I will show presently. I_3 which ends the term $(I_4\!:\!I_3)$ of i_1 begins no term except $(I_3\!:\!A)$ in the value of i_3. Hence i_1i_3 contains the term $(I_4\!:\!A)$. I now pass to the value of i_2, the value of which contains no term but $(I_3\!:\!A)$ hence i_2 as a multiplier always gives 0 as the product. I pass to the value of i_3. It contains no less than 8 terms. The first is of course $I_3\!:\!A$ which as multiplier gives no product. The second term of i_3 is $(I_2\!:\!I_1)$ and I_1 as already seen, is the first capital of $(I_1\!:\!A)$ in the value of i, and of $(I_1\!:\!I_3)$ in the value of i_3. Hence a term of i_3i_1 is $(I_2\!:\!A)$ which I duly set down in my blank multiplication-table and a term of i_3^2 $(I_2\!:\!I_3)$ which I set down in its proper place. The 3rd term of i_3 is $r(I_4\!:\!I_1)$ and I_1 as we have twice seen already begins $(I_1\!:\!A)$ in i_1 and $(I_1\!:\!I_3)$ in i_3. Hence $r(I_4\!:\!A)$ is to be set down as a term of i_3i_1 and $r(I_4\!:\!I_3)$ as a term of i_3^2. The 4th term of i_3 is $(I_1\!:\!I_3)$ and I_3 begins $(I_3\!:\!A)$ a term of i_3 so that $(I_1\!:\!A)$ is to be set down as a term of i_3^2, and I_3 begins nothing else. The 5th term of i_3 is $(I_5\!:\!I_3)$ and since I_3 begins only the term $(I_3\!:\!A)$ of i_3, the term $(I_5\!:\!A)$ is to be set down as a term of i_3^2. The sixth term is $r(I_2\!:\!I_4)$ and I_4 begins 5 terms, 1st $(I_4\!:\!I_3)$ of i_1 so that $r(I_2\!:\!I_3)$ is a term of i_3i_1; 2nd $r(I_4\!:\!I_1)$ of i_3 so that $r^2(I_2\!:\!I_1)$ is a term of i_3^2; 3rd $-r(I_4\!:\!I_5)$ of i_3, so that $-r^2(I_2\!:\!I_5)$ is a term of i_3^2; 4th $(I_4\!:\!A)$ of i_4; so that $(I_2\!:\!A)$ is a term of i_3i_4 and 5th and last $-(I_4\!:\!I_3)$ of i_5 so that $-(I_2\!:\!I_3)$ is a term of i_3i_5. The seventh term is $-(I_2\!:\!I_5)$ and I_5 begins two terms; 1st $(I_5\!:\!I_3)$ of i_3, so that $-(I_2\!:\!I_3)$ is a term of i_3^2, and 2nd $(I_5\!:\!A)$ of I_5, so that $-(I_2\!:\!A)$ is a term of i_3i_5. The eighth term is $-r(I_4\!:\!I_5)$ and as just seen this makes $-r(I_4\!:\!I_3)$ a term of i_3^2 and makes $-r(I_4\!:\!A)$ a term of i_3i_5. We now pass to the terms of i_4. Of course $(I_4\!:\!A)$ leads to nothing. The only other term is $(I_2\!:\!I_3)$ and i_3 begins no term but $(I_3\!:\!A)$ making $(I_2\!:\!A)$ a term of i_4i_3. We pass to the terms of i_5; skipping $(I_5\!:\!A)$, of course, the second term is $(r^2-1)\,(I_2\!:\!I_1)$ and I_1 begins two terms $(I_1\!:\!A)$ making $(r^2-1)\,(I_2\!:\!A)$ a term of i_5i_1; and $(I_1\!:\!I_3)$ of i_3 making $(r^2-1)\,(I_2\!:\!I_3)$ a term of i_5i_3. The third term is $-(I_4\!:\!I_3)$ and I_3 begins $(I_3\!:\!A)$ along, making $-(I_4\!:\!A)$ a term of i_5i_3. The fourth and last term is $-r^2(I_2\!:\!I_5)$ and I_5 begins two terms, the first being $(I_5\!:\!I_3)$ of i_3 making $-r^2(I_2\!:\!I_3)$ a term of i_5i_3; and the second being $(I_5\!:\!A)$ making $-r^2(I_2\!:\!A)$ a term of i_5^2. See the filled-up blank multiplication-table below.

Beginning with $i_3 i_1$, we first write down the whole development of i_1, and then under each term those terms of i_3 whose last capital is the same as the first capital if i_1; and then we draw a line under them and below it place the products, *being careful to take the terms on the lower of the two lines as the multipliers and the terms severally above them as the multiplicands.* Here is the work:

$$\begin{array}{lll}
\textit{Multiplicands} & (I_1{:}A) + (I_2{:}I_1) & + (I_4{:}I_3) \\
\hline
\textit{Multipliers} & \left\{\begin{array}{l} (I_2{:}I_1) \\ +r(I_4{:}I_1) \end{array}\right. & +r(I_2{:}I_4) \\
\hline
\textit{Products} & \left\{\begin{array}{l} (I_2{:}A) \\ +r(I_4{:}A) \end{array}\right. & +r(I_2{:}I_3) = i_2 + ri_4
\end{array}$$

We next treat $i_3 i_5$ in the same way:

$$\begin{array}{lll}
\textit{Multiplicands} & \left\{\begin{array}{l} (I_5{:}A) + (r^2-1)(I_2{:}I_1) - \\ \qquad -r^2(I_2{:}I_5) \end{array}\right. & (I_4{:}I_3) \\
\hline
\textit{Multipliers} & \left\{\begin{array}{l} -\ (I_2{:}I_5) \\ -r(I_4{:}I_5) \end{array}\right. & +r(I_2{:}I_4) \\
\hline
\textit{Products} & \left\{\begin{array}{l} -\ (I_2{:}A) \\ -r(I_4{:}A) \end{array}\right. & -r(I_2{:}I_3) = -i_2 - ri_4
\end{array}$$

Finally, we treat i_3^2 in the same way, thus:

$$\begin{array}{llllll}
\textit{Multiplicands} & \left\{\begin{array}{l} + (I_1{:}I_3) \\ \\ \\ \end{array}\right. & \begin{array}{l} + (I_2{:}I_1) \\ +r(I_2{:}I_4) \\ -\ (I{:}I_5) \end{array} & + (I_3{:}A) & \begin{array}{l} +r(I_4{:}I_1) \\ -r(I_4{:}I_5) \\ \\ \end{array} & + (I_5{:}I_3) \\
\hline
\textit{Multipliers} & \left\{\begin{array}{l} +\ (I_2{:}I_3) \\ +r(I_4{:}I_1) \end{array}\right. & & \begin{array}{l} + (I_1{:}I_3) \\ + (I_5{:}I_3) \end{array} & +r(I_2{:}I_4) & \begin{array}{l} -\ (I_2{:}I_5) \\ -r(I_4{:}I_5) \end{array} \\
\hline
\textit{Products} & \left\{\begin{array}{l} +\ (I_2{:}I_3) \\ +r(I_4{:}I_3) \end{array}\right. & & \begin{array}{l} + (I_1{:}A \\ + (I_5{:}A) \end{array} & \begin{array}{l} +r^2(I_2{:}I_1) \\ -r^2(I_2{:}I_5) \end{array} & \begin{array}{l} -\ (I_2{:}I_3) \\ -r(I_4{:}I_3) \end{array}
\end{array}$$

The first and last columns of products cancel each other. The terms whose last capitals are A suggest that the product equals $i_1 + i_5$. Let us see

$$\begin{array}{rl}
i_1 = & + (I_1{:}A) + (I_2{:}I_1) \qquad + (I_4{:}I_3) \\
i_5 = & \left\{\begin{array}{l} + (I_5{:}A) + (r^2-1)(I_2{:}I_1) - (I_4{:}I_3) \\ \qquad -r^2(I_2{:}I_5) \end{array}\right. \\
\hline
i_1 + i_5 = & \left\{\begin{array}{l} + (I_1{:}A) + r^2(I_2{:}I_1) \\ + (I_5{:}A) - r^2(I_2{:}I_5) \end{array}\right.
\end{array}$$

One important addition to this theorem, I failed to publish. Though it is an obvious corollary from what I did publish, I suppose I was so taken up with the interest of what I did publish that I overlooked it. It is that *second* form in which any such system of units may be expressed. The form I did publish amounted to this. The form I gave is got by taking for each unit i_1, i_2, i_3 etc. the corresponding capital letter followed by $:A$, — so as to make $(I_1:A)$, $(I_2:A)$, $(I_3:A)$ — as the first term and then running the eye along the rank of the multiplication table that belongs, for example to i_1 as multiplier, set down as a term of i_1 for every square of that rank that is occupied, every term in that square, only changing every i into the corresponding capital letter, and then adding a colon followed by the heading of the file thus heading being changed into a capital. That is the expression in ordinary language of the equation

$$i_m = (I_m:A) + \Sigma_l\Sigma_n[m, n, l]\,(I_n:I_l)$$

But obviously the ranks and files have the same sort of relation to the multiplication table. Only each rank belongs to one multiplier and each file to one multiplicand. Hence it is obvious that since the rule just given holds, it will hold equally well instead of giving each unit i_m the term $(I_m:A)$ to give each unit i_n the term $(\Omega:J_n)$ and then run down the file headed by i_n in the multiplication-table and copy as other terms of i_n every term found in every square only changing every i into a J, and putting before this J a colon preceded by the i at the left of the file in which the term occurs, this i also being changed into J. Thus, in the example I chose for illustration, of which I here repeat the multiplication table, [see the table, Fig. 7], the form I got was

$$\begin{aligned}
i_1 &= (I_1:A) + (I_2:I_1) \qquad + (I_4:I_3) \\
i_2 &= (I_2:A) \\
i_3 &= (I_3:A) + (I_2:I_1) \qquad + (I_1:I_3) + r(I_2:I_4) - (I_2:I_5) \\
&\qquad + r(I_4:I_1) \qquad + (I_5:I_3) \qquad - r(I_4:I_5) \\
i_4 &= (I_4:A) \qquad + (I_2:I_3) \\
i_5 &= (I_5:A) + (r^2-1)(I_2:I_1) - (I_4:I_3) \qquad - r^2(I_1:I_5)
\end{aligned}$$

I write the new form up the side of the multiplication table. [It is]

$$\begin{aligned}
i_1 = (\Omega:J_1) + (r^2-1)(J_5:J_2) \qquad &+ (J_3:J_2) + (J_1:J_2) \\
&+ r(J_3:J_4)
\end{aligned}$$

$$
\begin{aligned}
i_2 &= (\Omega:J_2) \\
i_3 &= (\Omega:J_3) - \quad (J_5:J_4) + (J_4:J_2) + (J_3:J_1) + (J_1:J_4) \\
& \qquad\qquad\qquad\qquad\qquad\qquad + (J_3:J_5) \\
i_4 &= (\Omega:J_4) \qquad\qquad\qquad\qquad\quad + r(J_3:J_2) \\
i_5 &= (\Omega:J_5) \quad - r^2(J_5:J_2) \qquad - (J_3:J_2) \\
& \qquad\qquad\qquad\qquad\qquad\qquad - r(J_3:J_4)
\end{aligned}
$$

Whence we easily make out that

$$
\begin{aligned}
i_1^2 &= (\Omega:J_2) = i_2 \qquad i_1 i_2 = 0 \qquad i_1 i_3 = (\Omega:J_4) + r(J_3:J_2) = i_4 \\
i_2 i_1 &= i_2 i_2 = i_2 i_3 = i_2 i_4 = i_2 i_5 = 0 \\
i_3 i_1 &= (\Omega:J_2) + r(\Omega:I_4) + \qquad (J_3:J_2) = \\
& \qquad\qquad\qquad - r(r^2-1)(J_3:J_2) \\
&= (\Omega:J_2) + r(\Omega:J_4) + r^2(J_3:J_2) = iJ + ri_4 \\
i_3 i_2 &= 0 \\
i_3^2 &= + (\Omega:J_1) - (J_5:J_2) + (J_3:J_4) + (J_1:J_2) = \\
& \quad + (\Omega:J_5) \qquad\qquad - (J_3:J_4) \\
& \qquad\qquad\qquad\qquad\qquad + (J_3:J_2) \\
&= (\Omega:J_1) + (r^2-1)(J_5:J_2) + r(J_3:J_4) + (J_1:J_2) = i_1 + i_5 \\
& \quad (\Omega:J_5) - \qquad r^2(J_5:J_2) - (J_3:J_2) \\
& \qquad\qquad\qquad\qquad\qquad - r(J_3:J_4) \\
i_3 i_4 &= r(\Omega:J_2) = [r]i_2 \\
i_3 i_5 &= -(\Omega:J_3) - r(\Omega:J_4) - r^2(J_3:J_2) = -i_2 - ri_4 \\
i_4 i_1 &= i_4 i_2 = 0 \\
i_4 i_3 &= (\Omega:J_2)[= i_2] \\
i_4^2 &= i_4 i_5 = 0 \\
i_5 i_1 &= (r^2-1)(\Omega:J_2) = (r^2-1)i_2 \\
i_5 i_2 &= 0 \\
i_5 i_3 &= -(\Omega:J_4) - r(J_3:J_2) = -i_4 \\
i_5 i_4 &= 0 \\
i_5^2 &= -r^2(\Omega:J_2) = -r^2 i_2
\end{aligned}
$$

In my notes on my father's Linear Associative Algebra, p. 91, I have given still another form of this algebra which is a little simpler than either of those deducible from my theorem, as is often the case. In this form there is one capital that occurs only as the first of a pair like Ω, and there is another that occurs only as the second like A and both occur in the expressions for every i. But, generally, those forms which contain no analogue either of A or of Ω are the simplest.

As it may seem disloyal in me to speak of Sylvester as I here do, he having been an old friend, I will mention that that friendship was entirely broken off under the following circumstances. He sent me a proof-sheet and requested me to insert a proper statement as to what I had proved concerning some subject, and further requested that I should send the proof to the printer. The matter in question was very likely this very theorem. I do not remember. I might easily look it up; but it is of no consequence. I appended the proper statement to the proof as he had requested; but instead of sending it to the printer, I sent proof and addendum, which was on a separate sheet to him, declining to send it myself to the printer, as it was proper he should see whether he approved of the insertion, and I particularly requested him to examine it. The thing was printed as I had requested [in] about a month (as I remember it, but my memory is treacherous on such points about 27 years ago, I am pretty sure it was several weeks, and I am not sure it was not several months). He never mentioned the matter to me, nor I to him; when suddenly he appeared in print saying that I had inserted the passage entirely without his knowledge and consent and had sent it to the printer as coming from him, and further that as far as he knew my statement was false. I replied to this and showed plainly that *his* statement was erroneous but he never made any apology and went on saying that, as far as he knew, there was no truth in the statement that I had proved the proposition in question.[4] Under these circumstances, I never had any further communication with him. I think it seems as if he simply *wouldn't* see a very easy demonstration, because it would be like him. But I admit that inasmuch as he never, I believe understood Gordan's symbolic method of dealing with invariants, though that was his special stamping-ground, it is possible that he could not see that two operations entirely undefined except by their multiplication-tables were the same if their multiplication-tables were the same.

However, I go on to something more interesting.

(But before I go any farther, I will just say that I have been looking in the books and have, I think, found out why Sylvester could not understand my theorem. It was, I have pretty good evidence to show, because he never could get it through his head what my capital-pairs $(I_1:I_2)$ and the like were. They are simply the representation of the *quality* of a factor, or *quantum* that might be multiplier and might be

[4] This controversy was carried on in the *Johns Hopkins University Circulars* beginning February 1883 and ending in April 1883. Manuscript 431a (3, 10d) carries another version of the story.

multiplicand, apart from its *quantity*. For I found a passage in one of his "Lectures on Universal Algebra" in which he says that the notion of regarding a matrix, or square array of numbers as a *sum* of numbers each multiplied by a symbol of its *place* in the array was one he got from me, though he believes I had been familiar with it for many years. But certainly in those days I knew nothing of there being any calculus of matrices. So I could not have regarded my capital pairs, which are used in my memoir of 1870, as denoting places in matrices. In fact I had printed at my own expense an account of my point pairs, and it was not until the copies were all but ready for delivery that finding Cayley's Memoir on Matrices had anticipated me, I never sent the copies out. That date was 1882 Jan 16. Cayley's Memoir was of 1858.)

Sylvester gave some lectures on what he called "Universal Algebra" by which he seemed to mean the Calculus of Matrices which is the same thing *mathematically*, though very different *logically*, as my "General Algebra of Logic." In those lectures, he made a splurge about what he ought to have called the *modulus* of a matrix, i.e. its measure of *how much* it amounts to, though he very oddly called it the "Nullity" of the Matrix. The idea itself was one of those that the instant it is started and the *determinant* of the matrix is only so much as *suggested*, strikes one as obviously correct. But somehow it never had been suggested before, as far as I know. I have forgotten almost entirely about those lectures. But I remember, or think I do that he made much of the *reciprocal* of a matrix, as though that were the *ipsissimum* of the doctrine and especially of the modulus; and perhaps that was why he called the modulus the "nullity." At any rate, I asked what if the modulus is zero? A zero cannot have a reciprocal. Oh, he said, that is a special case; I am considering only what is true *in general*. He quite forgot that what mathematicians mean by saying that something is so and so "in general" may, for all that, hardly ever be true. For what do they mean, I will tell you. All the elementary algebras say that an equation of the 5th or higher degree cannot be solved by means of radicals. What they ought to say is that it cannot be so solved "in general." ...

Milford Pa 1909 Dec 25

This is Vol. III of the letter of which the first two tomes have already gone.

Since I promised to say something about my System of Logic, I will do so. It is to regard Logic as the theory of signs in general; and will consist of 3 books. Book I, treats of the essential nature of a sign, and of the different main classes of *possible* signs. If I were making a natural history of such signs as exist, I should have to recognize irregularity in the divisions, just as I do, in some measure in classifying the Sciences. My classification of the Sciences is however intended to be useful in the future, and therefore is not absolutely confined to what exists. Indeed, I found it quite impossible to state the relations between the sciences without on the one hand, relying exclusively upon what the members of the different groups have said, — in which case the ideas governing my classification would be antiquated beyond what one would suppose before he tried that method, or else one must speculate upon what seems to be in the atmosphere of science but has perhaps never yet been uttered. Scientific *ideas* do not get uttered until long after they have been influential. The consequence is that I found I could make a useful classification only by adopting as a skeleton of it my own notions of how the sciences *ought* to be related. My classification of signs, however, is intended to be a classification of *possible* signs and therefore observation of existing signs is only of use in suggesting and reminding one of varieties that one might otherwise overlook. And our notions of what is possible are necessarily drawn largely from our innate ideas. In this first book I shall analyze and define and classify all the concepts of Logic so far as I know them. I start by defining what I mean by a sign. It is something determined by something else its object and itself influencing some person in such a way that that person becomes thereby mediately influenced or determined in some respect by that Object. This being what I mean by a Sign, I must classify signs according 1st to their natures in themselves, 2nd in relation to their Objects and 3rd in their relations to their *Interpretants*, i.e. the effects on the interpreter. You see here, the triplet is absolutely forced upon me by the nature of my concept of a Sign. Then in dividing Signs in the 1st respect, I must recognize that some are actual occurrences or are definite existing things. But other signs, such as the word "the," in the sense in which "the" is a single word, consist, each of them, in something being possible. I call such things (whether they be signs or not) "May-bes," perhaps better "can-be." Still other signs are neither Actuals nor May-bes. For example Greek syntax is a sign, — a medium by which my knowledge is determined by things that Plato or Sextus Empiricus has written; and yet it is neither an actual individual occurrence or thing nor does it consist in a mere possibility. It consists rather more truly in *impossibilities*, for example in the impossibility that a Greek would express certain ideas in other than in certain ways. I call things of this kind (whether Signs or not) "Would-bes." A Sign then may be a Would-be. Now it is plain that every conceivable thing is either a May-be, an Actual, or a Would-be. If it is not at once plain, it will become so on considering that "may be," "can be" and "might be" are but various applications of one idea; "may be" suggesting insufficient knowledge, "can be" insufficient action of the Subject (in grammar), and "might be" insufficiency of the circumstances. Now applying these ideas to the copula of a proposition, if "*S* may be *P*" and *no more*, then certainly "*S* may be non-*P*" so that nothing prevents both *P* and not-*P* being predicated under this Mode. On the other hand, both "*S* would be *P*" and "*S* would be non-*P*" may both be false; so that the principle of Excluded Middle does not apply under the Mode of Would be, which both apply under the Mode of Actuality. Thus, the only fourth Mode there could be would be one in which neither the principle of Contradiction nor that of Excluded Middle would apply. A state in which one neither knows that *S* is *P* nor that *S* is not *P* nor that *S* is not both nor that *S* is one or the other is a total absence of knowledge as to *S* being *P*; and similarly a state of *being* that neither makes *S P* nor prevents it, nor lets it be one or the other nor prevents it from being one or the other is not a state of being at all as to *S* being *P*. I make ten distinctions, and they are all trichotomies *because of my* making a classification of Maybes or Canbes. I might have drawn more than ten distinctions; but these 10 exhibit all the distinctions that are generally required in logic; and since my investigations of these involved my consideration, — virtually at least, — of 59049 questions, still leaving me on the portico of logic, I thought it wise to stop with these.

Book II, On Critic, discusses the warrant for each of the different kinds of reasoning. Throughout this Book and Book I, I do not allow myself to accept any discovery of "Psychology Proper," by which I mean the Empirical Science of the Modes of Functioning of finite Minds. For example, the modes of Association, its formation, suggestions through it, etc. Fatigue, — and in short the Physiology of the Mind. For in my opinion, excepting Metaphysics there is no science that is more in need of the science of Logic than Psychology proper is. On the other hand, I found Logic largely on a study which I call Phaneroscopy, which is the keen observation of and generalization from the direct Perception of what we are immediately aware of. I find there are three kinds of possible warrant for a Belief. Here is this distressing number 3 again, against which you seem to have sworn eternal enmity, but which *will* turn up again and again. I don't think you are willing to believe that space has three dimensions, are you? None of the 3 warrants is Positively Infallible although one of them is so in a Pickwickian sense. The first kind of warrant consists in the reasoner's being *disposed to believe* in his proposition. This goes toward warranting the belief, since the very undertaking to find out a truth one does not directly perceive assumes that things conform in a measure to what our reason thinks they should. In other words our Reason is akin to the Reason that governs the universe, we must assume that or despair of finding out anything. Now despair is always illogical; and we are warranted in thinking so, since otherwise all reasoning will be in vain. If it be so, a strong inward impulse to Believe a given proposition tends to show that proposition to be true; and if it be not so, we never can discover what we don't directly perceive, do what we may.

Dec. 28. I have been suffering horribly for 2 days and am now like a drowned rat. Juliette was also ill all last night and it sometimes seems all but hopeless to keep her alive through the winter if it is going to keep on as it has begun with 50°F only attainable for a couple of hours. My ink will freeze in a few days I hear and then what shall I do, I wonder. The second warrant is in case one's inference is from some state of things capable of expression in a proposition (generally a copulative proposition of some complexity) and when every state of things not denied by this proposition is a state of things in which the conclusion is true. Such inference is Deduction, or Necessary Inference. There are two kinds of Deduction; and it is truly significant that it should have been left for me to discover this. I first found, and subsequently *proved*, that every Deduction involves the observation of a Diagram (whether Optical, Tactical, or Acoustic) and having drawn the diagram (for I myself always work with Optical Diagrams) one finds the conclusion to be represented by it. Of course, a diagram is required to comprehend any assertion. My two genera of Deductions are 1st those in which any Diagram of a state of things in which the premisses are true represents the conclusion to be true and such reasoning I call *Corollarial* because all the corollaries that different editors have added to Euclid's Elements are of this nature. 2nd Kind. To the Diagram of the truth of the Premisses something else has to be added, which is usually a mere May-be and then the conclusion appears. I call this *Theorematic* reasoning because all the most important theorems are of this nature.

A very good example of this is the Ten Point Theorem whose diagram I have here drawn [Fig. 1]. If rays be drawn on a plane (or great circles on a hemisphere) so that three rays meet in one point, *O*, of the plane and on each of these rays two points, *A* and *a* on one, *B* and *b*, on another and *C* and *c* on the third; and if then from each of these six points two rays be drawn to 2 and the points on different rays through *O*; so that the new rays may be lettered *AB*, *AC*, *BC*, *ab*, *bc*, *ca*, there will be for each of the 3 rays through *O*, two of the six rays last drawn that do not pass

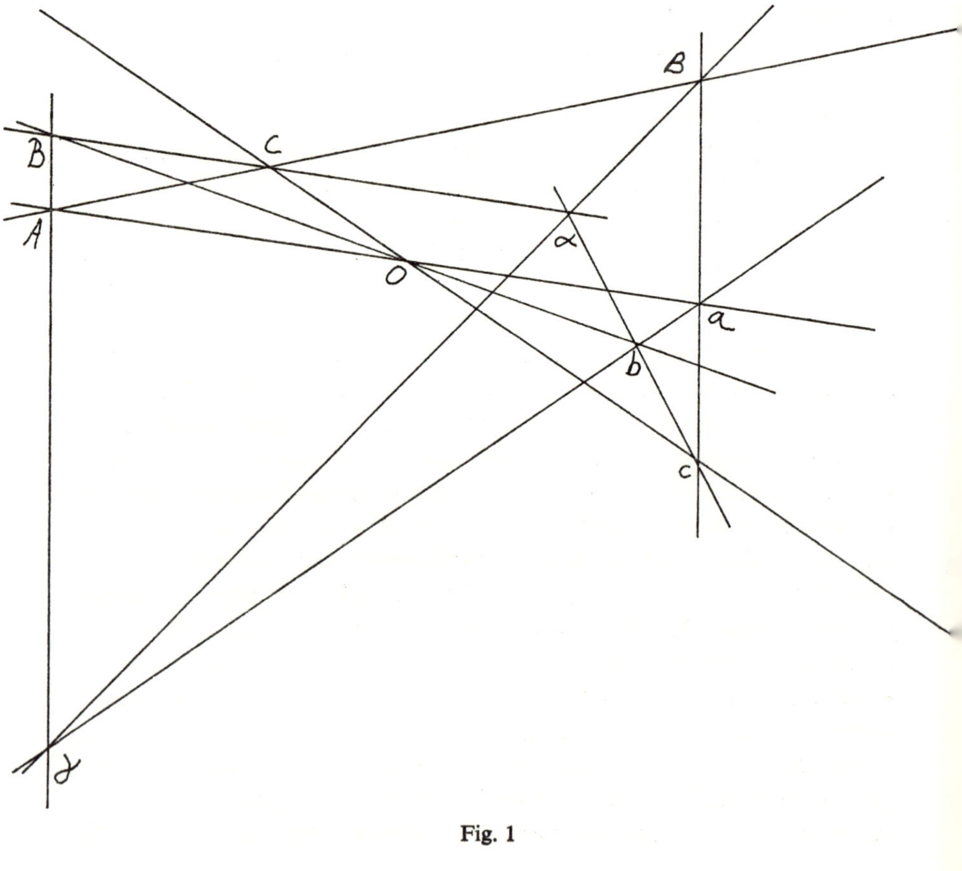

Fig. 1

through lettered points on that rays, but these two rays will cut each other and their three intersections; that is the intersection of *BC* and *bc*, which I mark, α, and the intersection of *CA* and *ca* which I mark, β, and the intersection of *AB* and *ab* which I mark γ, *will lie on one ray*! Now for more than two centuries, if not for three in all, the greatest mathematicians have tried to prove *that* by the diagram required to exhibit it *alone*, and have tried in vain. But it is readily proved in several ways by additions to the diagram; as for example by imagining the ray *OBb* to be the shadow on projection of a ray through *O* but not on the plane of *AOCac*. It then becomes evident, from the fact that two flats of dimensions *M* and *N* lying in in a flat of dimensions *P* not greater than $M + N$ necessarily have a flat of dimensions $M + N - P$ in common. Thus 2 planes in a flat of 3 dimensions have $2 + 2 - 3 = 1$ or a ray in common.

The 3rd kind of warrant is that which justifies the use of a method of inference provided it be carried out to the end consistently. There are 3 kinds of inference of this kind. They are all inferences from random samples. The strongest is that which is a sample (that is, a collection)of units. In that case, the theory of errors is applicable. The second kind is where there are no definite multitudes but where, as the sample is enlarged, the inference becomes stronger and stronger. The third kind which is the weakest of all forms of Induction is where the only defence is that if the conclusion is false, its falsity will *sometime* be detected if the method of inference be persisted in long enough. For example, if we infer from the fact that so closely as we have been able to measure, the sum of the angles of a triangle is 180°, therefore it is exactly so, the only warrant for this inference is that if we go on making our errors of measurement less and less, then if the sum be not exactly 180°, we shall ultimately find out that it is more or less if we persist in making the measurements. No inductive inference can be weaker than that and have any warrant at all.

My Book III treats of methods of research. Owing to my trying to get my book for Carus into as small dimensions as possible while still stating accurately my doctrine of Pragmatism, I prefer to postpone reading your two books until I have got my own ideas expressed. After I have done that I shall read your books and state just why. I do not think that my pragmatism calls upon me to renounce entirely all ideas of the Absolute. I certainly do not, for every mathematician knows very well that ideas of the Absolute have worked almost miracles in his own science, and let me tell you that when the unbelievers tell you that the Absolute of Mathematics is not that of Metaphysics, they are blinded by their own disinclination to believe in the Absolute. They *are* the same; only the method of the mathematicians and their use of diagrams enable them to get distincter notions of the matter than the Metaphysicians have. The mathematical Absolute is confined it is true to Space, but it bears the same relation to attainable space as the Metaphysical Absolute *ought* to bear to concepts of the Relative. At any rate, it is approximately so.

I thought your report upon the Hodgson manifestations was very just and admirable. Yet, after all, it seems to me to be a terrible blow to Psychical Research. What, this man who had been studying for years how he was going to communicate to some purpose could he do no more than imitate the external manner of his first greetings! It would seem that the spirits that communicate to us have lost their intellect, not to say their intelligence; and after a year or two are heard from no more. They do not gradually learn to communicate more, but on the contrary, *less*. Do they become extinct? I seem to read this lesson: that we ought to learn that there is infinitely more comfort and solace in reflecting that God, the Perfect Intelligence, governs the Universe than there can possibly be in any picture of what another life may be. Let us draw our consolation and joy and inspiration from that, and never mind what becomes of the personality. What can the Personality be but so much shortcoming. You tell me I don't deserve Juliette. But so far as there is any real meaning in deserve, I know but too bitterly, — and it is a bitterness that I love, — that I fall further short than any man if possible of deserving any good at all; and I cannot protest against any condemnation that may be visited upon me. I could tell you something on that head. I am accustomed to hear and read upon countenances that when I speak of religion people say I am a sham. I am supposed to be a coldblooded hypocrite of the lowest and most disgusting order; but I hardly expected as great a student of psychology and of Religious experiences as you are to class me so. I think there is too much about *deserve* in the New Testament. In that respect I can't help thinking that the mother of christianity, Buddhism, is superior to our own religion. This is what one of my selves, my intellectual self says. But enough, I will keep my religion to myself and to One that does not scoff at it.

I do not agree with you that my papers about the evolution of the Laws of Nature are the best things I have done. To begin with, let me say that I don't think there is any need of humility in regard to intellectual powers and achievements, for the reason that such things [are] truly *talents* in the sense of the parable and have nothing to do with character or merit. They are simply a load of responsibility which I have enjoyed, as one generally enjoys responsibilities, yet I have groaned under too. I admit, then, that the *idea* of the gradual evolution of the laws of nature is a magnificent one, and that it is, as I always intended to show in a final article, but Carus would only give me a thousand words for such an article, so that I had to give it up, and since he now *denies* flatly that he did not allow me to finish what I had to say, I mean to hunt up and send you his letter, and I want it to be printed after he and I are both dead and there can be no squabble about it, because I think it an essential part of the theory which ought to be brought in as a supplement. Because an event occurring by Absolute Chance ought, as a new importation of Variety into the Universe, to be regarded as a Creative act. And in addition, I think the logic of the argument against Necessitarianism, was strong. But beyond that, my development of the theory was poor and too hasty. It needed more years than I gave months to it, since my previous study of the matter was along a line I did not so much as touch upon. I think unquestionably my best work has been my Logic. Firstly, my proof that the warrant of Hypothesis is mainly that unless the human intellect is of the kin of the Creative Spirit, it is useless to attempt any discovery of positive truth — I mean empirical truth beyond direct perception. (I think *that*, in conjunction with the considerations in my Evolution series of papers make a strong argument for a Deity, though inferior, in *religious value*, to the one I wrote about in the Hibberts, which, of course, I did not claim as original. But such arguments, and good ones too, are plenty.) Secondly, my distinction of Corollarial and Theorematic Deduction, which, once stated, is as obvious as Columbus's Egg, is of enormous importance. Then my work in the Logic of Relatives simply revolutionizes Logic. It shows the nonsense of supposing that inferences from one premiss are so different from Syllogisms, — so that Kant referred them to different mansions of the realm of the Mind. Indeed you have only to put an "and" to connect the two premisses and they *become one*. Not that this insertion of the "and" would be at all an unimportant step if it were not virtually and implicitly there in the ordinary formula. Then to pass over my analysis of the syllogism inference into no less than *seven* distinct steps of inference. Then the showing that, from every proposition a strictly endless succession of different necessary consequences may be drawn, so that it is absurd to talk of a logical machine's being able to do what Reason does. And the equal absurdity of talking of THE conclusion from a pair of syllogistic premisses. Then my Memoir of 1870 in the Am. Acad. of Arts & Sci. Memoirs (4to) besides a host of other interesting points, contained a virtual rediscovery of Cayley's celebrated Doctrine of Matrices with which I was entirely unacquainted and showed that that was the Algebra of Relations in general, as I afterward, in my edition of my father's Linear Associative Algebra, showed very simply too was also identical with Linear Associative Algebra. And my second Memoir published in Vol. III of the Am. Jour. of Math. *omitted*, as too obvious for insertion, the proof of a theorem which proof Schroeder undertook afterward to "prove" was impossible. But subsequently, I came across the original MS. of my proof, that had seemed to me so obvious, and it has been printed as a foot-note to one of E. V. Huntington's papers in the publications of the Am. Math. Soc. It is odd it should have seemed to me so obvious; yet I can perfectly see how it did. Then another memoir in which I introduced into my algebra terms of second intention and other abstractions. Then my Algebra of Dyadic Relatives with which Schroeder fell in love and which is certainly very pretty, though he exaggerated its value (not so unjustly however as Whitehead's positively silly objection to it). Then my Universal Algebra of Logic, which I consider the most convenient to use as a calculus of any, and though not particularly pretty, very powerful and pretty deep. Finally my triumph in that line, my Existential Graphs, by which all deduction is reduced to *insertions* and *erasures*, and in which there are no connecting signs except the writing of terms on the same area enclosed in an oval or parentheses (which will do instead of ovals) and also heavy lines to express the identity of the individual objects whose signs are connected by such lines. This ought to be the Logic of the Future. Then my treatment of Induction not yet printed, though I have possessed it for many years. I divide Induction (as I said above I believe) into three kinds. One of these is that to which in the volume entitled "Studies in Logic by Members of the JH Univ" I restricted the word Induction, which I now call Quantitative Induction, the most perfect kind of Induction for which I gave the two Rules. 2nd what I now call Qualitative Induction, called "Hypothesis" in that volume, and 3rd what I now call *Crude Induction*, a most important kind that I entirely overlooked in my essay in that volume because I had not accurately defined the nature of the warrant for Induction. This is the weakest kind of Induction, and yet it is a kind with which we cannot dispense in science and still less in practical matters (— in science because practical considerations enter into scientific reasonings, unavoidably). I have done a lot of work in Methodeutic that is valuable and very little of it is printed. This will be the most widely useful part of my Big Book. A few scraps in my

"Amazing Mazes" have brought me several admiring letters from mathematicians of some power.

Then again, I think my Essay on Pragmatism in its new reduction in the volume Carus is to print has more value than my series of papers on the Evolution of Laws of Nature because the refutation of Laplace's theory of Induction given in all the text-books of Probability is completely proved; and so I think is my position that the Universe could not be less Uniform than it is. It is a mere remark that I made in those papers, yet I think a good one, that our notion that Nature is so uniform is simply due to the fact that its uniformities are important knowledge while its want of uniformity is without importance to us.

Now I will tell you why I set down all this boastful stuff. It is because I am not only over 70 years of age, but my powers are waning, and my health presents alarming features as to the continuance of sanity: and if I should be cut off before I had done anything to repay all the money that your genius for sympathy and unspeakable goodness has collected for us, — although if Juliette should pull through this winter and complete her work on this house, I firmly believe that that will suffice ultimately to refund the money, — yet it is a reason for wanting my System of Logic to pay and as Mill's went through 9 editions (though with the advantage of containing no special novelty), there is enough prospect of my book paying for me to wish that there should be something for my friends to say especially if I am dead that will help its sale. That is why I want to point out the things I have proved; and though I never was gifted in the way of constructing proofs, yet I will mention a few of my proofs that have had some decided merit; and they may help to neutralize the distrust that always confronts decided originality on that suspicious symptom that is called "genius." There are several that I have not mentioned in this letter, — I mean several fine proofs. Logic and exact reasoning are a good deal more useful than you think. The reason is that all that is exact and the science of logic especially deal with *possibilities* while the turn of your mind is toward *actualities*, and you do not fully appreciate possibility. Mathematics, for example, deals with nothing but hypothetical states of things, which far more often than not are either known to be false or extremely dubious. And the foundation, mother, and essence of possibility is subjective, in us, dreams. Your force is a penetrating judgment of what *is* the case in empirical matters.

Dec. 30.

Part of the above was written today, but it just occurs to me to put a new date to it. I am frightfully done up not having had a wink, not a wink of sleep the last two nights, couldn't lie still, bursting with intense disappointment at the results of a sale of books, — books that I loved — but whose sale I hoped was to facilitate Juliette's task of getting the house in disposable condition. It makes me sick, it drives me crazy! I can't dismiss it for an instant. Worst of all, after I had got together all the books I could possibly spare, my poor little girl put in some that I had given her in palmy days, French literature with extravagant bindings and as fresh as the day I gave them to her fresh from the famous binderies. And then — Hell! — she suggested that *she* could sell them to friends and I with my cursed fatuity negatived that saying New York was such a great mart for books. Well those books I spoke of brought 35 cents a volume. In 1867 when some paper of mine was getting printed at the University Press, I picked out of a waste paper basket there some old sheets of the Biglow Papers, rather dirty and the last signature wanting. I had a 30 cent binding put on it. That was the only thing at this sale that fetched a sixth of its value. It brought $12 and odd cents! I owe Carus an article I am writing, and I have got to get my Pop Sci articles of 1877-8 ready for him for which (the copyright) he paid $200, the most princely thing he ever did and he will also put them in 18 months' *Monists* and he won't want any thing else from me till that time expires 18 months from next July. As a rule he is by no means punctual in the small payments he does make. I think he would gladly be more liberal, but it is his wife who holds the purse-strings. It was her check that paid that $200. But as for Carus — what you say he is he is — he is a good fellow but his father was head of the Prussian Lutheran church. Paul was brought up in the episcopal palace in Königsburg. Now you know what that particular church is and how Lutheranized with Prussia breeds the most antiscientific, *personal* animus in philosophy!

I have a microscope bought of Nachet Fils in 1860 for about $300. I can give the exact price of the microscope itself. It is in perfect condition in essentials although in coming here the brass base was slightly bent. One skillful blow of a sledge will make it right. It has not only all sorts of attachments but a large lot of interesting slides. Of course microscope lenses in 1860 are not these of today. It would be judged deficient in illumination. It shows the last test but one of Pleurosigma subtillissima. If I could sell it cheap, I should rejoice to make up a little for my characteristic stupidity about the books. Will you mention it to some persons. I will send a full description of it.

I hate to suggest additional trouble to you but there is no particular hurry. Very little work on the house can be done in the dead of winter up here where I can't get the thermometer in the warmest room up to 50° F for long at a time. I remember your grippe too. I am just getting over one myself — But I am the most utter fool about selling anything or in conducting any negotiation or anything connected with human being *in actu*. Everybody too takes advantage of my poverty and I don't get any approach to the prices for what I write that I used to. Indeed I have no employment in writing at all and shan't have for 2 years. So it is no wonder that Juliette heartily agrees with you in the matter of my not deserving her. She says that if she has to pick out another husband, it will be *surtout pas un logicien.* And she adds that in this country it ought to be against the law for a young man to take up such a *folâterie de profession* unless he can satisfy the directors of the Society for the Prevention of Cruelty to Donkeys that he has money enough to hold out till he is 90 and at any rate he should be debarred from wedlock and be "*exilé à une léproserie jusqu'à le rétablissement de sa santé morale!*"

My dear Willie. We both bless you and yours with all our souls and wish you all the Happiest of New Years

and dear me! I have missed the mail!

1910 New Year's Night

The mail was missed. I have got Juliette so that I can leave her in bed and come and finish up this. But it will no longer be Juliette's delicate head and heart that does the writing — of course, it wouldn't be her fingers anyway. But I must record the fact that I went out to the mail and while waiting for it was taken with premonitions of one of my attacks and coming in found Juliette as pale as a ghost — and why women don't say what they want to convey I don't comprehend. But they will say one thing and be in a storm if one has not understood them as meaning exactly the contrary and still worse will understand us others as having said or as *meaning to convey* the exact opposite of the only thing we have had in our minds. It is not to be cured; so it must be endured. For now I have to write in place of her things she is too ill to dictate and she might as well call our rooster to do so! How can I convey a woman's message to a woman? However when I got back after finding I could not safely stand waiting for the mail wagon, I found

Juliette very pale and faint and as she said nothing at all was the matter I could not blunder so far as even to see that she had had a hæmorrhage. She had been thoroughly and thoroughly used up with all she had been doing before I left her and when I found her in such a condition! Then in the evening came the truly delightful letter from Mrs. James. What a faculty of comforting people, of which the secret is that they can see the sincerity that dictates it which Juliette says that it is more warming to body and soul than *many* hot bottles. But she says Mrs. James is not to send one because having found a hot flatiron in her bed objectionable she had "extravagantly indulged herself with a new rubber bottle." But, she says, it is so characteristic of Mrs. James always to be thinking of the comfort of others so keenly. "A lovely thought!" As for the jams she says "It is one of my greatest pleasures to prepare them for her. The raspberries this year were a failure because there were no red raspberries to be had nor currants neither." But Mrs. James is not to think of sending back the bottles since they are cheap, transportation dear, and the jars always handy (for the cook, I suppose. I venture on this commentary. It is probably a gross blunder.) But Juliette who is much altered today is much engrossed with Mrs. James's delicacy of heart.

But this must go now! Goodbye. Happy New Years in long succession is the ardent wish of both of us! I will soon write more about the microscope.

C.S.P.

F. P. E. B. JOURDAIN (L 230a)

P. O. Milford Pa. 1908 Dec. 5.

Dear Sir:[1]

I am very glad to receive your letter, and will answer it as well as I can, and send you some of my papers. But it seems to me unfortunate, — though possibly it may make no difference, — that I cannot, for the life of me, make out the 7th word of your first question, which I will here copy as well as I can: "Do you maintain, with Schröder, that a class (the 'aggregate' or 'Menge' of G. Cantor's work) is a *particular case* of a binary relative?" I reply, Certainly I do not, and had forgotten that Schroeder did. I will set down my view of it. But permit me, first, to protest against calling it 'aggregate' in English, or 'ensemble' in French, or 'insieme' in Italian, when it is a term of logic for which a familiar word is in well-established use, — the word *collection*. But should a student prefer to follow the analogy of grammar and call it an *Essential Plural*, I should not much object though grammarians would not call it a plural and it is the concern not of grammar but of logic. The historical Julius Caesar, the pillow with which Shakespeare makes Othello smother Desdemona, and the procession of the equinoxes make up a *collection* of three objects; and as DeMorgan well said, the members of every collection possess a common character which is peculiar to them. In the case I have instanced, it is the character of having been taken by me on this occasion to illustrate the meaning of the term 'collection.' In like manner in place of Cantor's word *Mächtigkeit*, I would prefer *multitude*. It further seems to me preferable to simplify Bolzano's definition of 'being more multitudinous than,' by saying that a collection, which we may call the *M*s, is more multitudinous than a collection which we may call the *N*s, if, and only if, no relation is possible in which every *M* stands to an *N* to which no other *M* is in the same relation; thus disregarding his other condition that there should be a relation in which every *N* stands to an

[1] To P. E. B. Jourdain (identification made by Max Fisch).

M to which no other *N* stands in the same relation. I am told that Borel has found a perfect demonstration that no two collections should be each more multitudinous than the other, in my sense. I have never had an opportunity of seeing Borel's proof, and I am rather sceptical about it for the reason that some years ago I myself invented what seemed to me a perfect proof, and submitted it to all the subtlest reasoners I knew, who all pronounced it sound; and yet after some months I detected, or thought I detected, a fallacy in it. I am sorry that I have now forgotten what it was. For after all, it may have been sound; just as I admitted that Schröder was right in saying that I had been wrong about a demonstration of one half the distributive principle of logical algebra; and yet I subsequently came across my memorandum of the demonstration, and it has been published in a paper by E. V. Huntington (Trans. Am. Math. Soc. Vol. V, pp. 300 *et seq.*)[2] since July 1904, without anybody's finding

[2] In a footnote to this paper Huntington wrote: "For the possibility of this simplification I am especially indebted to Mr. C. S. Peirce, who has kindly communicated to me a proof of the second part of the distributive law (22a, b) on the basis of this postulate 9." Peirce had succeeded in showing by a very complicated method that postulate 9 is independent of postulates 1-7, omitting postulate 8.

It has already been pointed out in the general introduction that Peirce, Borel, and Cantor are cited in a footnote in the 1917 edition of Huntington's *Continuum*. MS. 203 also refers to the Borel proof. The reference is found in an addition to the First Curiosity in *The Monist* and was "to be added to the long note provided Dr. Carus can find room" for it. The manuscript reads as follows:

> "*Addition*, 1908 May 24. In reading the proofs of this article, which was written nearly a year ago, I find myself in a condition to take, as it seems to me, a long stride toward the solution of this important and dubious question of whether Cantor and Dedekind, ... followed by the general body of mathematicians are right in holding the collective system of irrational and rational quantity to constitute a *continuum*, as I understand they do, or whether I have been right in maintaining that it should be called a *pseudo-continuum*. I still think, for the reasons given in *The Monist*, Vol. VII, pp. 205 *et* seqq., that there is room on a line for a collection of points of *any* multitude whatsoever, and not merely for a multitude equal to that of the different irrational values, which is, excepting one, the smallest of all infinite multitudes, while there is a denumeral multitude of distinctly greater multitudes, as is now, on all hands, admitted. I am obliged to grant, however, that the reasons to which I have just referred, being of the nature of logical analysis and not of mathematical demonstration, leave us, in the present state of the science of Logic, not fully satisfied. I should, therefore, if I had been able to do so, have resorted to a proof by Borel, of the proposition that any two unequal collections stand in the same relation that any two unequal finite collections do, since if they do, it seems to me clear that their units are inherently capable of being put into a linear arrangement in every order of succession and if this be quite satisfactorily proved, I should be satisfied that there is room on a line for a collection of points of any multitude. But to my vexation, I have never been able to procure a copy of Borel's paper; and seeing that my reasons based on logical analysis seem to preclude the possibility of any mathematical demonstration, and knowing by my

any fallacy in it during those $4\frac{1}{2}$ years. So much for the terminology. Allow me to call your attention to a little screed of mine, which I will mail to you, on the *Ethics of Terminology*. Now that England has capped the most stupendous folly of history, which began with first reform of Parliament in 1830, by a reform of the House of Lords, you will have to remember that the only possible salvation of a democracy is to clutch at morality as your last straw. It is all that keeps the United States afloat.

Now, approaching your question, I note that in many cases things that are really collections are given in perception; as the star-cluster Praesepe is to the naked eye, and as all things that are composed of atoms are. Only they cannot at once appear as single objects and as collections. They are not *mere* collections if their members have reciprocal relations to one another, such as that of being relatively near together spatially. The concept of a collection as such is that of an *ens rationis*, or creation of thought. An *ens rationis* may be Real, if we understand by the Real that which possesses such attributes as it does possess, independently of any person or definite existent group of persons thinking that it possesses them. Thus Hamlet is not Real, since his sanity depends on whether or not Shakespeare thought him sane. But that opinion which sufficient discussion *would* render unanimous, or even that opinion which, though it never would become unanimous, should be the only one that were not the only one that would not be

---

own experience the extreme difficulty of either avoiding or detecting a vicious circle in attempting to demonstrate that proposition, I still remain somewhat dubious about that.

But I wish to say now that while the view of Dedekind and Cantor seems inconsistent with the hypothesis that after a point has been [inserted] to denote each rational value between any two positive integers, or between zero or infinity and one such value, or between zero and infinity, the order of succession of the points on the line being the same as the order of values that they severally represent (which would be easily enough done, if one could accelerate his rate of working according to the proper law, and could mark a mathematical point at all, and possessed the means of magnifying the line indefinitely); and if after that a point were inserted in the proper order of succession to denote each irrational value (but how this could be done, I cannot in the least imagine)."

In his *Elements of non-Euclidean Geometry* (1909) J. L. Coolidge on p. 34 in speaking of the idea of *greater than* and *comparable* in the case of two non-re-entrant angles writes: "As for the *a priori* question of comparableness, we have perfectly clear definitions of greater than, less than, and equal as applied to infinite assemblages, but are entirely in the dark as to whether when two such assemblages are given, one of these relations must necessarily hold. (Cf. Borel, Leçons sur la théorie des fonctions, Paris, 1898, pp. 102-3.)" In Borel we find *Note I* (La notion des puissances, *102-110*, and L'égalité et l'inégalité des puissances as the first topic thereunder).

some time maintained for the last time (just as the probability of an event is that ratio of frequency that growing statistics would not ultimately support for the last time), could not be said to depend upon what any particular group of persons think. It seems to me that a collection stands upon the same footing as to reality that does the object of an abstract noun such as *probability* or *weight.* Weight is real, though levity is not. Probability is real in case statistics would approximate to an ultimate limit in the curious way which I have just defined; but in other cases it is not real. Now as to what should be the criterion of the reality of a collection, I think that its taking a definite value under definite circumstances is the correct choice; and from this point of view *Nothing* is a real collection. There is of course but a single nothing: it is the only thing of which every predicate is true. This I believe to be the only logical position. Those who object to calling it *real* are simply using that term in some sense different from my own which it is incumbent upon them to define. I congratulate myself upon its not being incumbent upon me; for I think they have a difficult task before them. It would certainly not do to confound Reality with *existence*, which merely consists in its subject's reacting against all other existents. The next question which naturally arises is whether a collection whose multitude is *one* differs from the sole object it contains. Is the Great Pyramid the same as the collection of Greatest Pyramids? No I say; for one is an *ens rationis* and the other is not. But what of the collection which embraces only the collection whose sole member is the Great Pyramid? They too are diverse. For though both are *entia rationis,* they are not created by the same thought.

I think I have thus said enough to show how I should go to work to answer any general question on the logic of collections.

So I turn to your second question. Undoubtedly De Morgan's paper on the Logic of Relations influenced me much when I came to know it; but a paper of mine in the Proceedings of the American Academy of Arts and Sciences (of Boston, Mass.) Vol. VII has a passage on p. 281 which shows I had been thinking of the matter, though it also shows that I had not advanced but very little way in it. This was read 1867 April 9, when it is clear that I had not seen DeMorgan's paper. Moreover, there is internal evidence that a paper by me of 1870 published in Vol. IX of the Memoirs of the same Academy (I will send you a copy) was nearly complete before I had much acquaintance with DeMorgan's paper. For having used the notation l^w to mean lover of every woman, it was only as an afterthought that I introduced ^lW to signify lover of nothing but women, which a reading of DeMorgan's paper would have shown me to be necessary. As far as my recollection goes, I was in London in 1870 for some months and called on DeMorgan and carried him my paper and he then presented me with his; and I should say from memory unchecked, that almost all my acquaintance with DeMorgan's system was derived from that and his Syllabus which he gave me the same day. But I suppose I must have been more influenced by him at first than this would imply. It was Boole whom I was chiefly thinking of in those days. My point of view remained quite opposed to some chief features of DeMorgan's such as that a proposition implies the existence of its subject, which is bed-rock truth for him. All I admit is that the interpreter of the proposition must have a previous acquaintance with its subject.

As to my definition of the distinction between a finite and an infinite collection, which always seemed and still seems to me to be quite obviously the same that Dedekind (to whom, by the way, I sent my paper, because I was so much struck with his Dirichlet's Zahlentheorie) brought out some half a dozen years later, notwithstanding Schroeder's having got some distinction from proving them to be the same, until long after that; years after, I was entirely unacquainted with Cantor's work, which I first learned from the Acta Mathematica; and as I was abroad all the summer of 1883, I probably did not see that (which I well remember I first saw at the Astor Library in New York, to which I had recommended the purchase of the Acta Mathematica).[3] I say I cannot well have made the least acquaintance with Cantor's work until the winter of 1883-4 or later. The paper in which I defined the distinction plainly shows my ignorance.

The only letter that I remember having written to Cantor contained a proof which I will here give that the collection of possible collections of members of any collection is more multitudinous than the collection itself. In giving this proof I shall use the following two equivalent forms

[3] Peirce had acted as a consultant to the Library in mathematical acquisitions at that time. A letter from J. M. Markoe in the archives of the Astor Library (now the New York Public Library) dated 4 June 1890 reads as follows: "I write to thank you on behalf of the Astor Library for your very full and valuable list of works on mathematical subjects which you deem worthy of a place in our Collection. It is a great help to us in selecting books to have such a careful and thorough Examination made for us by an expert, and we would be glad at any time to receive suggestions from you in the future. As for comparing your list with our catalogue, we will do that work, and you need give yourself no further trouble about the matter. Thanking you again for your kindness."

of expression, where *R* is a relation. '*A* is *R* to *B*,' '*B* is *R*'d by *A*' meaning *A* stands in the relation *R* to *B*. My proof will be a *reductio ad absurdum*. To say that there is a collection, that of the *X*s, which is such that the multitude of possible collections of *X*s is not more multitudinous than the *X*s themselves, is to say that there is a relation, say *R*, in which every possible collection of *X*s stands to some *X* to which no second collection of *X*s is likewise *R*. To reduce this to absurdity, I shall actually describe a collection of *X*s, and show that that collection is not *R* to any *X* to which no second collection of *X*s is *R*. It will facilitate my description of the refuting collection of *X*s to begin by dividing the *X*s into three genera, and one of these genera into two species as follows:

Genus A of *X*s consists of all those *X*s none of which is *R*'d by any collection of *X*s at all.

Genus B of *X*s consists of all those *X*s each of which is *R*'d by one collection of *X*s and by no other.

Species B1 consists of those *X*s of genus **B** each of which (say X_p) is *R*'d by a collection of *X*s that contains that very *X* (i.e. X_p).

Species B2 consists of those *X*s of genus B, each of which is excluded from that sole collection of *X*s which is *R* to it.

Genus C of *X*s consists of those *X*s each of which is *R*'d by more than one collection of *X*s.

Of course, if every collection of *X*s were *R* to an *X* not *R*'d by any other collection of *X*s, Genus B would necessarily be as multitudinous as the entire collection of collections of *X*s. But I proceed to describe a collection of *X*s which is not *R* to any *X* not *R*'d by another collection of *X*s. Namely it is that collection (which I will call the ξs) which contains every *X* of species B2 and contains no *X* of species B1. What *X*s of Genera A and C it may contain I do not care. The collection of ξs is not *R* to any *X* of genus A, by the definition of that genus. Nor is it *R* to any *X* of genus C that is not *R*'d by any other collection of *X*s, by the definition of that genus. Nor is it *R* to any *X* of species B1 since it contains no *X* of that species while no *X* of that species is *R*'d by any collection of *X*s that does not contain it. Nor is it *R* to any *X* of species B2 since it contains every *X* of that species, and no *X* of that species is *R*'d by a collection of *X*s that contains it. It is therefore a collection of *X*s that is not *R* to any *X* un-*R*'d by any other collection of *X*s. But *R* may be any relation whatever, and therefore there can be no relation in which every collection of *X*s stands to an *X* to which no other collection of *X*s is *R*; or in other words, no matter what collection the *X*s constitute, the collection of all possible collections of *X*s is more multitudinous than the collection of *X*s itself. It follows that calling the multitude of all integer numbers the *denumeral* multitude, the multitude of all possible collections of integers which is obviously that of all numerical expressions carried out endlessly into the decimal places, which again is the multitude of all analytical quantities rational and irrational is greater than the denumeral multitude and may be called the *first abnumerable* multitude; and there will be a denumeral series of abnumerable multitudes each greater than the one before it. But there will be no infinity-eth abnumerable multitude, since if there were the theorem just proved would be false of it.

I have made no other contribution to Cantor's theory, from lack of mathematical ingenuity, my forte consisting in logical analysis. I have a complete theory of this process, including its methodeutic, which I base upon my *existential graphs* which is my *chef d'oeuvre.* Yet I consider my theories of scientific reasoning to be of high importance, and also my idea (obtained by logical analysis), that the division of all logical terms into those of valencies 1, 2, and > 2, where 'valency' refers to the fact that, in existential graphs, every predicate has either a single connexion with one subject (as in "it rains" where the predicate is the present phenomenon and the subject is *rain* or *pluviation*); or secondly, it is a dyadic relative between two subjects and has valency $= 2$, as Napoleon was mortal, where Napoleon and Mortality are the two subjects, or finally, it connects more than two subjects, as the word *and* does when expressing as is usual *coidentity*, as in 'Napoleon was mortal and mendacious.' I adopt this trichotomy because no logically indecomposable relative has a valency > 3,  being *demonstrably* invariably capable of decomposition into 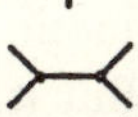. The concepts of valencies 1, 2, and 3 are important *categories.* A proposition can be separated into a predicate and subjects in more ways than one. But the proper way in logic is to take as the subject whatever there is of which sufficient knowledge cannot be conveyed in the proposition itself, but collateral experience on the part of its interpreter is requisite. Thus, if you say 'This rose is red,' a color-blind person will not apprehend your meaning, which shows that the subject has modality, being a *possible* when it is a quality, *de inesse* when it is an existent or actual fact, and *necessary* when it follows from a law or tendency. The result is that everything in a proposition that possibly can should be thrown into the subjects, leaving the *pure* predicate a mere form of connection, such as 'is,' 'possesses (as a character),' 'stands in the dyadic relation ____ to ____,' 'and' = 'is at once ____ and ____,' etc. A *Pure Predicate* should be '*continuous*' or 'self-containing' as '*A* is coexistent with B' = 'A coexists with something that coexists with B,' and 'A possesses the character ϱ' = 'A possesses the character of possessing the character ϱ' and 'A stands in the relation λ to B' = 'A stands in the relation of being in the relation of standing in it to the relation λ to the relation of a relation to its correlate to B,' etc. The function of reason is to trace out in the real world analogues of logical relations. Thus, corresponding to subject and predicate, or that to which a predication refers, and the *predicate* the substance of the predication, reason supposes there is an element of *matter* which gives being, i.e. is that which is, and an element of *form* which is *how* it is. Corresponding to Major Premiss, Minor Premiss and Conclusion, are 1st a law of nature, 2nd a cause which subsumes this or that occasion under that law, and 3rd the phenomenal effect. This is a version, of course, of Kant's Transcendental Logic.

We think in signs; and indeed meditation takes the form of a dialogue in which one makes constant appeal to his self of a subsequent moment for ratification of his meaning in respect to his thought = signs really representing the objects they profess to represent. Logic therefore is almost a branch of ethics, being the theory of the control of signs in respect to their relation to their objects. My idea of a sign has been so generalized that I have at length despaired of making anybody comprehend it, so that for the sake of being understood, I now limit it, so as to define a sign as anything which is on the one hand so determined (or specialized) by an object and on the other hand so determines the mind of an interpreter of it that the latter is thereby determined mediately, or indirectly, by that real object that determines the sign. Even this may well be thought an excessively generalized definition. The determination of the Interpreter's mind I term the *Interpretant* of the sign. But it is necessary to distinguish the Immediate Object of a sign which is the Object as it is represented in the sign from the real thing which is its Dynamical Object in its own Being. And in like manner it is requisite to distinguish the Immediate Interpretant, or the Interpretant as the Sign itself represents it, the Dynamical Interpretant, which is the effectual Interpretation the sign does as a matter of fact produce in the mind of the Interpreter, and the Rational Interpretant, which is the effect it ought to have upon the mind of the Interpreter, as such. I divide Signs by ten trichotomies governed by the 3 categories of Valency in respect 1st to the Nature of the Sign itself as being either a possible Quality, or an Occurrence (i.e. either an Existent thing or an actual fact), or a general form, such as the word 'the' which is one and the same word every time it is written, printed, spoken, or telegaphed; 2nd, according to the Nature of their Immediate Objects, 3rd according to the nature of their Dynamical Objects, 4th according to the relation of the Sign to its Dynamical Object. (And since this is the trichotomy I have most studied, and consequently most frequently mention, I will here say that the division is into, 1st, *Icons*, which represent their Objects by virtue of resembling them, as a geometrical figure in a geometry-book, or as any Diagram, or Array of letters in algebra, where the resemblance is not sensual but is intellectual; 2nd into *Indices*, which represent their Objects by virtue of being in fact modified by them, as a clinical thermometer may represent fever, or a letter attached to a figure of a triangle may from its position represent an angle of the triangle; and 3rd into *Symbols*, which represent their Objects by virtue merely of the certainty (or probability) that they will be so interpreted; as any noun represents the thing for which it stands.) I fear this definition may be open to criticism. Symbols are either conventional signs or they are proper names (or involve in meaning proper names). We have to distinguish Symbols, which are not themselves existent things from *Instances* of them, which are Icons of them. Just as if the word 'the' occurs 20 times on my copy of a certain page of a certain book, those are 20 'Instances' of a single Symbol. In the 5th to 10th places I divide Signs according to their Interpretants.

I go into all this as a preparation for a slight sketch of my theory of the nature of numbers. Although cardinal numbers represent the values of multitudes, yet they are specialized ordinals, since they represent positions in the series of multitudes. Still, they have a peculiar interest since they suggest peculiar, and highly mathematical, definitions of sums, products, and exponentials of multitudes, altogether different from the more abstract ordinal definitions; and their yielding the same rules of arithmetic is a highly interesting fact for the logician. So that I will glance at them. Since a multitude is a *whole*, a definition of *part* and *whole* is in order. A *Part* is a member of a collection in the Being of whose members together with their relationship the Being of another thing, their *Whole*, consists. But an *Ingredient* is a member of a collection of certain quantities of certain kinds of things whose composition in a certain way is the condition of the Being of a certain quantity of another kind of thing. Thus, the Ingredients of a color of a certain hue, a certain quantity of chroma, and a certain quantity of luminosity are a certain quantity of light of the wave-length of C, a certain quantity of light of the wave-length of E, a certain quantity of light of the wave-length of about G, whose being simultaneously emitted or otherwise proceeding from a surface without any regularity of phase is the condition of that surface appearing to a normal eye undistinguishable in hue, chroma, and luminosity from the color of the light first mentioned. ... In various cases what are called 'parts' are, properly speaking, conditions of the Being, or production of that result which is called their 'Whole.' This is important to the sketch of my view of numbers (though there is nothing very novel about this) since that is wholly a work of logical analysis; and the *parts* into which such analysis decomposes concepts are "parts" in the sense of being conditions.

G. C. J. KEYSER (L 233)

P. O. Milford Pa. [1-7 Oct. 1908]

My dear Professor Keyser:

I intimated to you, before you went away on your vacation, that I desired to propose to Mr. Jacks a series of four articles to be written by me which should consider how our fundamental hypotheses of physics and psychology ought to be modified in case it should be accepted as proved that there have been and continue to be occasional communications with the dead.

When the American branch of the Psychical Research Society was started I refused to join it, on the ground that I did not think science was prepared to grapple with these problems. I did not then think, and do not now, that the psychologists since Fechner have shown much power. I do not mean to say that I could show them how to do better; but I do not think they have produced a transcendent genius who could show them how to go to work. In part I must acknowledge that I was mistaken.

I should show in my article how unlikely it would be that they should reach any positive results in working in the only way they saw their way to work. The evidence is of the most execrable kind. But one thing they certainly have done: namely, they have brought forward an enormous mass of utterly unscientific observation, which certainly staggers one. In my first article I should criticize the psychologists and the psychical researchers, not that they could have done better, but that the world is waiting for the genius who shall show us how to attack such problems. I shall take the theory of telepathy as an example of incompetence. Myer enunciated most vaguely and in mere negative terms. That sort of thing will not serve for a scientific hypothesis, which should boldly propose a proposition from which *predictions of observations* can be made.

However, if we are staggered by the mass of unscientific observations that the Psychical Researchers have adduced, as I confess I am, then what *I* can contribute to the study, by the aid of my logical studies, I will do in these articles. And the first step is to ask, supposing the phenomena of spiritualism are "veridicent," what do they disclose as to the character of the other life? (For my criticism of telepathy will go further than I can explain in this letter, and I think spiritualism makes the proper hypothesis in the premises.) The answer to that question will be given in some detail; but it may be summed up in saying that it would appear that the other world has a great resemblance to self-deception. And yet I would not infer that it is so. On the contrary, those characters may be sufficiently explained otherwise. It is too shocking to suppose the very respectable and intelligent persons who make the sacrifice to truth of telling of experiences so suspicious were temporarily insane etc. etc.

But, I shall say, regardless of those manifestations our fundamental hypotheses require modification anyway. Logic requires it. I shall then make a few remarks about logic, — the insufficiency of all modern logic. The logic of Aristotle answered very well the purposes of medieval thought. (It did not help ancient science, but the several branches of ancient science had their several logics which did help them. The logic of mathematics used by Euclid was far superior to Aristotle's in recognizing the absolute need of a "construction" παρασκευή. The logic of the Epicureans had a *close* analogy with the system of J. S. Mill, and was about as serviceable.) But no system of logic has approached adequacy for modern thought (except my own which remains unpublished because of my narrow means and the absence of all encouragement. However, I won't *say* that. I am very near 70 years of age, and I am doing my best, as well as starvation and penury permit, to give it to the world. It will do me no good, because I have no faculty for business. I only turn from it to give an illustration of it in the proposed series of articles, provided the honorarium is sufficient to enable me to take a year out of my more theoretical work, and from the *Monist*.) Mill had a pretty good genius for logic; but he was dragged down by that wretched metaphysics of his father and of Bentham. When I say it was wretched, I mean the effects of its errors on Mill's Logic were so, for in spite of its errors, it was *relatively* exact thought, and did the world immense good. Mill's Logic is, I suppose, still held in favor, not to say in awe, in England. I shall only show that all the scientific work he adduced as admirable in his first edition turned out pretty poor stuff, showing that his logic does not distinguish good from bad reasoning. Aside from Mill's there is no system of logic very helpful to science. Jevons saw some of the weaknesses of Mill, but produced a poor book with no logic in it, to speak of. However, I shall hurry over those writers and show very briefly what logic really is. All our thinking takes place in signs of some kind. The premisses are signs of the conclusion. The nature of signs is what logic really ought to study and there are ten principal points of view from which signs ought to be regarded and from each of these points of view, signs may be divided into classes. There is none of the resulting genera which is not employed in reasoning somewhere. However, I shall not go into that but shall show, what the function of a scientific hypothesis is, and that one of the rules that it ought to obey should be: Whenever it is proved that a given form of idea is important in science, the question arises, How far is it important? And the way to reach the answer to that question is to begin by carrying that form of idea as far as possible, until there is positive evidence that some other opposing Idea prevents its being carried further. Such an Idea is that of Evolution. Its logical value is much greater than Spencer suspected, who most illogically wished to rest it upon the antagonistic concept of *persistence*. When we see the enormous importance of evolution, both in the moral and in the physical universes, how the whole world seems to have been designed, not to be perfect, but to rise, and grow, and ameliorate, I declare that it is urgent that the idea of evolution should be extended far beyond Spencer's conceptions, both as to the Physical and as to the psychical universes. Now this I have done and have traced its consequences, and although the physical world has, in regard to its laws, reached a nearly stationary condition, a condition of maturity of growth, yet I have been able to deduce from the hypothesis of the Laws having been due to evolution, not merely their principal known characters, but one or two others that were not known at the time I worked them out, but have since been discovered. Of course, there is a limit to the introduction into our hypotheses of evolution in time. Namely the law of time itself cannot be so explained; although on comparing physical with moral causation we do come upon a relic of evolution. Yet time itself forces us to a wider view of evolution. For time is the image of logical sequence and a logical, Hegelian-like evolution must be recognized as logically preceding the temporal evolution.

Evolution looks backward toward homogeneity; and thus the hypothesis requires us to suppose an affinity between the three Universes of Ideas, of Matter and of Soul. Now there must in fact be such an affinity, since the two latter mutually react, which would be impossible if they had no common nature; and Ideas act on the other two. And although the Universe of Ideas seems to have reached Maturity *logically before* Time, yet the Universe of Ideas appears to show some traces of a common origin with the others. But this seems a difficult point.

That is a sketch of what I should try to express in my first article; or at least to get into it as much of this as I could render clear.

In my second article I should consider the psychical, the pneumatic world; and here I should at first seem to be very materialistic. Naturally; for as has often been remarked, spiritualism is a sort of materialism. It seems to me that psychology since Lotze has not been pursued with very much ability. The old associationalism of Hartley (and Gay) and Berkeley and Hume seems to have been begun with very great ability. But as for the present state of the science, look at the way in which many of the psychologists talk of consciousness. Why do they not try to find out something of the constitution of the soul, just as physicists had to begin by getting some notion of the constitution of matter? Is it because we know nothing of the *soul* except by its phenomena? The very same thing is true of matter but it does not deter physicists from investigating the constitution of matter with brilliant success. It looks to me as if, in their secret hearts, the psychologists thought that the only substance beneath manifestations of mind was certain special varieties of albuminoids or beta-amino-acids, but hadn't the courage to confess to themselves what they thought. It is a rule of the logic of hypothesis that whatever one finds an impulse to believe one should develop into so definite a form that experiment and observation may have a fair opportunity to refute it, if it be not true. The proposition so developed should thereupon be adopted provisionally and students should at once go hard to work to develop its consequences and compare them with the facts. To me it is difficult to believe that either the alpha or the beta amino acids produce Feeling, by mere physical action without any different sort of [cause]. But I would much sooner admit *that* provisionally than to rest upon consciousness, as many do, who even go so far as to *define* psychology as the science of consciousness. For it seems to me that their "consciousness," as they talk of it, is a close parallel to the substantial forms of the medieval schoolmen. For instance, what a rumpus they kick up over the proposition that consciousness is an *epiphenomenon*! Now it is very easy to trace the genesis of the concept of consciousness. We start from any phenomenon of perception, we abstract from certain characters of it, — and especially from all those of dynamic activity, — to prescind pure awareness. Of course it cannot be dynamically active, because we have *abstracted* from that. Consequently to say it is an epiphenomenon is a mere analytic proposition. Then why all this flurry? Because psychologists think that consciousness is *substantial*. It is, in short, a scholastic *substantial form*. What is the subject of consciousness; what is it that is conscious? If I were a psychologist, I swear I would not go to bed until I had picked out some definite verifiable hypothesis on the subject which I would devote my whole being to developing, until it was disproved, a result that I would pursue with all the ardour of *my* "subject of consciousness." To say that consciousness has *no* subject is to erect it into a substantial form. To say "we don't know," is sloth! *The business of a man of science is to guess*, and disprove guess after guess, being guided by the particular way the last guess failed in forming the next one. A scientific genius has seldom had to guess as many times as Keppler did. (I spell him with 2 pees because so he always signed himself *in German*. Usually writing Latin, he was "Kepler.") I shall give in my article very strong reasons for disbelieving that ordinary matter can be conscious except possibly in an almost infinitesimal degree. One of them is that the subject of consciousness takes habits, which matter can only do under certain circumstances very well worth analyzing. Namely water will wear a bed for itself and shows all the characteristics of habits in doing so; and it is very instructive as showing what is requisite for habit-forming and habit-breaking. Namely it needs a body of considerable resistance and stiffness which can be overcome by the force of a more yielding body. Now I grant that the amino acids may be the former of these. But what is the latter. Habit is not the only character that is significant. There is Fechner's Law, a greatly significant fact which the psychologists absurdly reject for the most part because forsooth sensation cannot be measured! This shows that they have not studied the nature of measurement as it has been elucidated by the two brilliant mathematicians Felix Klein and William Clifford. Nor do they know that Number is *in all cases* purely ordinal at bottom. But Fechner showed by his own experiments and Weber's, and he would have done well to adduce Ptolemy's star magnitudes to which all the operations of arithmetic can be applied *in all their significance*, — he showed that sensation, as naturally measured by us, just as Ptolemy did (or Hipparchus more likely) is proportional to the *logarithm* of the excitation. Now the logarithm is one of the most distinctly significant of all functions; and here is a great aid in forming our hypothesis which the ageometric psychologists have stupidly neglected. It is one of their most striking stupidities. The most fundamental way of showing what the relation between a logarithm and an antilogarithm or exponential is, is to say that if we have a number of compartments, or headings, equal to the base of the system, and a plural of objects equal to the logarithm, then the antilogarithm is the multitude of ways in which the objects of that collection may be distributed under those headings. Thus if 2 is the base, and we call the heads A and B, *three* things can be distributed in 2^3 or 8 ways

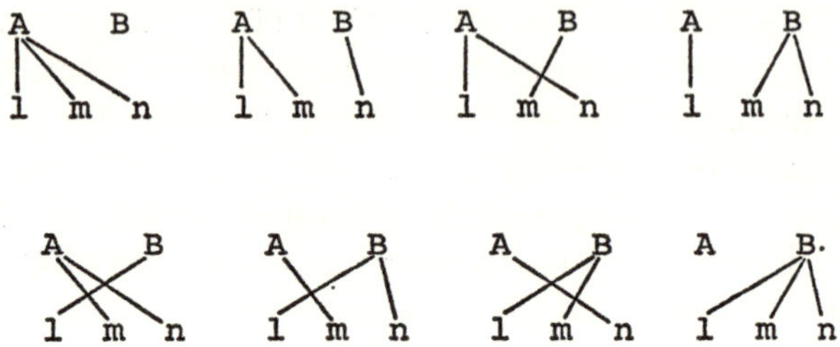

(Of course you know most of these as well as I do, but I don't know whether Mr. Jacks, to whom I hope you will think it worthwhile to submit this letter does so or not. I will show in a P.S. how I show the ordinal character of rational numbers.) Now state the relation as you please, so long as it exhibits its true nature it will come to the same results, the first of which is that the logarithm expresses the value of the more fundamental, elementary, and less *evolute* thing. We are thus assured that the *sensation* is nearer the very substance of the world than the physical excitation, or force, is.

I will make the following tentative hypothesis, which I may very likely improve before the time comes to write this second paper. If not, it will be better than none as I now state it. For to penetrate into secrets one must go on some definite, or tolerably definite, hypothesis. I will suppose that in the infinite past, after the purely logical development, at the time of the creation of Time, the Universe of Ideas was already tolerably distinct from the others, but that the Universes of soul and of matter were not distinct, there being throughout space (not necessarily restricted to three dimensions) a sort of fluid under no very definite laws, which was somewhat conscious but without distinction into separate personalities — a sort of *soul-stuff*. I will suppose that, in the absence of laws, this fluid had various motions in different parts. These motions, from the empirical point of view, i.e. as they would have appeared, had there been any person to observe and study them, would have appeared to be without any law, and therefore to be absolutely fortuitous and arbitrary; while from a higher point of view, they could only be regarded as effects of direct creative energy. The fluid, originally homogeneous, in consequence of these various motions, mixed with a capricious (i.e. creative) viscosity set up vortices. These increased in number as time went on, until these vortices became the atoms of a secondary fluid, — just as vortices are atoms of a fluid in the vortex-theory of atoms. In this fluid, in which the viscosity, or internal friction, was somewhat greater, new vortices were set up, just as, according to the vortex-atom theory, there are vortices in rapidly and variously moving water, while the water itself consists of a collection of vortices in an underlying fluid. I will suppose that in this second fluid, governed by habits altogether independent of consciousness, there was a secondary consciousness far less intense than in the underlying fluid. I will suppose that this second fluid came to have such a multitude of vortices that they constituted a third fluid. And I will suppose that this process went on until there were an endless (or nearly endless) series of fluids each composed of vortices of the next underlying fluid. All this is easy to conceive. In the final, more composite, fluids, Feeling would have almost disappeared, and they would be regulated by formed habits. For the formation of habit always lowers the intensity i.e. vivacity of Feeling. Finally the last of these fluids, or rather the totality of its vortices, is what we recognize as matter.

Of course, all this complication of vortices of vortices amounts, after all, to nothing but a very intricate multiple vortex in the fontal fluid; but one set of vortices being billions of times smaller than the next, so that there is a vast series of fluids, each almost homogeneous, although in the last resort it amounts to but one fluid with a multiple vortex motion, yet that series of phenomenal fluids (i.e. what would be phenomenal, if men were to carry science so far as to discover them empirically) have, each of them, a special nature of its own, quite distinct from each of the others, and a special mode of being of its own. When I speak of "fluids," I use that term for simplicity, without meaning to deny that parts of them may approximate to rigidity, so as to appear as solids.

We are so placed in this perhaps both-ways endless series of fluids that we find our ordinary matter to consist, apparently, of a pair of vortex-fluids, the vortices of the two members of the pair being very near the same size, — that is, the larger not having radii perhaps a hundred times greater than the smaller. The larger we call atoms the smaller electrons. If there are but about 1000 electrons to an atom of hydrogen, the linear dimensions of the electron are presumably about a tenth that of the H-atom. Now the Uranium atom being only about 6 times as large as the Hydrogen atom, — more likely less than more, it is true, — one would expect an electron to differ no more from a hydrogen atom than a Uranium atom does, the other way, which I can hardly believe. Perhaps, for some reason, a vortex more than 1/60 of the size of another will be much more independent in its motions than if it were smaller. An electron would be 1/62 of an atom of U, but 1/59 of an atom of Pb. The Pb then has thrown off all its inherent electrons while a U atom has not quite. However, I am by no means sure that the hypothesis of vortex-motion will answer the purpose. I shall probably have decided that question in a few months. If not, it will be necessary to suppose motions of small range in a fourth dimension. But should I not have made sure of this point at the time I write my article (which is just possible) I don't think it will be ruinous to my general idea to leave this point uncertain.

It is I think distinctly less repugnant to good sense to suppose electricity to have Feeling than to attach this idea to ordinary matter. Nevertheless, I would not suppose that; but would suppose that there were numbers of intermediate fluids, — and perhaps an endless series of them, — before we reach the quasi-fluid whose motions are feelings. And I am inclined to think that the more "subtle" of them, as I will call those that are nearer to the "soul-stuff" have motions in higher numbers of dimensions than 3. It would be contrary to the principles of logic to suppose this without reason, that is, unless it would positively improve the hypothesis in some way. But there certainly *is* reason, though I will not now go into it since I have not decided whether it is sufficient to overcome the objection to complicating the hypothesis or not. If it be adopted, matter will be thereby supposed to be a phenomenon on the relative or quasi *surface* of the soul.

My hypothesis of vortices of vortices, — for which, by the way, can you suggest a good name? How would the *Introvortical* theory do? — my hypothesis accounts perfectly well for the effects of matter upon the mind. For I must suppose, and indeed, it is *inherent* in the nature of the hypothesis that all these fluids have a slight viscosity so that the motion of one vortex is easily converted, in small part, into a motion of sub-vortices. It will be much like production of heat by friction, or by stirring water so as to throw it into vortices, and destroy them again. But the explanation of the action of the mind upon matter is by no means so easy, — The analysis of the phenomena of volition and self-control — and what is called free-will is nothing but automatic regulation, of a very complicated nature, perhaps indefinitely complicated, — this analysis shows, apart from all dubious hypotheses, two things clearly: 1st, the dynamical action of the soul on the body is very slight indeed, comparable to pressing a telegraph-key to explode tons of nitroglycerine or iodide of nitrogen; and 2nd, that there certainly *is some* energy imparted to the matter. It must be remembered that all attempts to measure the work done in labor and account for it by physiological changes are much too rough to strain out any minute quantity. The non-scientific mind has the most ridiculous ideas of the precision of laboratory-work, and would be much surprised to learn that, excepting electrical measurements, the bulk of it does not exceed the precision of an upholsterer who comes to measure a window for a pair of curtains. The measurement of ten feet to a quarter of an inch is more precise than ordinary chemical analysis, the standard usually aimed at being 1/3 of one percent. Each atomic weight is the result of the expenditure (on the average) of many hundred times as much industry and pains as an ordinary analysis, so that they must be several tens of times as accurate. Yet the International Atomic weights for 1904 which give every significant figure that is supposed to have any significance, give in

3 cases 5 significant figures
32 cases 4 significant figures
36 cases 3 significant figures
6 cases 2 significant figures
1 case 1 significant figure

The soul then *certainly* does act dynamically on matter. It does not follow that it acts *directly* upon matter, because there may be involved an[1] *endless* series of transformations of energy from that of a motion of one fluid into that of motion of another, all these fluids being of a spiritual nature, followed by a *beginningless* series of transformations of energy from that of one fluid into that of another, all *these* latter fluids of a material nature. To get a clear idea of this, consider the spiral whose equation is

$$\theta = \frac{1}{\log\left(\frac{1-r/a}{1-b/a}\cdot\frac{b/c-1}{r/c-1}\right)}$$

where $a > b > c$

[1] There is another version of this page with the special illustration beginning at this point. The spiral given there is labelled Fig. 2 here. The page reads as follows: "*endless* series of transformations of energy from motion of one fluid to motion of another, all these fluids being spiritual, followed by a *beginningless* series of transformations of energy from motion in one fluid to motion in another, all *these* fluids being material. It is just as a spiral within a circumference *A*, — it is easy to write the equation of it, — may make an endless series of turns before it reaches an inner circumference *B*, and may then keep right on making a beginningless series of turns before it reaches a third circumference *C*. $\Theta = \log(r-c)$ or even $\Theta = \frac{1}{r^2-c}$ is a sufficient illustration."

Incidentally, at Peirce's request, Keyser returned on 12 November 1908 the letter printed in this collection.

Then obviously if $r > a$	θ is imaginary
if $a > r > c$	θ is real
if $a > r > b$	θ is negative
if $r = b$	θ is infinite
if $b > r > c$	θ is positive
if $c > r$	θ is imaginary

Consequently the spiral begins with $r = a$ and ends with $r = c$. At those values of r, $\theta = 0$. It winds round in a negative direction and takes an infinite series of turns before $r = b$. It then keeps right on in the same direction θ diminishing with r and taking another infinite but beginningless series of turns before r gets fairly away from $r = b$. This illustrates a beginningless series following an endless one [Figs. 1, 2].

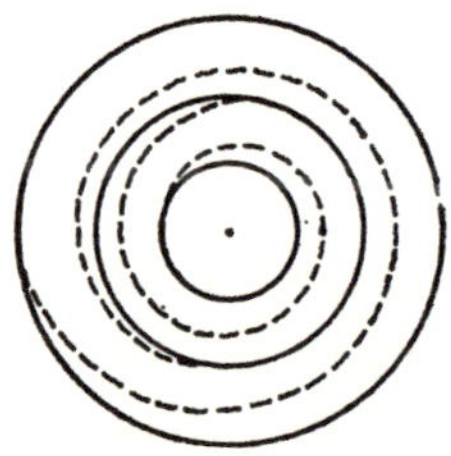

Fig. 1

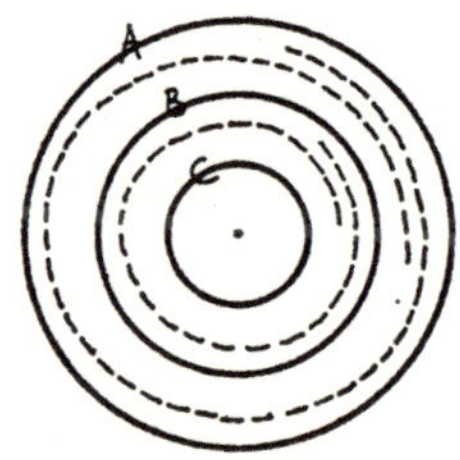

Fig. 2

If we suppose the series of fluids thus to consist of an infinite series of spiritual and an infinite series of material ones and that the diameter of the vortices of each is a million times greater than that of the next less material one, the diameter of the soul-stuff atoms will be (writing M for a million) $RM^{-\infty}$ which is an infinitesimal of the *infinite order*, which some may think appropriate for soul-stuff.

For myself, I shall not suppose that the fluids are infinite but only very numerous. At least, not until I have some good reason for this.

I might say a great deal more about the soul; but I don't want to give my readers too much at one meal.

This vortical theory, or something like it (I don't feel satisfied that I have quite the right hypothesis) ought to be accepted "anyway," regardless of the question of another world. Ought, I mean to be accepted as a hypothesis, i.e. as what we intend to believe, look forward to believing, in case its consequences should be verified. When scientific men have once done something like that, and have recognized that we already have ample evidence that the body acts on the soul, and that the soul acts dynamically on the body, they will cease to view with scorn the idea of communicating with the dead. I confess it is hard to believe evidence of so bad a kind as we have of that, though it is at least equally hard not to believe such a great mass of evidence, bad as it is. I may have a paragraph upon how any amount of bad evidence can be worth anything. They may think positive evidence is in defect; but they can no longer think it a thing not worth looking into. On the contrary, it will seem a plausible thing but needing good evidence to support it.

H. E. H. MOORE (L 299)

Milford Pa 1904 Jan. 2

My dear Professor Moore:

I wish you the happiest and most successful possible of new years, and I write again concerning that proposition concerning multitudes.

The principle that if each one of a class of propositions is possible in every possible determination of the others in respect to truth and falsity, all are possible simultaneously, holds for classes of existent and destined facts; but it does not hold for the possibilities of possibilities of which an ultranumerable collection must be composed; and therefore my proposed demonstration is fallacious. Having discovered this, it seemed to me that the bases of the whole theory must be carefully reconsidered. This work I have accomplished, and as a result of it, I find it necessary to define a *collection* (Menge).

Cantor gives no definition of it at all, although he thinks he does so. He uses a false method of definition much affected by Kant. But the process never can yield a definition in any proper sense of the word. This general question I will not now discuss. Suffice it here to say that Cantor's supposed definition is not a mathematical definition. A mathematical definition is either purely verbal or it analyzes a conception. In either case, it must be useful in mathematical demonstration. If it analyzes a conception, it involves hypotheses which must in some way be made explicit before the mathematical theory can be developed.

What I say is that the hypotheses of the theory of multitude are not made explicit hitherto. I proceed to show what ought to be stated. For that purpose, I must first explain a few terms used by me. By a *rheme* I mean a blank form resulting from removing certain parts of a proposition and leaving a *blank* in place of each, these blanks being of such a nature that if every one is filled with a proper name a proposition will result. A *rheme* might therefore be called a relative predicate, that is, one requiring several subjects.

By a *lexis* I mean a rheme which in a statement of a proposition is left unanalyzed, like a single word (*Lexis* is the Greek for *word*).

By a *rheme of otherness* I mean such a rheme that every lexis it involves is one or other of these two

____ is not ____
____ is at once ____ and ____

Let us now consider what we mean, in the doctrine of multitude, by a collection, or *Menge*. In a wider sense, a collection is an individual object such that the truth of any proposition asserting any predicate of it consists in the truth of a relative predicate of certain individuals called its members, or, what is the same thing consists in the truth of a proposition all the blanks in whose rheme are filled by proper names or designations of certain individuals called the members of the collection. In this sense, a regiment is a collection. For whatever is true of the regiment is so by virtue of some state of facts about its officers and men. But the collection of the theory of numbers is a particular species of the collection in that general sense.

Whatever lot of individuals you please have some character that is common to all of them and is possessed by nothing else in the universe. For every lot of individuals is capable of being exactly described, and that description is a character common and peculiar to them. If I draw up a list of things most heterogeneous, the fact that they are on this list is such a character.

Now if we say that *every* character defines a collection, we must in the first place recognize *nothing* as a collection. The collection, then, is an *ens rationis*; and consequently the collection which has but a single member is a distinct object from that member, since the latter may be a self-subsisting individual. Moreover, under this understanding, the identity of the collection seems to consist in the identity of its common and peculiar character. We might call a collection in that sense a *sam*, as a provisional designation.

But if we speak of a *universe*, meaning an object in which individuals present themselves to us which individuals are the principal subjects of our discourse, and if we form another word *gath* (first syllable of gather) and say that a *gath* is any individual object whose existence is neither more nor less than the presence in the universe of whatever objects are present there of which a general predicate is true, so that it is not the character of the predicate but the existence of the objects that constitute the gath, then its identity is the identity of those objects regardless of their character (although this has to be referred to in order to describe the gath) and a gath of a single member is that very member, and a gath of no member is nonexistent.

I think we must say that in the doctrine of multitude there is supposed to be a collection of all the objects of any description that applies to any object of the universe.

In the next place if two sams be such that every member of either is a member of the other, those two are one individual collection in the sense of the theory of multitude.

To say that a collection has such and such a multitude can only mean (in advance of theorems about the properties of multitude) that certain rhemes of otherness are true of members of it and certain ones are not. (I do not mean to say here that a collection may not be such that every rheme of otherness is true of its members nor that it may not be such that no rheme of otherness applies to its members; that is a trifling matter.) When we suppose a collection in the theory of multitude, we suppose that there are some rhemes of otherness that are true of members of it and presumably that there are some that are not true of any members of it.

It is not expressly supposed that one collection is different from another in any other general properties than that.

It is, on the contrary, vaguely and implicitly supposed that all collections are alike in their general properties except those of multitude, — I mean *those* general properties which concern the doctrine of multitude. But how to make this explicit and distinct I hardly know.

Why not adopt this postulate?

If a collection, the *m*s, consists of *m*s that have the character *a* and *m*s that want this character, there is a one-to-one relation, ϱ, in which every *m* stands to an *m* while to every *m* an *m* stands in that relation, and every *m* that is *a* is ϱ to an *m* that has a certain character *b* and every *m* that is not *a* is ϱ to an *m* that is not *b*, and this relation may always be such that either no *m* that is *a* is *b* or else every *m* that is not *a* is *b*.

In some way or other, the homogeneousness of every collection for the purpose of the theory of multitudes must be expressly stated or else we must be allowed to suppose it in each special case on the ground that it does not conflict with any express hypothesis.

In either way, the above proposition must practically be allowed. But from this the proposition in question follows in the most obvious manner, as a mere corollary.

For if the *p*s are not in one-to-one relation to the *q*s, let us call the collection of all that is either *p* or *q* the collection of *r*s. Then, on the assumption just made, there are two collections the *p*s and *q*s which make up the *m*s such that all the *p*s are in one-to-one relation to all the *p*s and the *q*s to all the *q*s and either every p is a q' or else every q is a p'. But the former alternative being excluded we have only the latter.

So that is the state of the case. If you allow me to assume anything to be possible that does not conflict with an express hypothesis, the proposition is at once true, since it only expresses a possibility of that description as is easily shown.

If, however, this is not allowed the hypotheses of the theory have to be enlarged so as in some way to express the homogeneousness of collections; and when this is done, the proposition must follow as an easy corollary from the additional postulate or clause of a definition that is wanted.

P. O. Milford Pa. 1904 Nov. 21

My dear Professor Moore:

§1. This letter being a long one, I shall divide it into sections numbered so that §*m*,*n* is the *n*th division of the *m*th section, etc. I will append a table of contents. In this table I mark with a point hand (thus: ☞) certain parts that I should like to have you read even if you haven't time to read the rest, though I shall not insert anything that I do not think will prove interesting to you, unless it is something that I particularly desire to bring to your attention and therefore mark ☞.

§2. As you urged me some time ago to draw up some mathematico-logical paper, I now propose to sketch the contents of a paper I gave the Academy last week. The whole thing is completely worked out and I will explain the obstacles to my writing it out and printing it, with a suggestion of how you might perhaps help me to do so.

§2,2. The paper contains two definitions of the system of positive integer numbers.

Definition of *Ordinal.* By an *ordinal* I mean a positive integer number. I do not defend the designation. But having nothing to do in this paper with Cantor's ordinal numbers, it is convenient here to use this word in this sense.

Definition of *Definition.* By defining a system of objects that have no other being than as creations of thought, and therefore have no characters, whether positive or negative, than such as thought has endowed them with (so that the principle of excluded middle does not apply to them) I mean stating certain relations subsisting between them, from which every relation subsisting between any of them is deducible.

But while every such statement is a definition of the system, yet the purpose of a definition is to make the different properties of the system distinct, and the further the analysis is carried the better. The aim should be to make as many distinct clauses stating independent relations as possible; though in some cases there may [be] a limit to this. Of course, if the objects had any other being than as creations of thought, and were N in number, there would be N independent characters, and many of these would be of no present interest, and their statement could only be confusing. But with creations of thought this is not usually, if ever, the case. It should be the intention to make the different clauses of the definition as independent as possible; but I do not propose to take any elaborate pains to avoid implying anything in one clause that is implied in another, but leave the definitions to be revised so as to effect this object.

§2.2.2. As a preparation for the First Definition of Ordinals I have to make a division of characters into the *Ordinary* (or continuous) and the *Singular* (or discrete); and I have to define the relation of being *intermediate* as subsisting between combinations of characters.[1] ...

§2.2.3. As a preparation for the Second Definition of Ordinals I explain a locution and define three terms. (One definition I insert in writing this, but it occurs in earlier printed writings of mine.)

The Locution. Instead of the clumsy phrase "P stands in the relation r to Q," I say indifferently, as convenience dictates, either

P is r to Q,
or Q is r'd by P.

Definition of *Appurtenance.* By an appurtance, I mean any relation, r, whose essential nature is repugnant to any one subject P being r to two different individual objects or to two mutually exclusive generals.

Definition of *Comparative Fulfillment.* I term any relation, r, a *comparative fulfillment* if, and only if, to say that A is r to B, is by the very nature of the relation, the same thing as to say that nothing can be r of

[1] A whole page of MS is missing.

A without being *r* of *B*.

Corollary. Since whatever is *r* of *A* is identically *r* of *A*, it follows that if any relation, *r*, is a comparative fulfillment it is a *suilation*, that is, is a relation in which every individual stands to itself. (See my Nomenclature and Divisions of Dyadic Relations, 1903, p. 7.)

Definition of an *Exclusively Existential Relation*. In the work just referred to, I remark that my writings concerning the logic of relations are confined to existential relations, which are there defined as relations which can subsist between two existent objects. I there inadvertently define an existential relation in that way. But my writings alluded to are restricted to relations which *only* subsist between existent objects. Not to change a definition of a term once printed, I will call such relations *exclusively* existential. By *existent* is here meant occurring as individuals in some universe, which may, however, be ideal.

§2.2.3.2. *Second Definition of Ordinals.*

Clause I. There is a certain appurtenance, *N* (called 'being next higher than'), such that, whatever exclusively existential relation between ordinals *r* may be, and whatever ordinals *P* and *Q* may be, either from the facts that certain ordinals are *N* to certain ordinals and that certain ordinals are not *N* to certain ordinals, taken in conjunction with the clauses of this definition, it follows by logical necessity that *P* is in such a relation to *Q* that, by the definition of *r*, *P* is *r* to *Q*, or else *P* is not *r* to *Q*.

Clause II. Every ordinal is *N*'d by an ordinal.

Clause III. There is an ordinal that is not *N* to any ordinal; (and such an ordinal is called a zero, or 0).

Clause IV. Whatever ordinal is *N* to an ordinal is also in a certain relation, *g*, of comparative fulfillment to that ordinal (called 'being at least as high as' that ordinal).

Clause V. Whatever different ordinals *P* and *Q* may be, either *P* is *g* to *Q* or is *g*'d by *Q*.

§2.2.4. I prove that either of these definitions considered as a series of theorems, can be deduced from the other, and that all the universal exclusively existential relations between ordinals can be deduced from either definition.

§2.2.5. It is not without interest to define Addition and Multiplication and to deduce their properties from the definitions. I here set down the definitions with some remarks. One is embarrassed to choose between the innumerable modes of possible definition. From the point of view of the first definition of ordinals which considers the relations of characters of individual ordinals we may define the two operations by means of the relations between relations between ordinals, somewhat as follows.

First Pair of Ordinal Definitions.

Any ordinal, *S*, is said to be the *sum* of an ordinal, *P*, as *addend*, and an ordinal, *Q*, is *augend*, if, and only if, *S* stands in such a triadic relation, *s*, to *P* for *Q*, that

1st, Whatever ordinals *I* and *J* may be, there is an ordinal that is *s* to *I* for *J*.

2nd, Whatever ordinals *K*, *I*, and *J′* may be,

either *K* does not possess every character of the system possessed by *I*, and whatever ordinal *J* may be, *K* is not *s* to *I* for *J*;

or *K* does not possess every character of the system possessed by *J′* and whatever ordinal *I′* may be *K* is not *s* to *I′* for *J′*

or *I* possesses every character of the system possessed by *J′*, when there is an ordinal *I′* and an ordinal *J*, such that
I′ possesses every character of the system possessed by *J*, and *K* is *s* to *I* for *J* and is *s* to *I′* for *J′*

or *J′* possesses every character of the system possessed by *I*, when there is an ordinal *I′* and an ordinal *J*, such that
J possesses every character of the system possessed by *I′*, and *K* is *s* to *I* for *J* and is *s* to *I′* for *J′*.

I think no other clause is needed, but not having examined the matter very closely I may very likely be mistaken. [An]other definition consonant with the first definition of ordinals is as follows:

Any ordinal, *S*, is a *sum* of an ordinal, *A*, as *addend* and an ordinal, *B*, *augend*, if, and only if, there is a dyadic relation, *t*, such that

1st, every character, λ, of the system that is possessed by a sum, *S*, but not by the augend, *B*, is *t* to a character, α of the system possessed by the addend, *A*, α being such that it is *dependent upon*, that is, is not possessed by any ordinal that does not possess, every character of the system possessed by *A* that is *t*'d by some character of the system possessed by *S* upon which character λ is dependent, and α further being such that no character of *A* that is dependent upon α is *t'd* by λ or by any character upon which λ is dependent.

2nd, every character, β, of the system that is possessed by the addend, *A*, is *t*'d by a character, μ, of the system that is possessed by *S* but not by the augend, *B*, this character μ being such that it is dependent upon every character of the system possessed by *S* but not by *B* and that is *t* to a character of the system possessed by *A* and having β dependent upon it, and further μ being such that no character of the system possessed by S but not by B which character is dependent upon μ is t either to β or to any character upon which β is dependent.

Such things are much more lucidly expressed in my old universal algebra of logic, which I will briefly explain. Letters on the line have mostly subjacent indices. Except for the two letters Π and Σ (which after all are not truly exceptions) every letter on the line expresses a predicate whose *individual* subject or subjects (for I call the direct and indirect *objects* of grammar "subjects" and I employ exclusively the 'exemplar' form of expression) are indicated by the subjacent indices. In particular q means 'has the character of,' its first index, a Greek letter indicating the individual character, the second index, an Italic minuscule, or l.c. letter, indicating the subject; and the letter r on the line means 'is in the relation,' the first Index, or subjacent letter, a Greek letter, indicating the mode of relation, or general relation, and the other subjacent letters indicating the correlates. These are Italic lower case letters. The last is the 'relate' or grammatical individual subject. The subjacent index next to the last is the direct correlate, or direct grammatical individual object, the antepenultimate subjacent letter indicates the indirect object. Thus $r_{\alpha ijk}$ means 'k is in the relation α to j for i.' Π_x means 'Take any individual you like, and call it x,' Σ_y means 'There is an individual which I will call y.' The operator precedes the operand. Thus $\Pi_\alpha \Sigma_i q_{\alpha i}$ means, 'Take any character, α, you please, and there is an individual object of the universe of discourse i, such that i has the character α; while $\Sigma_i \Pi_\alpha q_{\alpha i}$ means 'There is an individual of the universe, i, such that whatever character α may be i has the character α.' An *obelus*, or heavy line over an expression of predication negatives it. Thus $\Sigma_\alpha \Pi_i \bar{q}_{\alpha i}$ means 'There is a character α such that whatever individual of the universe of discourse i may be, i has not the character α.' The sign ψ means 'or,' and the sign $\cdot$ means 'and.' Thus, $\Sigma_\alpha \Pi_i \Sigma_j \Pi_k (r_{\alpha ji} \cdot \bar{r}_{\alpha ij} \psi\ \bar{r}_{\alpha ik} \psi\ r_{\alpha kj})$ means, 'There is a relation α such that, no matter what individual i may denote, an individual j exists such that whatever individual k may be, either i is α to j; while j is not α to i, or else k is not α to i or else j is α to k.' In order to express subjectal abstraction, I write for example (I forgot to say that I_{ij} means i and j are identical, T_{ij} means i is not j, ♈$_{ij}$ (sign of Aries) means i and j are in the same universe, O_{ij} means i and j are not in the same universe.)

$(\Pi_\alpha \Sigma_\beta \Pi_i \Sigma_j \bar{r}_{\alpha ij} \psi\ r_{\beta ji}) = x \Sigma_\gamma q_{\gamma x}$ which means 'Take any relation α you please, and using x to denote the hypothesis that "there is a relation β such that whatever individual there may be that [is] α'd by everything is β to something," then there is a character which the state of things x possesses.' Using this simple system of expression, the last definition may be expressed as follows:

Q, there is an ordinal, P, such that S is s to P for Q.

2nd, if Q possesses every character of the system possessed by S, and P possesses no character of the system, S is s to P for Q.

3rd, if Q possesses no character of the system, S is s to S for Q. S_{ijk} means 'k is a sum of j as addend and i as augend.'

The definition in its thoroughly analytical form is bewildering. Namely it is as follows

$$\Pi_s\Pi_a\Pi_b\Sigma_\tau\Pi_\lambda\Sigma_\alpha\Pi_\xi\Pi_\iota\Pi_\psi\Sigma_\iota\Pi_\phi\Sigma_l\Pi_\omega\Sigma_k\Pi_\beta\Sigma_\mu\Pi_\sigma\Sigma_p\Pi_\rho\Sigma_q\Pi_\upsilon\Pi_m\Pi_\chi\Sigma_n$$
$$\Pi_{\tau'}\Sigma_{\lambda'}\Pi_{\alpha'}\Sigma_{\xi'}\Sigma_{\iota'}\Sigma_{\psi'}\Pi_{j'}\Sigma_{\phi'}\Pi_{l'}\Sigma_{\omega'}\Pi_{k'}\Sigma_{\beta'}\Pi_{\mu'}\Sigma_{\sigma'}\Pi_{p'}\Sigma_{\rho'}\Pi_{q'}\Sigma_{\upsilon'}\Sigma_{m'}\Sigma_{\chi'}\Pi_{n'}$$
$$\{\bar{s}_{bas} \psi [\bar{q}_{\lambda s} \psi q_{\lambda b} \psi r_{\tau\alpha\lambda} \cdot q_{\alpha a} \cdot (\bar{q}_{\alpha i} \psi q_{\xi i} \psi \bar{q}_{\xi a} \psi \bar{r}_{\tau\xi\psi} \psi \bar{q}_{\psi s} \psi q_{\lambda j} \cdot \bar{q}_{\psi j}) \cdot$$
$$(\bar{r}_{\tau\phi\lambda} \cdot \bar{r}_{\tau\phi\omega} \psi q_{\lambda k} \cdot \bar{q}_{\phi k} \psi \bar{q}_{\psi a} \psi q_{\phi l} \cdot \bar{q}_{\alpha l})] \cdot$$
$$[\bar{q}_{\beta\sigma} \psi r_{\tau\beta\mu} \cdot q_{\mu s} \cdot \bar{q}_{\mu b} \cdot (\bar{q}_{\sigma s} \psi q_{\sigma b} \psi q_{\sigma p} \cdot \bar{q}_{\mu p} \psi \bar{r}_{\tau\beta\sigma} \cdot (\bar{r}_{\tau\rho\sigma} \psi q_{\beta q} \cdot \bar{q}_{\rho q})) \cdot$$
$$(\bar{q}_{\mu m} \psi q_{\upsilon m} \psi \bar{q}_{\upsilon s} \psi q_{\upsilon b} \psi \bar{r}_{\tau\chi\upsilon} \psi \bar{q}_{\chi a} \psi q_{\beta n} \cdot \bar{q}_{\chi n})]\} \cdot$$
$$\{s_{bas} \psi q_{\lambda' s} \cdot \bar{q}_{\lambda' b} \cdot [\bar{r}_{\tau'\alpha'\lambda'} \psi \bar{q}_{\alpha' a} \psi q_{\alpha'\iota'} \cdot \bar{q}_{\xi'\iota'} \cdot q_{\xi' a} \cdot r_{\tau'\xi'\psi'} \cdot q_{\psi' s} \cdot (\bar{q}_{\lambda' j'} \psi q_{\psi' j'}) \psi$$
$$(r_{\tau'\phi'\lambda'} \psi r_{\tau'\phi'\omega'}) \cdot (\bar{q}_{\lambda' k} \psi q_{\phi' k'}) \cdot q_{\psi' a} \cdot (\bar{q}_{\phi' l'} \psi q_{\alpha' l'})] \psi$$
$$q_{\beta' a'}[\bar{r}_{\tau'\beta'\mu'} \psi \bar{q}_{\mu' s} \psi q_{\mu' b} \psi q_{\sigma' s} \cdot \bar{q}_{\sigma' b} \cdot (\bar{q}_{\sigma' p'} \psi q_{\mu' p'}) \cdot (r_{\tau'\beta'\sigma'} \psi r_{\tau'\rho'\sigma'} \cdot$$
$$(\bar{q}_{\beta' q'} \psi q_{\rho' q'}) \psi q_{\mu' m'} \cdot \bar{q}_{\upsilon' m'} \cdot q_{\upsilon' s} \cdot \bar{q}_{\upsilon' b} \cdot r_{\tau'\chi'\upsilon'} \cdot q_{\chi' a} \cdot (\bar{q}_{\beta' n'} \psi q_{\chi' n'})]\}$$

But it would be easy to simplify this by denoting compound relations by single letters which, of course, would be defined; and this would have to be done were the paper ever printed. At present, I have no time for such work; but as an example of it we might put

$$d_{\eta\theta} = \Pi_h \bar{q}_{\eta h} \psi q_{\theta h} \quad \text{when} \quad \bar{d}_{\eta\theta} = \Sigma_h q_{\eta h} \cdot \bar{q}_{\theta h}$$

Multiplication may be defined in the spirit of the first definition of ordinals as follows:

Whatever ordinals p, m, and n may be, p is the *product* of m, as *multiplier*, and n, as *multiplicand*, if, and only if, there is a triadic relation, π, and a triadic relation ω, such that

1st, μ, ν, σ, σ' being any characters of the system whatever,
$\Pi_\mu\Pi_\nu\Pi_\sigma\Pi_{\sigma'}\, \bar{q}_{\mu m} \psi \bar{q}_{\nu n} \psi \bar{r}_{\pi\nu\mu\sigma} \psi \bar{r}_{\pi\nu\mu\sigma'} \psi 1_{\sigma'\sigma}$

2nd, $\Pi_\sigma\Sigma_\mu\Sigma_\nu\, \bar{q}_{\sigma p} \psi r_{\pi\nu\mu\sigma} \cdot q_{\mu m} \cdot q_{\nu n}$

3rd, $\Pi_\mu\Pi_{\mu'}\Pi_\nu\Pi_{\nu'}\Pi_\sigma\, \bar{q}_{\sigma p} \psi \bar{r}_{\omega\nu\mu\sigma} \psi \bar{r}_{\omega\nu'\mu'\sigma} \psi 1_{\mu\mu'} \cdot 1_{\nu\nu'}$

4th, $\Pi_\mu\Pi_\nu\Sigma_\sigma\, \bar{q}_{\mu m} \psi \bar{q}_{\nu n} \psi r_{\omega\nu\mu\sigma} \cdot q_{\sigma p}$

Corresponding to the second definition of ordinals, we have the following definitions of addition and multiplication.

Whatever ordinals a, b, s may be, s is a *sum* of a as *addend* and b as *augend*, if and only if,

1st, either a or b is N to some ordinal or s is not N to any ordinal.

2nd, either a is not N to a' or s is N to a sum of a' as addend and b as augend.

3rd, either b is not N to b', or s is N to a sum of a as addend and b' as augend.

4th, No ordinal is a sum of ordinals unless it is logically necessitated to be so by this definition.

Multiplication will be defined by

$$\begin{aligned} 0 \cdot 0 &= 0 \\ m \cdot Nn &= m \cdot n + m \\ Nm \cdot n &= n + m \cdot n. \end{aligned}$$

Addition and Multiplication may also be multitudinally defined, as follows:

A multitude, s, is a *sum* of a multitude, m, as *addend*, and a multitude, n, as *augend*, if, and only if, the Ss, the Ms, and the Ns, being any three collections whose multitudes are x, m, n respectively, there are, by logical necessity, two relations, μ and ν, such that every S is either μ to some M to which no other S is μ, or is ν to some N to which no other S is ν; while every M is μ'd by S by which no other M is μ'd nor any N is ν'd, and every N is ν'd by some S by which no M is μ'd nor any other N is ν'd.

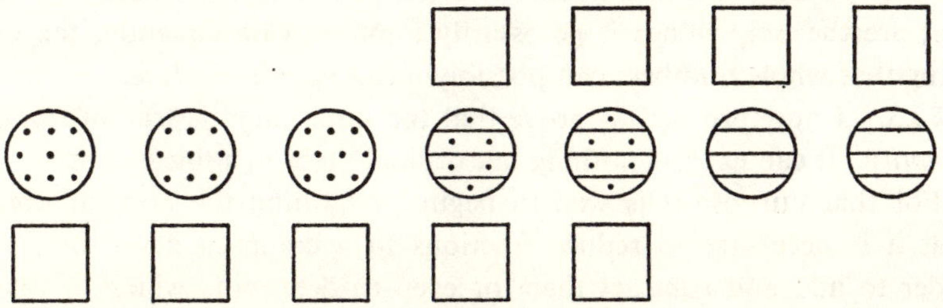

Fig. 1

Thus if m is the multitude of dotted disks above [Fig. 1] and n is the multitude of barred disks, s is the multitude of squares, μ is the relation of being under in the figure and ν is the relation of being over.

A product $m \cdot n$ where m is the multitude of Ms and n the multitude of Ns may either be defined as the multitude of individuals in the Ms if each M is a collection of n individuals or it may be defined as the multitude of different ways of joining a single M to a single N.

Note that m^n, where m is the multitude Ms and n the multitude of Ns, may be defined as the multitude of ways in which each N may be joined to a single M.

Thus 2^3 [in Fig. 2]

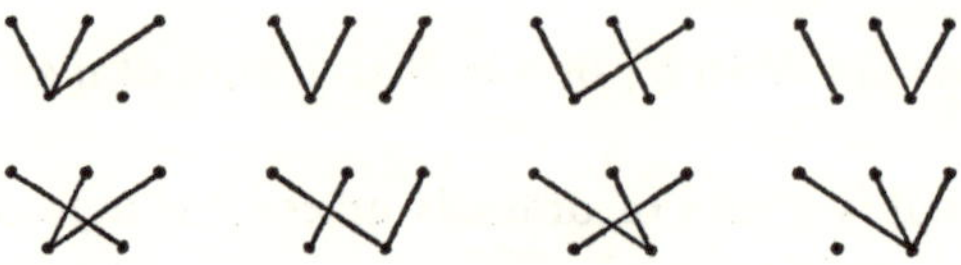

Fig. 2

while 3^2 [in Fig. 3]

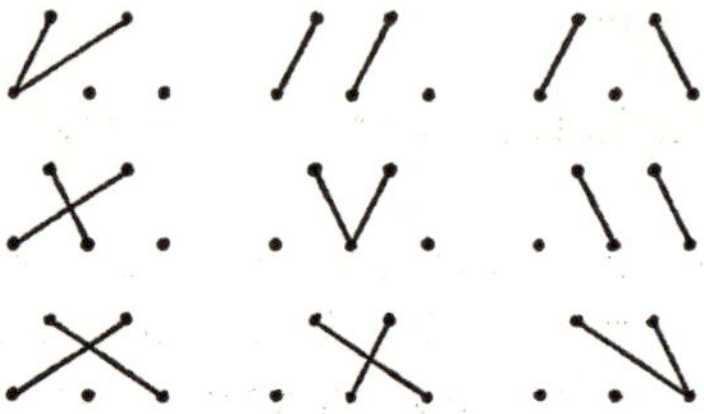

Fig. 3

§2.2.6. I prove that in every case in which a whole number can be used to express a fact, it implies that that to which it is applied has a certain ordinal place in a linear series; and that and what follows from that are the only things it necessarily implies. Consequently, the only thing that whole numbers can possibly mean is *ordinal place*.

§2.3. I now proceed to prove that the same thing is true of *rational quantity*. It can express nothing but ordinal place in a linear series.

For that purpose it is well to begin by refuting the popular notion that it is necessary to reduce fractions to a common denominator in order to add and subtract them or even to determine which of two is the greatest.

It is easy to set down the *fractions* (i.e. fractional *expressions*) in the order of their magnitudes without any other operation of arithmetic than those of numeration, by the following rule:

1st, set down the following row

$$\tfrac{0}{1} < \tfrac{1}{1} < \tfrac{2}{1} < \tfrac{3}{1} < \tfrac{4}{1} < \text{etc.}$$

2nd, Now insert the other fractions in the following way. No fraction of denominator n (whatever integer n may be) is to be inserted until all (up to the limit of magnitude) of lesser denominators have been inserted.

3rd, Of fractions of any one denominator n, no one as $\dfrac{m}{n}$ is to be inserted until all of denominator n and lesser numerator have been inserted.

4th, The succession of fractions $\frac{0}{n+1}, \frac{1}{n+1}, \frac{2}{n+1}$ etc. are to be inserted into the row after the insertion of all those of denominator n by placing them successively from left to right in every nth interval between a fraction and a sign in the row as it stands after the fractions of denominator n have been inserted.

5th, Every time a fraction is inserted between a sign on its left and a fraction on its right, that sign on the left (whatever it may be) is to be repeated at the right. But every time a fraction is inserted between a fraction on its left and a sign on its right, the sign $=$ is to be set down on the left of it.

By this rule we get successively,

$$\tfrac{0}{1} \quad < \quad \tfrac{1}{1}$$

$$\tfrac{0}{1} = \tfrac{0}{2} < \tfrac{1}{2} < \tfrac{1}{1}$$

$$\tfrac{0}{1} = \tfrac{0}{3}| = \tfrac{0}{2} <| \tfrac{1}{3} < \tfrac{1}{2}| < \tfrac{2}{3} <| \tfrac{1}{1}$$

$$\tfrac{0}{1} = \tfrac{0}{3} =| \tfrac{0}{4} = \tfrac{0}{2} <| \tfrac{1}{4} < \tfrac{1}{3} <| \tfrac{1}{2} = \tfrac{2}{4} <| \tfrac{2}{3} < \tfrac{3}{4} <| \tfrac{1}{1}$$

$$\tfrac{0}{1} = \tfrac{0}{3} = \tfrac{0}{5}| = \tfrac{0}{4} = \tfrac{0}{2} <| \tfrac{1}{5} < \tfrac{1}{4} < \tfrac{1}{3}| < \tfrac{2}{5} < \tfrac{1}{2} =| \tfrac{2}{4} < \tfrac{3}{5} < \tfrac{2}{3}| < \tfrac{3}{4}| < \tfrac{4}{5} <| \tfrac{1}{1}$$

$$\tfrac{0}{1} = \tfrac{0}{3} = \tfrac{0}{5} = \tfrac{0}{6} = \tfrac{0}{4} = \tfrac{0}{2} < \tfrac{1}{6} < \tfrac{1}{5} < \tfrac{1}{4} < \tfrac{1}{3} = \tfrac{2}{6} < \tfrac{2}{5} < \tfrac{1}{2} = \tfrac{3}{6} = \tfrac{2}{4} < \tfrac{3}{5} < \tfrac{2}{3} = \tfrac{4}{6} < \tfrac{3}{4} < \tfrac{4}{5} < \tfrac{5}{6} < \tfrac{1}{1}$$

§2.3.2. By the use of addition only we can set down all rational positive quantities in their order of magnitude and expressed in their lowest terms, by beginning with $\frac{0}{1}\ \frac{1}{0}$ and then going through the row over and over setting down in each interval a fraction whose numerator and denominator are respectively the sums of those to the immediate right and left hand of the interval. Thus

$$\tfrac{0}{1}$$

$$\tfrac{0}{1} \qquad \tfrac{1}{0}$$

$$\tfrac{1}{1}$$

$$\tfrac{1}{2} \qquad \tfrac{2}{1}$$

$$\tfrac{1}{3} \quad \tfrac{2}{3} \quad \tfrac{3}{2} \quad \tfrac{3}{1}$$

$$\tfrac{1}{4} \quad \tfrac{2}{5} \quad \tfrac{3}{5} \quad \tfrac{3}{4} \quad \tfrac{4}{3} \quad \tfrac{5}{3} \quad \tfrac{5}{2} \quad \tfrac{4}{1}$$

$$\tfrac{1}{5} \ \tfrac{2}{7} \ \tfrac{3}{8} \ \tfrac{3}{7} \ \tfrac{4}{7} \ \tfrac{5}{8} \ \tfrac{5}{7} \ \tfrac{4}{5} \ \tfrac{5}{4} \ \tfrac{7}{5} \ \tfrac{8}{5} \ \tfrac{7}{4} \ \tfrac{7}{3} \ \tfrac{8}{3} \ \tfrac{7}{2} \ \tfrac{5}{1}$$

$$\tfrac{1}{6} \ \tfrac{2}{9} \ \tfrac{3}{11} \ \tfrac{3}{10} \ \tfrac{4}{11} \ \tfrac{5}{13} \ \tfrac{5}{12} \ \tfrac{4}{9} \ \tfrac{5}{9} \ \tfrac{7}{12} \ \tfrac{8}{13} \ \tfrac{7}{11} \ \tfrac{7}{10} \ \tfrac{8}{11} \ \tfrac{7}{9} \ \tfrac{5}{6} \ \tfrac{6}{5} \ \tfrac{9}{7} \ \tfrac{11}{8} \ \tfrac{10}{7} \ \tfrac{11}{7} \ \tfrac{13}{8} \ \tfrac{12}{7} \ \tfrac{9}{5} \ \tfrac{9}{4} \ \tfrac{12}{5} \ \tfrac{13}{5} \ \tfrac{11}{4} \ \tfrac{10}{3} \ \tfrac{11}{3} \ \tfrac{9}{2} \ \tfrac{6}{1}$$

Of course, I prove the properties of this series, which is easy enough.

A description of this process in abstract terms will constitute a definition of the system of relations of the rational quantities from which all their properties will be deducible. So in order to define the sum of fractions we have only to describe in abstract terms the process of addition, which will be obvious from an example. Required the sum of $\frac{2}{5}$ and $\frac{3}{7}$

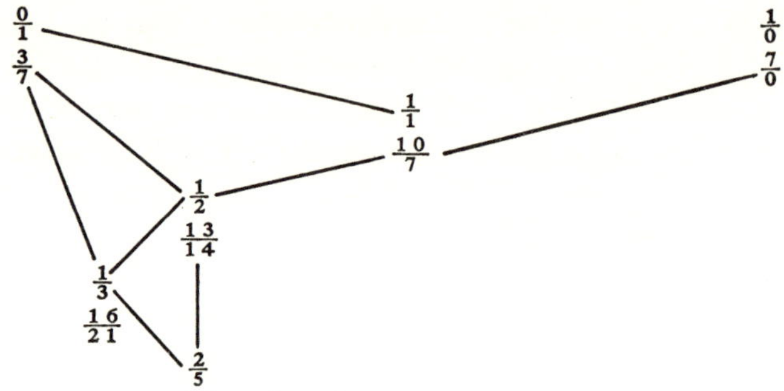

$$\frac{29}{35} = \frac{3}{7} + \frac{2}{5}$$

This obviously will not always give the answer in the lowest terms. There must be a better way that will, I think.

§2.3.3. I here show that no kind of quantity, rational or irrational, can possibly express anything but an order place in a linear series, — an important principle.

§2.4. At this point, the course of my argument compels me to develop the entire doctrine of multitude with the exception of one proposition of which I have no need. (I am in a condition to prove it, had I occasion for it.) Here my logical researches stand me in good stead. The first problem, and indeed the main problem of multitude, which is of a logical nature, may be said to consist in defining a collection; for this cannot be defined until the main difficulties are overcome. (The definition of Cantor is of that abstractive kind which, though it was such a favorite of Kant's, is no definition at all; and in most cases, as in this case, is mere rubbish, affording no useful information of any description.) But this way of stating the problem gives no hint of the fundamental character of the work required. In this letter, I shall pass over the logical parts of my paper lightly, because they are, I suppose, somewhat out of your line of interest, although they are by very much the most valuable part of the work from every point of view. I may say, however, that the investigation required as a propedeutic to the doctrine of multitude, involves a deep-ploughing criticism of the principles of contradiction and of excluded middle. It is true that these principles may correctly be regarded as constituting a definition of negation; but it does not follow that the criticism of them is a mere question of definitions, or that it is not of great avail in this borderland between mathematics and logic.

One of the points I establish is almost or quite axiomatic. However, I prove it, or bring it into connection with other truth. It is that existence of individual cases can never exhaust the possible varieties under a general term.

Another most important conception which clears up matters very much is that of *domains of actuality*, which I so name on account of their analogy to Galois's domains of rationality. The analogy is easily discerned, after the first stages are passed. We have in the first place objects which can be distinctly and fully described (when I say in the *first* place, that is only relative to what I am saying here). Then we have objects like irrational quantities which cannot be fully described but of which any one can be described with an indefinite approach to fullness. In fact they cannot exist in the same full sense. Thus, the past has an absolute definiteness. It was as it was, absolutely. It is not so with the future. You may entertain the theory of necessity. But still the future is not *ipso facto* definite as the past is. It has not the kind of actuality. It is not the *fait accompli*. Accordingly, in describing a general, we often use the future tense.

It is indispensable in these matters to avoid all confusion between what is *general* and what is *vague*. It might seem almost impossible to confuse the two concepts, which are truly as wide apart as the poles. Yet we all do so continually. The typical general is the substance of a wish or purpose. I wish for a young horse. I do not wish for a stallion, a mare, or a gelding, — I just wish for a young horse. The characteristic of the general is that the principle of excluded middle does not apply to it. The type of the vague is the state of things at the present instant. It is as settled a fact as the past. Yet all that we can change is the present instant. The only thing we are quite certain of is what is this instant experienced. Yet nothing is so hopelessly occult. The characteristic of the vague is that the principle of contradiction does not apply to it. It is this or that; one or other and neither. Or, better, it is either not this or not that, and it is both, like a point where a continuous variable makes a *saltus*. Of course, the vague is mere fiction. Yet the only thing thoroughly real is the present state of things and that is vague. The merely possible is vague, since the only mode of being it has consists in its being possibly so, and possibly not so.

I am giving you no idea whatever of the mode of reasoning by which I establish all this, and cannot undertake to do so in this letter. But one thing I can say; that we are not in the least degree considering what is to be found in the physical universe, and we are millions of miles away from all psychology. We are considering objects that are as purely hypothetical as those the pure mathematician deals with; but we are not considering them in a merely mathematical way. I make a survey of all the conceptions that ever have been recorded, and false conceptions are just as valuable as true ones, or more so, for this purpose; — and adding to these all the conceptions they suggest or that I can in any way create; and this done I find they fall into a certain system. In that way, I get firm ground under my feet in going further.

Therefore, you see, if when I talk of "actuality" and "existence," of these concepts, if you were to understand these terms in their usual senses, you would misapprehend me quite as completely as you would if you were to gather that all my talk related to the latest fashions in millinery.

To attempt to make myself understood by anybody but a mathematician would be futile. Nor could the minds the best adapted to understanding these things, — say Clifford and J. E. Oliver, — possibly go through the process of accurately apprehending all the necessary conceptions in less than, — well, if I say several weeks, I am attributing to them powers marvellously above my own. As for the whole existing race of philosophers, — say John Dewey, to mention a relatively superior man whom you see, — why, they are the sort of trash who are puzzled by Achilles and the Tortoise! Think of trying to drive any exact thought through such skulls! Royce is the only philosopher I know of real power of thought now living.

To return to my "domains of actuality," the reasons why, although I am speaking of purely hypothetical objects, the kind of objects the pure mathematician deals with, that I designate these as domains of *actuality*, using a word expressive of a mode of being instead of a word expressive of a mode of thought, are, in the first place that the worst misapprehension I have to guard against is being supposed to be talking psychology, and then in the second place because the reason why we cannot *conceive* of a denumeral (*abzählbar*) collection with the same definiteness that we can a finite collection is precisely the reason why such a collection cannot have been brought into the same mode of being that a finite collection can, — the reason why it cannot *have been brought about*. Therefore, I require a word referring as much to being as to being thought, — and this, of course, must be a word expressive of a mode of being. You cannot describe a denumeral collection unless there happens to be a law governing it that you can state. You don't then describe the collection itself. You only describe a general description of it. Now when you come to a first abnumerable collection, — say the collection of irrational real positive numbers, — you cannot even in that sense describe a description of it, you can only describe a *description of a description* of it. And this indirectness equally affects its mode of being. With the next abnumerable you cannot even do that much. Finally, when you rise to the idea of the totality of all possible collections, every vestigate of definiteness disappears. The objects, the ultimate parts of the whole that is contemplated, not only can no longer be distinct, however indirectly, in thought, but they equally and for the same reason cannot be distinct from one another in being; they melt into one another like single ordinary characters. I am in possession of several distinct ways of proving this.

If you admit that it is impossible that existence should ever exhaust all the possible varieties for which any general term gives room, it is obvious that my proposition is a corollary of this.

§2.4.2. *Terminology*. *Multitude* is the obvious word for what Cantor calls *Mächtigkeit* (*puissance*, *potenza*). What the devil is the sense of taking such inexpressive terms when there is so familiar a one? As for "cardinal number," I cannot assent at all to his use of that term. The cardinal numbers are vocables used in the performance of that experiment called counting. It is quite a secondary use of them as adjectives of multitude. We have occasion in this study to speak of *cardinal numbers* proper and the term is needed in its proper and familiar sense, while it is a pure superfluity in the sense of multitude. As for his balance between *ordinal number* (to his use of which I fully assent, although I here for the sake of brevity call the integer numbers ordinals, in which perhaps I do wrong), — but as to that balance, I doubt its justice. Multitude appears to me to be a much more special conception than an ordinal. I propose to call his *wohlgeordnet* series a *Cantorian series*, and perhaps I should prefer to call his ordinal numbers *Cantorials*. His *Menge* (*ensemble*, *insieme*) is just what the common word *collection* means, this word not at all implying that the objects are brought together otherwise than in thought. The whole doctrine needs a name. Names of sciences are usually formed from the Greek; but that language is peculiar in not having any word for *many* or *how many*, etc. in contradistinction to *much*. The plural πολλά is used, and the *ordinal* word πλῆθος πόστο[ν] means a throng, but also magnitude, and more commonly this in mathematics. Not caring for the quality of our Greek we might call the doctrine *pollote'tics*, but I doubt its being understood at sight. Since Latin has *quot*, I am tempted to imitate *sutrina* and call the doctrine *quótitrine* or *quotitri'na*.

I call the finite multitudes *enumerable*

The $\aleph_0$ multitude, the *simply endless*, or *denumeral* multitude (I vary the ending *-al* instead of *-able*, because this is a single multitude derived differently from any other.)

The $\aleph_1$ or $2^{\aleph_0}$ I call the *simply-abnumerable* multitude

The $\aleph_2$ or $2^{\aleph_1}$ I call the *doubly-abnumerable*

These run up through the finite orders of abnumerable; and I prove that these are all the multitudes that are possible.

§2.4.3. The first step in the exposition of any doctrine is to define the purpose of it. *It is a principle of definition that whatever has a purpose or intention is to be defined by means of its purpose or intention.*

The purpose of quotitrina is to ascertain the general conditions under which one collection is g or $\bar{g}$ and $\breve{g}$ or $\check{\bar{g}}$ to another (g meaning as *great as*), $\bar{g}$ not as great as, $\breve{g}$ as small as, $\check{\bar{g}}$ not as small as.

In order to complete this definition we must define a *collection* and the *relations of multitude.*

The word *collection* (*Menge*, *ensemble*, *insieme*) is a term of art which *has never been defined.* I know that Cantor proposes a definition; but it is one of those "definitions by abstraction" which Kant (naturally a logician of great power, but very ignorant of logic) was particularly disposed to use, but which I show are not definitions at all. They are psychological experiments, usually ill-performed, and hardly ever of any importance. Cantor's proposition has nothing to do with quotitrina.

Under these circumstances, since it is settled that quotitrina is about the magnitudes of collections, the only question is what *collection* ought to mean in order to give the proper unity to the doctrine. It is obvious that a collection is an individual object whose *existence* consists simply and solely in the existence of certain objects called its *members.* The question, however, arises whether this statement is any part of the *definition* of a collection. For it does not usually belong to a definition to state the conditions of the *existence* of the definitum. It is true that some definitions virtually do so, but the purpose of a definition is to set forth the *essence*, and not to account for the *existence* of the definitum. On the other hand, not merely the *word* collection, but the *thing itself* is a creation of thought; and the question is whether, in stating in general terms, on what occasions thought does create a collection, we are not stating the essence, or idea, of it. There are two entirely different things that are often confused from no cause that I can see except that the words *abstract* and *abstraction* are applied to both. One is ἀφαίρεσις leaving something out of account in order to attend to something else. That is *precisive abstraction.* The other consists in making a subject out of a predicate. Instead of saying, Opium puts people to sleep, you say it has a dormitive virtue. This is an all important proceeding in mathematics. For example take all "symbolic" methods, in which operations are operated upon. That may be called *subjectal abstraction.* This use of the word abstract goes back to the beginning of the XIIIth century while the other use is earlier still. So both are of unquestionable respectability. But they have nothing in common. What I say in treating such subjects I am apt to mean. They have nothing in common. No doubt subjectal abstraction presupposes a certain considerable precisive abstraction in each case; but that was not introduced in making the subjectal abstraction, it was there before. Experience is first forced upon us in the form of a flow of images. Thereupon thought makes certain assertions. It professes to pick the image into pieces and to detect in it certain characters. This is not literally true. The image has no parts, least of all predicates. Thus predication involves precisive abstraction. Precisive abstraction creates predicates. Subjectal abstraction creates subjects. Both predicates and subjects are creations of thought. But this is hardly more than a phrase; for *creation* and *thought* have different meanings as applied to the two. Without precisive abstraction man would not be man; but I can well believe, — indeed, I do think it probable, — that a large fraction of the races of mankind, by no means necessarily very low in the arts, are entirely devoid of the power of subjectal abstraction. I have not closely examined into the facts of language that might determine this. I am neither an anthropologist nor a linguist, and therefore may be excused for not having done so; but certain facts about almost all those languages whose structure I have studied and that are neither Aryan nor Shemitic, — perhaps two dozen in all, seem to indicate a weakness in subjectal abstraction; and certainly thought enough for the staple of life may be had without it. However, it is subjectal abstraction that I have to speak of exclusively. Putting aside precisive abstraction altogether, it is necessary to consider a little what is meant by saying that the product of subjectal abstraction is a creation of thought. It is one of the stock objects of ridicule. That about the "dormitive virtue" of opium has made myriads merry; although after all when one says that opium has a dormitive virtue one means that there is something in opium which explains the fact that it puts people to sleep; and that is what must be said before the latter fact can get understood. That the abstract subject is an *ens rationis*, or creation of thought does not mean that it is a fiction. The popular ridicule of it is one of the manifestations of that stoical (and Epicurean, but more marked in stoicism) doctrine that existence is the only mode of being which came in shortly before Descartes, in consequence of the disgust and resentment which progressive minds felt for the Dunces, or Scotists. If one thinks of it, a *possibility* is a far more important fact than any *actuality* can be. For what do we mean by a fact being "important"? We mean that it has a bearing upon conduct; —conduct aiming at material well-being in case the importance is practical, conduct in reasoning if the importance is mathematical, etc. Now to say that a fact has a bearing on conduct, implies that in the *future* (which already is not exactly actual, in the narrowest sense) we shall have different alternative lines of conduct *possible* to us, and that the consideration of the fact should on due deliberation tend to lead us to choose one of those rather than the other. It is wholly a question of what is *possible* and what is not so; and to say that possibility is nothing in real fact is to wed such a *necessitarianism* as to make it irrational to behave rationally. What is possible is not *merely* a question of thinking; it is a question of the relation of thought to reality. In fact one may say that no actuality has any importance at all except so far as the actuality of one fact may argue the possibility or impossibility of another. An abstraction is a creation of thought; but the real fact which is important in this connection is not that actual thinking has caused the predicate to be converted into a subject, but that this is *possible*. The abstraction, in any important sense, is not an actual thought but a general type to which thought may conform. This general type is not a creation of actual thinking, but is an inevitable result of sufficient thought. Now that is the sort of stuff that truth is made of. Therefore while we must not forget that the abstraction is a creation of thought which we can shape exactly as we like (which I shall show that Schönflies, for example, sometimes forgets) yet for all that it is a question of real truth whether the abstraction, once it has been shaped, has or has not this or that character (which is more frequently forgotten). I enlarge on this matter in my memoir because the extreme sensationalism of modern (recent, Wundtian) psychologists has nearly atrophied their power of thinking about anything otherwise than as ...

My dear Professor Moore:

I feel bound to tell you that almost immediately after my second letter to you I found the principle that all of a class of propositions any one of which is possible in all possible distributions of the others as to truth and falsity are simultaneously possible. Though true of propositions expressing actual facts of existence or expressing what may be such a fact, does not hold for the possibilities of possibilities of which a collection greater than the collection of all whole numbers (an ultranumerable collection) must be composed. This led me to think it necessary to reëxamine the foundations of the theory of multitude. Now I find that there are states of things logically possible which the hypotheses of the theory, so far as they are explicit, neither exclude nor positively suppose possible. Since Euclid, mathematicians have reckoned among the fundamental hypotheses of theories some postulates asserting merely that this or that is possible. I do not see the necessity of that. Whatever is logically possible and does not conflict with any positive hypothesis is left by the hypotheses possible, just as it was before they were enunciated. If this view be admitted, I can prove the proposition that two collections cannot be greater than one another in various ways. But a difficulty arises in the application of the principle that whatever does not conflict with the hypotheses is possible. Namely, I may prove by means of that principle that for any two collections there is some one-to-one relation of all members of one of them to members of the other or else to all members of that other of members of the first; but an objector may say that the very same principle supports the proposition that there may be two collections such that there is no one-to-one relation of all the members of either to members of the other. To this I reply that when you suppose a collection, it is well understood that you suppose this collection to be distinguished from other things. That is, you suppose the well-recognized logical principle that the members of every collection have some character *common* and *peculiar* to them. From this it follows that there must be characters that are not realized at all, that is, that some propositions that are possible are not actually true. Thus, subjects and predicates stand upon an altogether different footing as to possibility. To say that two collections can be placed in one-to-one correspondence so that all the members of one or other of them correspond to members of the other, or, as I phrase it, to say that there is a one-to-one relation of all the members of one to members of the other is a statement of a mere possibility which cannot be denied without some arbitrary restriction of the possibilities of relations, which I would call a *transient* relation of it. It need not be a one-to-one relation. That member of the collection like A above I would call the origin of that transient for that collection.

I use the word *aggregate* etc. for the collection whose members are the members of two collections called its *aggregants.*

I use the word *product* (with *multiply* etc.) for the collection of all different pairs of which one member belongs to one *factor* collection, the other to the other *factor.*

I use the word *exponential* to mean the collection of all collections whose members are members of a collection called the *exponent* collection. Thus the $(m + n)$th abnumeral multitude is the nth exponential of the mth abnumeral multitude.

I have little concern with ordinal numbers. But I call a *wohlgeordnet* series a *Cantorian* succession or sequence.

As for the principle of my last letter, although it holds good so long as it refers to existent objects or objects that may exist, I find it does not apply to ultranumeral collections, and therefore my supposed demonstration falls to the ground. I, therefore, fall back on the next easiest of those in my possession, as follows:

(But I will first mention another point of my terminology. I never speak of the members of two collections "being brought into" one-to-one correspondence, or use any phrase like it. For an indesignate relation has no other being than possibility; so that if those members can "be brought into" one-to-one correspondence, they already *are* in a one-to-one relation to each other. So likewise all the members of a *gath,* actually *have* some character common and peculiar to them. This is a recognized principle of logic.)

I start with the proposition that every innumerable collection is an aggregate or collection of the members of mutually exclusive denumeral collections (and that of course in many ways) I will not stop to prove that, as I think it cannot be questioned. Every innumerable collection may be aggregated to another equal innumerable collection, the two being mutually exclusive; and there is a one-to-one relation in which every member of the resulting aggregate stands to a member of the collection first considered, while to every member of the collection first considered some member of the aggregate is in that same relation, and no other individual is so. I mean of course that such a relation there *is* (i.e. may be). So then this very relation separates the innumerable collection into two equal parts. Now we are only considering the pure mathematical relations of gaths, i.e. such as exist between its members by virtue of the hypothesis of the mathematician, in this case that certain gaths exist. But a gath consists simply in the existence of its members, and consequently since existence is common to all of them, all parts of the gath are alike except that a scheme of otherness may exist between the members of one part that does not exist between the members of another part. Hence, since every innumerable gath is an aggregate of two mutually exclusive exchange the gath of *B*s for the equal gath of *B*s were no *A* that any given part of it, or else the complementary part, shall be wholly a part of one of those halves.

Suppose then there are two innumerable gaths, the *A*s and the *B*s, such that there is no relation in which every *A* stands to a *B* to which no other *A* stands in the same relation. If these two gaths are not mutually exclusive exchange the gath of *B*s for the equal gath of *B*′s where no *A* is a *B*′. Form the aggregate of the *A*s and the *B*′s. Separate this aggregate into two equal halves such that the *B*s all are members of one half. Then every *B*′ is in a one-to-one relation (that of identity) to some member of the aggregate, and there is a one-to-[one] relation in which every member of the aggregate stands to a member of that half to which no *B*′ belongs, and every member of this half stands in a one-to-one relation (that of identity) to some *A*. Consequently, there is a one-to-one relation in which every *B* stands to an *A*. Q.E.D. This proves the proposition for innumerable collections. For enumerable gaths it is readily proved. For the *sam* of nothing, perhaps we should say that it is greater than itself.

Tell me, please, whether you can discover any flaw in this. If you do, I shall have to fall back on another apparent proof I have ready.

As to the remaining fundamental question, of whether the abnumeral multitudes are all the ultranumerable multitudes, I am at present perfectly balanced between *Yea* and *Nay*. I hope soon to reach a conclusion, but do not feel at all confident of it.

The connection of the doctrine with the logical maxim called *pragmatism* is interesting. All things that exist ought by pragmatism (as a regulative principle) to form an enumerable collection. But what may be *in futuro* forms a denumeral collection.

Now according to *tychism*, law determines some things (excludes some future contingencies) and leaves others indeterminate. In that case, the possible different courses of the future have a first abnumeral multitude.

The possibilities of such possibilities will be of the second abnumerable multitude and when we reach the infinitieth exponential, which is thoroughly potential with no relic of the arbitrary existential element left, we have a *true* continuity, such that on a line there are not only points at every value of the analytical variable (all such values forming a first abnumeral collection) but there is *room* for any multitude of points whatsoever.

I do not know whether or not you have seen my proof that whatever Collection M may be $2^M > M$. I therefore add it in ink of a different color.

If you think what I have worth printing and can find me a place to print it, I will write it out in form.

Here is the proof I allude to, that whatever the *M*s may be, there is no relation in which every *sam* of *M*s stands to an M to which no other *sam* of *M*s stands in the same relation:

For if there be such a relation let r be its corresponding concrete relative term; so that to say that X is r to Y or that Y is r'd by X is the same as to say that X stands to Y in the relation in question.

So then it is supposed that every *sam* of *M*s is r to some M to which no other sam of *M*s is r. But I will show that it is not so by describing a sam of *M*s that is not r to any M to which no other sam of *M*s is r. This sam of *M*s that I am going to describe I will call my "test sam." In order to describe it, I will first divide the *M*s into three classes as follows:

Class I shall comprise all *M*s (if any) to which no sam of *M*s is r, together with all *M*s to which more sams of M than one are r.

Class II shall be composed of all those *M*s and only of those *M*s of each one of which, say M', it is true that there is but one sam of *M*s that is r to it, and this sam of *M*s that is r to M' *includes* M'.

Class III shall be composed of all those *M*s (and of them only) to each one of which a single sam of *M*s is r this sam of *M*s *excluding* every M of class III to which it is alone r.

Now my "test sam" shall include all *M*s of class III, shall exclude all *M*s of class II, and whether it includes or excludes those of class I I do not care. What I say is that this test sam is not r to any M to which no other sam of *M*s is r.

It plainly is not the sole r of any M of class I, since no M of that class is r'd by one sole sam of *M*s.

It is not r to any M of class II because every M of that class is included in the sole sam of *M*s that is r to it, while the test sam excludes every M of class II.

It is not r to any M of class III since every M of that class is excluded from the sam of *M*s that is r to it, while the test sam includes every M

of class III.

Consequently, the test sam is not *r* to any *M* to which no other sam of *M*s is *r*. Q.E.D.

I need not say that it follows from the first proposition of this letter (which is also the first of the theory) that for every collection whatever there is such a relation ϱ that if *X* and *Y* are any two *different* individuals of the collection either *X* is ϱ to whatever *Y* is ϱ to and to something besides, or else conversely *Y* is ϱ to everything to which *X* is ϱ and to something besides. From this I presume it follows that for every collection there is some relation which constitutes that collection to be a Cantorian sequence. Only manifestly the Cantorian sequence must then be extended beyond what Cantor seems to have provided symbols for (as far as I remember).

There is a paper of Cantor's that I have never read (although he sent it to me) but which looks highly interesting in the Zeitschrift für philosophischen Kritik. I mention it because perhaps you too have overlooked it in such an out of the way journal.

Very truly
C. S. Peirce

P.S. You will perceive that this proof rests upon a proposition supposed to be involved in our idea of a collection. Cantor thinks he defines a collection; but he (like most Germans) does not know what a definition is. A definition should be a precise statement of what is involved in predicating a term. Cantor's definition (of a kind a favorite with German philosophers) only tells you what kind of psychological experiment to make in order to find the conception of a collection.

It is involved in our idea that for every collection there is a possible complication of otherness that is true of members of it. I call it a scheme of otherness. Whatever complication of otherness is true of a part is ipso facto true of the whole. But the converse is not generally true.

It is also involved in our idea of a collection that it is constituted by the mere individual existence of the subjects of schemes of otherness. Consequently, it is alike in all its parts. That is to say, if a collection is divided into two parts A and $\bar{A}$ and is also divided into two parts B and $\bar{B}$, then there is a division into two parts B' and $\bar{B}'$, such that whatever scheme of otherness is predicable of B is predicable of B', and conversely; and whatever scheme of otherness is predicable of $\bar{B}$ is predicable of $\bar{B}'$, and conversely; and furthermore either every member of A is a member of B' or else every member of $\bar{A}$ is a member of $\bar{B}'$. I see no simpler way of expressing the homogeneity which we implicitly suppose to hold of every collection in saying that it is constituted by the mere existence of its members.

Milford Pa. 1902 March 20

My dear Sir:

I am very much obliged to you for your kind remittance of all these papers and as I am laid up today and unfit for work, my time is at my disposal to write something to you about the subject of one of them the value of which I fully appreciate although my own line of thought is quite different. I speak of your paper on the Axioms of Projective Geometry. I am not well enough and it would be too long to go into the real course of my thought. I can only sketch them. The subject is not a mathematical one; and although the aid of mathematics is necessary to setting it right and cannot be dispensed with, it is not very deep mathematics as mathematics. It requires aid from Logic of a much more difficult kind. It takes the power of the logician to the very utmost; so that it is more a subject for a logician than for a mathematician.

Allow me to remonstrate on your calling these hypotheses Axioms. According to my notions of terminology when a term has been in *use as a term of science* since ancient times, it is proper to use it in the ancient sense, illuminated if necessary by modern understanding of the matter. Now the hypotheses of geometry were called *Postulates* by Euclid, and this entirely agrees with Aristotle's explanation of what a Postulate is, if we remember that Aristotle was only just mathematician enough to gain entry into the school of Plato. The German sense of the word Postulate (although it may be found in some Greek commentator) does not in historical continuity date back farther than Christian Wolf, and has, I am happy to say, not been accepted in England and France. Euclid's Axioms, — he calls them 'Common Knowledge,' but the word Axiom was used as a synonym in his time and earlier, — do not relate to Geometry but are merely propositions which the geometer borrows and it is not his business as a geometrician to discuss them. I should therefore prefer the word *Postulate.* But if you do not like that, it seems to me that *Hypothesis* is perfectly unobjectionable.

Now as to the Postulates of Projective Geometry, in my opinion before they can be intelligently considered, it is necessary to get those of Topical Geometry perfectly stated. Just as Metrical Geometry is at bottom only the Projective Geometry of a certain Absolute, so Projective Geometry is nothing but the Topical Geometry of a certain Family of Lines. Properly Metrical Geometry is not the science of space but of the laws of displacement of rigid bodies; and so, to my mind Projective Geometry is not the science of space but only of Optics, which happens to coincide with the path of a particle not acted on by any force. But with Topical Geometry it is different. That is Space itself. This may sound like verbiage but it is not so. I can't fully explain in this letter the exact idea under it. But at any rate, it is easy to see that Topical Geometry must come first.

The first thing necessary is to analyze *continuity*: The continuity of analysis won't answer here at all. It is necessary to say what that continuity is which common sense attributes to time. That time flows in one sense has nothing to do with it. That is a notion about causation in time and no more concerns time itself than the notion of a straight line, or ray, concerns space itself.

Now I will only say that starting with the logic of relatives, and *having nothing at all to do with any relation* except such as are *formally* defined, I begin by defining a *collection* (not an easy job for a logician) and a *part* of a collection, and a *part* in general, and what the relation of greater and less is, etc. I thus show, to begin with, that (*multitude* being that character of a collection by virtue of which it is greater or less, "Mächtigkeit" as Cantor calls it, with German awkwardness) there is no maximum multitude. Here we have a hint about continuity; for anybody will say that in a lapse of time there is room for any multitude of events however great. Consequently, *time* is not a collection of instants, nor space a collection of points. But although there is merely a simple endless series of multitudes, there is no contradiction in supposing something which begins by being a collection increasing through all these multitudes and surpassing them. But for that purpose it is necessary to suppose some form of relation to exist; otherwise when the collection ceases to be a collection[1] it will not be anything definite. It becomes a continuum. This continuum does not consist of indivisibles, of points, or instants, and does not contain any except so far as its continuity is ruptured. The continuum is a *General*. It is a General of relation. Every General is a continuum vaguely defined. I find that a perfect continuum

[1] The rest of the letter is a restoration from MS. s-90.

of one dimension, or a *line*, — having no points of discontinuity, must return into itself. For if any part is unlike another part it has something exceptional and is not perfectly general. Distance is metrical, even infinite distance. Therefore a line cannot run off to nowhere without returning into itself unless it is to be imperfectly continuous. A spiral that winds from an outer limiting circle to an inner one *stops* (i.e. supposing it does not exist outside the outer nor inside the inner) and that stopping is a discontinuity whether there be any means of determining at what points of the 2 circles it stops or not. If it don't stop, and don't return, it goes on [Fig. 1].

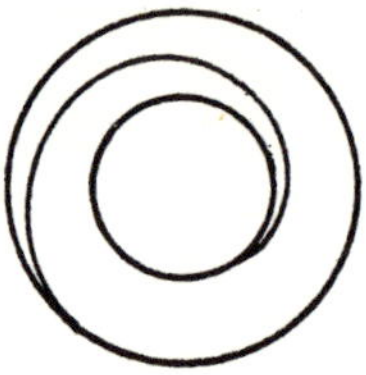

Fig. 1

Looked at from this point of view, it seems to me quite clear that two points do not determine a segment of a line. For which segment is it? If you choose to take a plane at infinity, that is simply the absolute of metric geometry you are taking. But I cannot see that Projective Geometry has the slightest occasion to distinguish the locus at infinity. The geometer ought to consider space itself to be perfectly continuous, because he has not the slightest reason to suppose it has anything exceptional about it, and because it is simpler to suppose it has nothing exceptional. Then why should projective geometry bother its head about infinity? It supposes of course that the plane is perissid, or one sided.

I hope to see you in Washington next month.

Very truly
C. S. Peirce

P.S. A particle is shot due north and is unacted on by any force. When it has travelled as long as years can measure, we have no reason to suppose that Time gives out and comes to a dead standstill; for years do not constitute time but only measure it. Neither have we reason to suppose that space comes to an end at the point the particle has then reached. What then will it do next? There is no reason to suppose it will behave in any exceptional way, simply because it has reached the Absolute, which is only like the Equator, — a conventional ellipsoid. True, it cannot get any *distance* away in any measurable time, but time will go on and space will extend just the same without a ripple on either, and the particle will move right on and come round again after the lapse of sufficient time. Perhaps it will by that time have made the circuit of time as well, so as to be back at its original place at the very moment when it was supposed to have started but at which it cannot have started since it is not of a nature to be acted on by any force.

Dear Moore:

I chanced in considering a point about the foundation of quantity to make the obvious remark that the simplest universe in which there is a relation that is a fourth root of unity is one of four individuals, and that between these there are three such relations with their converses. Wondering what connection there was between this fact and there being three dimensions in space, I naturally formed the group generated by these four relations; for I am so little acquainted with the theory of groups that it was new to me, as far as I recollect. I suppose it has been thoroughly worked out; but I will set down my results. In order to form an image of the relations, let the four individuals of the universe be the four diagonals of a cube. Then the group consists of unity with three rotations through 180° about the axes through middle points of the faces, and eight rotations through 120° about the four diagonals of the cube. These twelve may be called the *plus* relations. There are besides six quadrantal rotations about axes through the centres of the faces and six rotations through 180° about axes through the middles of the edges. These twelve may be called *minus* relations. The product of two relations is *plus* if both are *plus* or both *minus*, and is *minus* if one is *plus* and the other *minus*. Each of these two classes of relations is divided into three sets on the principle that every individual of the four is related to every other in one, and one only, of the relations of a set. The substitutions of the six sets are as follows:

Positive sets	*Negative sets*
Set P	Set L
1 = ABCD	ABDC
BADC	BACD
CDAB	DCAB
DCBA	CDBA

Set Q	Set M
ACDB	ADCB
DBAC	CBAD
BDCA	BCDA
CABD	DABC

Set R	Set N
ADBC	ACBD
CBDA	DBCA
DACB	CADB
BCAD	BDAC

The multiplication-table of the sets is this [Fig. 1].

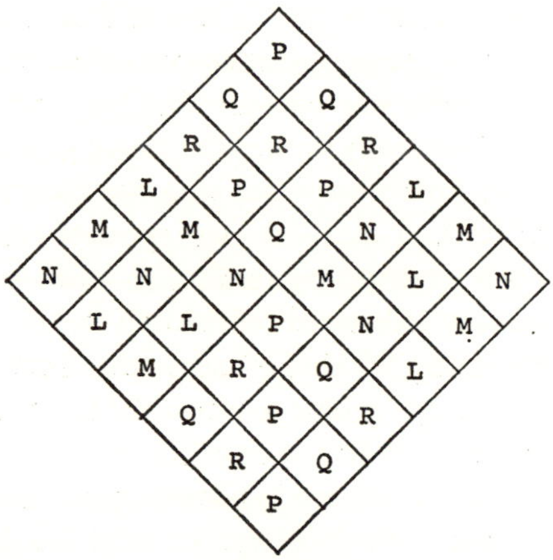

Fig. 1

Of course, this gives a suitable algebra for four dimensional space, and would be of good service in studying such space.

I. HOWES NORRIS, Jr. (L 321)

P. O. Milford, Pa. 1912 May 28.

Mr. Howes Norris, Jr.
Vineyard Haven, Mass.

Dear Sir:

You pay me the compliment of asking an autograph of me. You once before made the same request, and I intended to comply; but there was a difficulty. Namely, you desired that my answer should express a sentiment. Now I suppose the interest of an autograph depends upon its being peculiarly characteristic of the individual writer. Now I do not indulge in sentiments unless they are such as my reason approves, nor even then unless the particular instance of such a general kind equally commands my approval. Of course, they form a most indispensable part of our nature; but I have them in suspicion unless they are of sorts that are common to almost all men. It follows that my sentiments are of the most commonplace complexion, and are not at all idiosyncrasies such as will gratify a collector of autographs. You can see that this is so by what I have already said. The difficulty still subsists, and is now accompanied by another. For I am now, with exasperating slowness, recovering from a long and serious illness, which has weakened my energy and intellect to such a degree that it forces itself upon my own ordinary self-conceit and on many days renders my handwriting illegible to myself, and so disgusts me with myself as to indispose me to make an exhibition of myself even to my closest friends.

A few ideas, however, as old as myself, have kindly sauntered in, in their superannuated totter, to relieve the tedium of the sick-room, while exercising a most sedulous care to avoid every suggestion that could possibly startle one. One of these thoughts that came to me the other day was that when we are in possession of a simple method which is capable, of itself alone, of perfectly fulfilling a desirable purpose, it is best not to complicate our action by introducing a second method, entirely alien to the first, so as to give to the second method a part of the job that the first could do more simply and more elegantly without any aid. The proposition, I confess, seemed to my poor noddle an obvious dictate of common sense; but I perceive that it conflicts with most of the arrangements that are held to be the greatest triumphs of human intellect. For instance, such we are apt to regard the "arabic" system of numerical notation.

The one clear and all-sufficient mathematical idea in the system is that there is a row of places, extending indefinitely right and left of a central units' place in which a number n has its own value, n; while in the first, second, third, etc. place to the left of that it has the value, n^1, n^2, n^3, etc. (or in general in the pth place to the left it has the value n^p), and in the first, second, third, etc. place to *right* of the units place it has the value n^{-1}, n^{-2}, n^{-3} (or, in general, in the qth place to right of the units' place it has the value n^{-2}). In order to make this method denote every real finite number, for the base of numeration we must employ the number *two*. It would be requisite, then, to have new names for the numbers; but a system far superior to that we follow, could be contrived in so many different ways, that I need not dwell on that. The main things are the notation and the arithmetical operations. A place would be shown to be *filled* by having a vertical line in it, or to be *vacant* by having a dot in it. The units' or zero power place would be marked by its vertical line or dot being heavier there than in other places. For example what we call "*one*" would be written **|**

"two" would be written | •
"three" would be written | **|**
"four" would be written | . •
"five" would be written | . **|**
"six" would be written | | •
"seven" would be written | | **|**

and so on.

Objections are usually hurried up, without waiting for necessary explanations; because people think it a mark of intelligence to start objections, however speedily and conclusively their vacuity may be demonstrated. No doubt, therefore, at this point a whole Pandora's box of objections would be started, such as that since two to the tenth power is not much more than ten to the third power, in this way nearly $3\frac{1}{3}$ times as many figures would be required as in our present system. No doubt the number of characters used *would* be $3\frac{1}{3}$ (or, more exactly 3.321928) as great, if all our numbers and all our calculations were carried out to an infinitely great number of places, and would approximate to that. But as a matter of fact, the ratio proposed, 3.321928 is decidedly though not enormously too high for two reasons. The first is that the number of characters is not the same as an exponent or logarithm, but is one more than that number. Consequently even if 2^{10} were no greater than $(10)^3$, the ratio of the number of characters required would not be $\frac{10}{3}$, which is greater than 3, but $\frac{11}{4}$ which is less than 3. But this is not all; for, in the second place, small numbers occur much more frequently than large ones. In those astronomical and geodetical computing offices, where dozens of men are incessantly employed in making logarithmic calculations, it is the universal experience that an ordinary table of logarithms, gets completely worn to bits at the beginning before the end shows any decided sign of wear, while a table of antilogarithms, which contains substantially the same facts, except that in place of giving in the body of the table the exponents of powers of ten whose values are entered at equal intervals at the side, it gives in the body of the table what the other enters at equal intervals of magnitude at the side, and *vice versa*, — such a table of antilogarithms is invariably worn out, like the parson's one horse shay, at an equal rate in all its parts. This proves that numbers whose logarithms are taken from either form of table, which are what we may call "expressions of measurement," or of ratios of magnitude to those of arbitrary units occur with ratios of frequency which do not depend upon the absolute values [of] numbers expressing them, but only upon the real ratios of the magnitudes they express. That is, numbers ranging from 1 to 2 will occur with the same frequency as numbers ranging from one million to two million; because the ratios, which are the real facts which they express are identical. Consequently any one number between 1 and 2 expressed accurately to a given decimal point will occur a million times as frequently as any given number a million times as great, supposing both to be mere numbers of measurement. This consideration taken in connexion with the fact that the number of characters is not the same as the value of an exponent greatly increases the importance of the other consideration. For such "expressions of measurement," I have calculated the number of characters required, on the average of a great number of independent occasions, to express them on the "secundal" system of notation (that is the system in which a character in any given "place" denotes *double* what it would in the "place" next to the right, instead of *ten times* that as in the "decimal" system) supposing them to be expressed with the same accuracy as they would be by a given number of figures of decimals; and in the following table n is the number of figures required to write the "expression of measurement" in the usual way (not counting zeros or "ciphers" that haven't other figures *both* to their right and to their left), while r is the number by which n must be multiplied to give the corresponding number of characters required for the equally accurate "secundal" expression.

n	1	2	3	4	5	10
r	2.046	2.736	2.921	3.0381	3.09073	3.10702

1912 June 1

During the three days last past I have been by no means equal to writing. It results from my table on the last page that the number required to express a given quantity secundally is about three times that need to express it decimally. But by using a plain vertical line for a unit, and a simple dot for zero, the *space* occupied need not be much greater. As for the *time* and *labor*, if we use a - (hyphen) for *two* consecutive zeros, and [something] like a *w* or *u* for *three*, probably the advantage will incline toward the secundal side. One of the stupid complications of our usual way of writing numbers is the *decimal-point*. Since there is no peculiarity about the relation between any special pair of consecutive decimal places; while the decimal-point prevents the two between which it occurs from being, in print, at any rate as near to one another as other consecutive places are. What precisely is peculiar hereabouts is the units' place itself. Whatever the base of numeration, B, may be, a unit in this place has the same value B°. Accordingly, it ought to be marked by the heaviness of the figure that occupies it. For 3.5 we ought write | **|** | and for $4\frac{1}{4}$ | . . . |. But the greatest practical merit of the secundal system lies in its having several different methods of performing each operation, from which one can at sight select without hesitation the one most convenient for the case in hand. But owing to the prolixity to which my letter already reaches, I shall here show but one way, each, of adding and subtracting. Possibly I may make some slight suggestions on division and on extracting roots. Of course, the Addition-table and the Multiplication-table have to [be] mastered. The former consists of two facts. The first is that zero added to any number, or with any number added to it, gives that number as the sum. The second is that one added to itself gives zero with one to carry. On the other hand, zero multiplied by or into any number gives itself while one multiplied by one gives one.

(1912 June 4 — the first day I have been fit to write since my last page.)

The importance of the numbers which I denote by $F(n)$ or Fn arises from certain truths of which the foremost is that $\frac{1}{Fn}$ is expressed secundally by $n-1$ zeros immediately to the right of the units' place and followed by a single 1, these $n+1$ figures being repeated endlessly to the right, like a circulating decimal. I write the circulating part only once, putting over it a line that roofs it on. Thus for $\frac{\text{i}}{\text{ı i}}$ which is :. ı . ı . ı . ı . ı . ı etc. ad infinitum, I write :$\overline{\text{. i}}$, and for $\frac{\text{i}}{\text{ı ı i}}$ I write :$\overline{\text{. . i}}$ The other facts that aid division consist in certain simple rules for finding the least of the multipliers expressed secundally by a row of ones that is divisible by any given prime. By means of these rules we know that the least Fn that is divisible by ı . i is ı ı ı ı; so that since ı ı ı i ÷ ı . i = ı i, it follows that i/ı . i = :$\overline{\text{. . ı i}}$ So the least Fn divisible by ı . . . i is F ı . . i which is ı ı ı i × ı . . . i so that $\frac{\text{i}}{\text{ı . . . i}}$ = :$\overline{\text{. . . . ı ı ı i}}$

These rules are mostly very simple applications of one of the most historically famous propositions of mathematics called *Fermat's theorem* (or else of an easy extension of it). It should be stated in every encyclopedic dictionary and proved in every encyclopedia of any pretensions. The best on this fascinating subject is Dedekind's edition of Lejeune-Dirichlet's "Vorlesungen über Zahlentheorie." But most books give the reader more or less difficulty owing to their not precisely defining their technical terms and vigorously sticking to their definitions. Since definitions are in my line, I here define the principal terms needed.

To *count* a lot of single objects to pronounce upon all the latter singly all the different *integers*, or *whole numbers*, without repetition from the lowest, *one*, to the highest of those that have to be used, the highest of them being the *number of the lot*. But it will be necessary to affix some mark upon every object that has already been counted, so as to distinguish those that remain to be counted; and therefore the first step should be to make sure that *none* of them are marked as if they had been counted. In making sure of this, it would be a good plan to say "none" or "zero." Zero is thus *not* an "integer," in that it is not attached to a single object as having been already counted. Yet it is useful in counting, both as marking the first step and in showing that a category, or supposed lot of single objects, does not in fact contain any. It, along with the integers, may be called a *denumerand.*

The term "factor" etymologically implies a whole number that goes to make up *another* whole number by being multiplied into (or by) some whole number. A *prime*, or *prime number*, is an integer that has no factor among integers. By a *divisor* of an integer may most usefully be meant an integer which will give the former as the product on being multiplied by (or into) an integer not greater than itself. This makes every integer a divisor of itself, but makes *one* a divisor of nothing but itself.

June 4. *F*(*n*), then, or, more simply written, *Fn*, denotes a whole number which is secundally expressed by a row of *n* units, of which the last stands in the units' place. The importance of such numbers in dividing arises from a group of circumstances of which the foremost is that if we consider the secundal expression consisting of a unit in every *n*th place to the right of the units' place, all the other places being occupied by zeros, it is plain that the product of this by *Fn* (the secundal expression of which consists of a row of *n* units of which the last is in the units' place), will be expressed by an endless row of units filling all the places to the right of the units' place. Now this will evidently differ from unity by less than any quantity that could ever result from successive halvings; and *pragmatism* (or the doctrine that assertions mean no more than they can possibly come to) teaches that this is as much as to say that the product will differ from unity *not at all.*

I think I mentioned that I mean in the rest of this letter to try how it will work to mark the units' place by putting a dot over the figure (vertical line or dot) in that place, instead of by thickening that figure. I will also indicate a circulating secundal fraction by writing the circulant once only and drawing a sort of cover over it. We have just found then that $\imath\, i \times :\overset{\frown}{.\,\imath}$ that is $\imath\, i \times :.\imath.\imath.\imath.\imath.\imath.\imath.\imath.\imath.\imath$ and so on endlessly equals i, as does $\imath\,\imath\, i \times :\overset{\frown}{..\,\imath}$ and $\imath\,\imath\,\imath\, i \times :...\imath...\imath...\imath$ etc. So fifteen, which is secundally written $\imath\,\imath\,\imath\, i$, which is evidently the product of $\imath\, i \times \imath\,.\,i$ when multiplied by $:\overset{\frown}{...\,\imath}$ gives i as the product. Consequently one fifth of it, or $\frac{i}{\imath\, i}$ will be $\imath\,.\,i$ times that product, or $:\overset{\frown}{.\,\imath\,.\,\imath}$, which amounts to the same endless row of figures as $:\overset{\frown}{.\,\imath}$ If we had multiplied by $\imath\, i$ instead of by $\imath\,.\,i$, we should have found that $\frac{i}{\imath\,.\,i} = :\overset{\frown}{..\,\imath\,\imath}$. Now every whole number whatsoever except powers of two is an exact divisor of some set of numbers of the form that I have called *Fn*, so that the reciprocal of every number can be readily found, if one is acquainted with the properties of *Fn*. I need not say how much this facilitates division.

It will pay, then, to look a little into the divisibility of the *F*s; and these properties are not so obvious with any other system of numerical notation. Remark, to begin with that, *m* and *n* being whole numbers, every divisor of *Fn* and *Fm* is also a divisor of *F*(*mn*). This is obvious from the secundal expression. For *F*(*mn*) is expressed by a row of $m \times n$ units. If then any number, *p*, will divide *Fn* without a remainder, it will divide *m* such rows, each followed by the next; and so if *m* and *n* are relatively prime, $F(mn) = Fm \times Fn$.

Before going further, it seems to me convenient to agree upon a few verbal distinctions. First, between an "integer," or whole number, and a "non-fractional number." By an "integer" I mean a number equal to a unit or sum of units. By a non-fractional number I mean a number expressible secundally or decimally without using any place to the right of the units' place; so that *zero* is a non-fractional number, but is not an integer. Similarly, I would distinguish between a "factor" and a "divisor" of an integer. By a "factor" of an integer, I mean as the etymology implies, any one of a finite number of integers which *go to make up*, by being multiplied together, a *new* integer; and any integer that has no factors thus alone satisfies the definition of a *prime*. But a divisor of a number is any number which, multiplied by a non-fractional number, gives. ...[1]

(1912 June) By a *factor* of any integer, say of *N*, one would understand, from the derivation of the word, an integer which *produces N* as the *product* of two plural integers of which the factor is one. This would not be admitted, and *taken by itself*, ought to be admitted, as a good reason for adopting a definition of a term of science. At the same time, I see reason to apprehend future inconvenience and embarrassment from the willingness of scientific men to use Greek and Latin words in senses utterly unknown to those languages in manufacturing terms of science. For instance, from the Greek βίος, meaning *life-time* we have not only borrowed the old word "biography," where βίος has its proper meaning, but we have also fabricated *biology*, where the first-part represents an idea quite foreign to βίος, even if it ever entered the head of a Greek.

[1] A whole page seems to be missing from the letter.

Now what if one found occasion to compound those two terms and speak of *biobiography*, i.e. the account of the course of life of an animal or plant? Such considerations prompt me to take a certain degree of account of the original significations. In the case of the word "factor," this is supported by the convenience of defining the word *prime* as an integer without factors, etc. I shall therefore adopt this definition: A *factor* of any integer, N, is a plural integer which, multiplied by a plural integer, produces, or *makes*, N as the *product*.

There are three other arithmetical terms, seldom carefully defined, but often used as nearly synonymous with *factor*. They are "measure," "divisor," and "quotient." One is sometimes said to be the only common "measure" of all whole numbers; and every number (rational or irrational) is sometimes spoken of as a "measure" of itself. If we are to give "divisor" a different meaning from "factor," it would seem to me most appropriate and consonant with usage (or least in conflict with it) to make every number a divisor of itself, since that division will give an integral quotient, but not to regard 1 as a divisor of any *other* number. Unfortunately, I believe it is impossible to embody this signification in a definition that shall conform to what *ought* to be regarded as the requirements of logic in definition. For one of such rules should be that a disjunction cannot define a concept. Thus to say that the subject of biology is "whatever pertains to an animal *or* a plant" does not give any understanding of the *reason* of these two classes and nothing else being included. The definition of a class should state what single character belongs equally to every possible member of the class. But it so happens that the principal need of the term "divisor," in this undefinable sense, is to enable us to define a certain function of any integer, N, introduced (so far as I know) by Gauss in 1801, and denoted by him ΦN, without giving it any name (which is very inconvenient since ΦN is usually understood to be applicable to *any* quantity whose value depends, or is regarded as depending on the value of N) and I think Sylvester (who once brought an utterly mendacious accusation against me, but), who certainly invented many other useful mathematical terms, did well to call it the "totient" of N, and to denote it by τN, although he has not been generally followed in this. Now the *totient* of N is definable as the number of integers that in the undefinable sense are *non-divisors* of N not greater than N; though many writers, instead of "not greater" say inconsistently (since $\tau 1 = 1$) "smaller" and Gauss himself does so at first, and then changes to "not greater" a few pages further on. You see, therefore, that it is not *divisor* but *nondivisor* that is the important idea;

and this can be defined. Namely the non-divisors of N as small as N are all those integers not greater than N, that are not products of integers with factors of N. Thus the non-divisors of ninety are odd numbers, not divisible by any prime factor (2, 3, 5) of 90 nor greater than 90; that is, they are

1, 7, 11, 13, 17, 19, 23, 29,
31, 37, 41, 43, 47, 49, 53, 59
61, 67, 71, 73, 77, 79, 81, 89 and the totient of 90 or $\tau 90 = 24 = 90 \cdot \frac{1}{2} \cdot \frac{2}{3} \cdot \frac{4}{5}$ so $\tau 6 = 6 \times \frac{1}{2} \times \frac{2}{3} = [2]$

$$\tau 5 = 5 \times \tfrac{4}{5} = 4$$
$$\tau 4 = 4 \times \tfrac{1}{2} = 2$$
$$\tau 3 = 3 \times \tfrac{2}{3} = 2$$
$$\tau 2 = 2 \times \tfrac{1}{2} = 1$$
$$\tau 1 = 1 \times \tfrac{1-1}{1} = 0$$

In general, if the prime factors of N are $p_1\, p_2\, p_3$ etc. then $\tau N = N \cdot \frac{p_1-1}{p_1} \cdot \frac{p_2-1}{p_2} \cdot \frac{p_3-1}{p_3}$ etc. only when N is prime it must be considered, for this purpose, its own factor. For that reason I should have said that $p_1\, p_2$ etc. are the prime divisors of N. But I am getting fatigued.

The only thing among my "four truths" that you will find any difficulty in proving is that if p is an odd prime $F(p-1)$ is divisible by p. This is an immediate consequence of the celebrated proposition that generally goes by the name of "Fermat's theorem," being one of those that he jotted down on the margin of the first printed edition of the *Arithmetica* of Διόφαντος,[2] which is that if n is any integer not a multiple of the prime, p, then $n^p - 1$ is divisible by p. Of course $F(p) = 2^p - 1$; and 2 is not a multiple of any other prime. The easiest way to prove Fermat's theorem is to start with the binomial theorem $(x+y)^n = x^n + nx^{n-1}y + \frac{n(n-1)}{2}x^{n-2}y^2 + \text{etc.} + \frac{n(n-1)}{2}x^2y^{n-2} + nxy^{n-1} + y^n$. You see that the coefficients of all the terms except the 1st x^n and the y^n carry n as a factor of their numerators; so that if n is a prime, these coefficients are all divisible by n; and there will be $n-1$ of them. Then putting $x = y = 1$ we have $(1+1)^p = 2^p = 2 + p \times$ some even number. Dividing by 2, $2^{p-1} - 1 = p \times$ an integer

Thus
$$2^3 = 8 = 2 + 3 \times 2$$
$$2^5 = 32 = 2 + 5 \times 2 \times 3$$
$$2^7 = 128 = 2 + 7 \times 6 \times 3$$
$$2^{11} = 2048 = 2 + 11 \times 186$$

[2] Fermat owned a copy of the Bachet edition (1621), the first edition published which contained the Greek text.

This seems to be all that concerns secundal arithmetic. But by the same sort of reasoning from the binomial theorem it is easy to prove not only Fermat's full theorem, but also the extended theorem which extended theorem is that if x is any integer prime to m, i.e. not divisible by any prime factor of m (for instance, differing from a multiple of m by one less or one more than any prime factor of m except 2 and 3), then $x^{\tau m}-1$ is divisible by m. For example, the 4th power that does not end (when decimally expressed) in 5 nor in 0, will be one more than a multiple of 5, that is, will end in 6 or 1. Thus $2^4=16$ and $3^4=81$ etc. So putting $m=6$, we see that (since $\tau 6=2$) the square of any odd number not divisible by 3 will be 1 more than a multiple of 6 thus $5^2-1=4\times 6$, $7^2-1=8\times 6$, $11^2-1=20\times 6$. This is easily proved from the fact that $(a+b)^p$, p being a prime, when divided by p, leaves the same remainder as does a^p+b^p, which is obvious from the binomial theorem; and since a and b are sums of units, any integer to a prime power, leaves the same remainder when divided by p as does the number itself, i.e. n^p-n is divisible by p; and if n is prime to p, $\dfrac{n^p-n}{n-1}$ will be a whole number. ...

(1912 June 10. This is the first day on which I have been fit to write since the last on which I did write to you; but I only mean that this is true in case it turns out that I am fit to write today. I am making my letter a very long one; but I have never heard that made a MS any the worse as an autograph. It adds one characteristic of my years, — long-windedness or garrulity.)

I was dealing with the numbers that I denote by $F(n)$ or Fn; that is those whole numbers which are (each of them) expressed secundally by an uninterrupted row of n units of which the last is in the units' place. In other words $Fn=2^{n+1}-1$. Consequently, $F(mn)$, where m is a whole number, is always a multiple of Fn. The causes of the great aid that the numbers Fn afford in operations of division in secundal arithmetic are as follows:

1st, the reciprocal of any Fn can be written secundally at sight, since it is obviously the number which would be secundally expressed by putting a unit in every nth secundal place to the right of the units' place. I write a circulating secundal by simply writing the circulant, (or, circulating part) once, and drawing over it (calling this line a roof). Thus, seven being $F3$, or $\mathrm{I\,I\,\dot{I}}$, $\dfrac{\mathrm{I}}{\mathrm{I\,I\,\dot{I}}}$ will be $:\,.\,.\,\mathrm{I}\,.\,.\,\mathrm{I}\,.\,.\,\mathrm{I}\,.\,.\,\mathrm{I}\,.\,.\,\mathrm{I}$ and so on *ad infinitum*, which I write $:\overset{\frown}{.\,.\,\mathrm{I}}$. Of course it will sometimes be convenient to write it $: . . | \overparen{. . |}$ or $: . \overparen{. | .}$ or $: . . | . . | . . \overparen{| . .}$ etc.

2nd, if the number by which one wishes to divide, say q, exactly divides one of the Fn numbers, so that, r being a plural integer, $Fn = q \cdot r$ or course $\frac{1}{q} = \frac{r}{Fn}$. Thus one fifth will be $: \overparen{. . | |}$ since $F4$ is | | | i = | . i × | i.

So $\frac{|}{| . . i}$ will be $: \overparen{. . . | | |}$ because | | | | | i = | | i × | . . |.

3rd, every number is an exact divisor of *some Fn*. In fact, any number p that is an odd *prime* is a divisor of the number $F(p-1)$. Thus the prime | . i divides | | | | and the prime | | i divides | | | | | i. This fact is an immediate corollary from one of the most historically famous propositions of mathematics, known as "Fermat's theorem" *par* excellence being the first of a number of new mathematical propositions, difficult to prove, that the Toulouse lawyer, Pierre de Fermat, jotted down on the margins of his copy of the arithmetic of the Greek Diophantos. These jottings were printed in 1670, after Fermat's death, and remained for more than a century a complete enigma to all the mathematicians; though I believe that all (at any rate, all but one) have now been demonstrated. The one that usually goes by the name of "Fermat's theorem" was generalized in 1801 by Gauss, and perhaps I will state that extension of it, although it is proved in every good encyclopedia of any pretensions. I will mention at once that whatever whole number m may be $F(2m)$ is divisible by $F2$, or | i; that $F(3m)$ is divisible by $F3$, or | | i, that $F(4m)$ is divisible by | . i; $F(5m)$ by | . | i, $F(6m)$ by | . . i; $F(7m)$ by $F7$ or | | | | | | i; $F(8m)$ by | . . . i; $F(9m)$ by | . . | . . i, and (as I implied in the first sentence of what I have been writing about the 3rd cause of the usefulness of the *Fns*) F(| . | : n) by | . | i; and so on.

4th, the prime factors of *most Fn*s can be found by the above 3rd cause of their utility; but there are cases in which they cannot so be discovered. For instance, if you had to divide secundally by | . | | i or 23; and then, since the smallest factor of a composite number must be less than its square root, one sees at once that 23 is a prime, that divides $F22$. But $F22 = 4194303$, which is not a very convenient number to divide by, even secundally. There is a much more convenient way of dividing by 23, since F | . | i is divisible by 23. But no general rule for finding out the factors of numbers has ever been devised. I have got a secundal method which I believe to be less laborious than any hitherto published for finding the least and greatest two factors of any odd number.

(1912 June 13. If I find myself able to write today, it will be the first day on which I could since I wrote last.)

I was dealing with the $F(n)$s. The causes of their being so helpful in division are the following truths.

1st, the reciprocal of Fn, secundally expressed, consists of units in every nth secundal place to the right of the units' place, with zeros in all the other places. I write the circulant of any circulating secundal fraction with a line over it, like this ∩, just covering the circulant. For example $\frac{\text{i}}{\text{ııi}}$ = : . . ı . . ı . . ı etc. ad infinitum = $:\overset{\frown}{..\text{ı}}$.

2nd, if the number with which one wishes to divide, say m, is a divisor of an Fn, then writing q for $\frac{Fn}{m}$ of course $\frac{1}{m} = q \times \frac{1}{Fn}$. Thus since $F4 = 3 \times 5$, of course $\frac{1}{5} = \frac{3}{Fn}$ which is $:\overset{\frown}{..\text{ıı}}$

3rd, this furnishes a method of dividing by any odd number; and of course division by a power of two is merely a matter of shifting the units place by a number of places to the left equal to the exponent of that power. For that every odd number exactly divides some Fn is proved as follows. First, if the number in question is a prime, say p, then $F(p-1)$ is divisible by p. (Thus 5 and 7 being prime, $F4 = F(5-1) =$ ı . i × ı i and $F6 = F(7-1) =$ ıııııı = ııi × ı . . i). This is an immediate consequence of one of the most historically famous propositions of mathematics, which goes by the name of "Fermat's theorem." Pierre de Fermat was a lawyer of Toulouse who in reading the arithmetic of the Greek (Alexandrian) Διόφαντος, Diophantus, took occasion to jot down in the margins any additional propositions that might strike him. In 1670, after his death, they were printed, and for long, — certainly for more than a century, — proved an insoluble enigma to the mathematical world. In fact, when I was younger, and scientific discoveries did not rain upon as thick as now-a-days, there was still one of Fermat's statements that had not been proved either true or false. How it is now I don't know. All the others were true, I believe. The one that goes by the name of "Fermat's theorem" par excellence is that if p is any prime and n is any integer not divisible by p, then $n^{p-1}-1$ is divisible by p. For example, let $p = 5$. Then the 4th power of any whole number, written decimally, will have either 6 or 1 in the units' place; so that one less than that will be divisible by 5. If p is an *odd* prime and $n = 2$, $n^{p-1}-1$ will be $F(p-1)$ and will, by the theorem be divisible by p. I may mention that Fermat came nearer to the methods of the differential calculus than anybody else before Leibniz and Newton. Euler first *published* a demonstration, though I believe Leibniz left one in MS, of Fermat's theorem. In 1801 Gauss in his Disquisitiones Arithmeticae proved an extension of Fermat's theorem. Namely, that if m and n are two incommensurable integers, and though this is usually stated to be sufficient, yet it is not so; and it is best to take a prime for m, and τn is a number related to n in a way that I will define below (and τn is best called by the name Sylvester gave it, the *totient* of n), then $m^{\tau n} - 1$ is divisible by n. The totient had been noticed by Euler. Oh, I had better say at once that if p_1, p_2, p_3 etc. are the prime factors of n, then τn may be defined as $n \cdot \frac{p_1 - 1}{p_1} \cdot \frac{p_2 - 1}{p_2} \cdot \frac{p_3 - 1}{p_3} \cdot$ etc. For instance if p is a prime $\tau p = p\frac{p-1}{p} = p - 1$. Since 4 has no prime factor by 2, $\tau 4 = 4 \times \frac{1}{2} = 2$ and then 3 being prime to 4, $3^{\tau 4} - 1 = 9 - 1 = 8$ is divisible by 4. $\tau 6 = 6 \times \frac{1}{2} \times \frac{2}{3} = 2$ and taking a number incommensurable with 6, say 25, we shall have $(25)^2 - 1 = 624$ divisible by 6. The only prime factor of 9 is 3. Hence $\tau 9 = 9 \times \frac{2}{3} = 6$. Then $2^6 - 1 = 63$ and $5^6 - 1 = 15624$, each of which is divisible by 9. But $10^6 - 1 = 111111$ is *not* divisible by 9.

(1912 June 14. If it turns out that I am well enough to write today, it will be the first day on which I have been so since I wrote last. This shows I *try* to conform to your request.)

I was dealing with the $F(n)$s. The facilitation of division by these numbers is due to 4 truths, which I will first state without proving them, so that you may the more readily appreciate how they aid division, provided they are true, as I will then go on to show, after some needful preliminaries. ...

4th [Truth]. Division by 2^n is effected by shifting the units' place n places to the left.

Before going on to the demonstration of these truths and other propositions, it will be well to make two insertions. I will here only mention that the first is a general table of as many of the $F(n)$s from $F1$ up as there will be room for on the fourth page of this sheets with their prime factors. The first column shows n expressed secundally. All the numbers being integers, it is needless to mark the units' place. The 2nd column shows the ns decimally expressed. The 3rd gives the *Fns*. The secundal expressions each consisting of a row of n units, need not be shown. But all are given decimally. The 4th column contains a P against those *Fn*s that are prime, and less than a million. Since there are published tables of prime factors showing those under ten millions, but I haven't got them,

I have omitted the other *P*s. The other columns give the prime factors of each *Fn*, decimally expressed.

ı	1	1	P													
ı .	2	3	P	3												
ı ı	3	7	P		7											
ı . .	4	15		3		5										
ı . ı	5	31	P				31									
ı ı .	6	63		$3^2.7$												
ı ı ı	7	127	P					127								
ı . . .	8	255		3		5			17							
ı . . ı	9	511			7					73						
ı . ı .	10	1023		3			31				11					
ı . ı ı	11	2047							23, 89							
ı ı . .	12	4095		$3^2.7.5$								13				
ı ı . ı	13	8191	P										8191			
ı ı ı .	14	16382		3				127						43		
ı ı ı ı	15	32767			7		31								151	
ı	16	65535		3		5			17					257		
ı . . . ı	17	131071	P													
ı . . ı .	18	262143		$3^3.7$						73				19		
ı . . ı ı	19	524287	P													
ı . ı . .	20	1048575		3		$5^2.31$					11					41
ı . ı . ı	21	2097151			7^2			127								337
ı . ı ı .	22	4194303		3					23, 89							683
ı . ı ı ı	23	8368607						47								178481
ı ı : . . .	24	16777215		$3^2.7.5$					17			13				241
ı ı . . ı	25	33554431					31			601						1801
ı ı . ı .	26	67108863		3						2731			8191			

I will just mention a few of the properties of these numbers, *Fn*, that a computer will find useful. The first is that if *n* is an odd prime, $F(n-1)$ will be divisible by *n*, while $F(n-1)n^m$, whatever integer *m* may be [is] divisible by n^{m+1}. Of course, one sees at once that i ı ı ı ı i (I try the experiment of marking the units' place by a dot high over the line) is divisible by ı i and by ı ı i and that when divided by both these it yields another ı i as final quotient; so that it is also divisible by ı . . i and by ı . ı . i. Now since i divided by ı ı ı ı ı i will obviously yield for the quotient the circulating secundal :$\overset{\frown}{.....}$ı where I mark the circulant by covering it with a sort of roof, it follows that i/ı . . i will be : . . . ı ı i. A slightly different way of reasoning may be exemplified thus: it [is] plain that ı ı ı ı ı ı ı ı i is divisible by ı ı i giving as quotient ı . . ı . . i which in decimal form is $8^2+8+1 = 73$. Consequently i/ı . . ı . . i is : . . . . . . ı ı i. There is another way of reasoning which is occasionally useful. Namely, although it follows from what has been said that if *Fn* is a prime, then *n* itself must be so, yet the converse is not necessarily true. Let us see, then, whether ı ı ı ı ı ı ı ı ı ı i that is *F* ı . ı i, can be separated into two factors, or not. Since this is not a perfect square, one of the factors (if it be not prime) must be at most no greater than *F* ı . i, or ı ı ı ı i. Let us represent this problematical factor by the algebraic expression

$$x_{\text{ı.:}}(\text{ı :})^{\text{ı.:}} + x_{\text{ıi}}(\text{ı :})^{\text{ıi}} + x_{\text{ı:}}(\text{ı :})^{\text{ı:}} + x_{\text{ı}}(\text{ı :}) + x_{:}$$

where the five unknown coefficients must have, each of them, either the value *unity* or the value *zero*. We may represent the larger factor (if there be any) by a similar expression, having *y*s instead of *x*s for coefficients; and just as every one of the five *x*s must satisfy the equation $x(1-x) = 0$ whose roots are zero and unity, the two possible values of each x, so every one of the *y*s (whose number must be more than 5 probably) must satisfy the equation $y(1-y) = 0$. We thus have a problem in Boole's Calculus of Logic, for which see his "Laws of Thought." I will just illustrate here my way of treating it, by setting forth a few of my first steps. Multiplying (or rather conceiving to be multiplied) together the two polynomials representing the two factors, we see that the only term of their product that will not involve 1: as a factor will be $x_{:}y_{:}$, and even this would vanish if either $x_{:}$ or $y_{:}$ vanished. So, if there be two factors, it must be that $x_{:} = \text{i}$ and $y_{:} = \text{i}$. Next, the only term of the product of the two polynomials that involves no higher power of ı : than the first is $(x_{:}y_{\text{i}} + x_{\text{i}}y_{:})\,(\text{ı :})$. Substituting, in this, the values just found for $x_{:}$ and $y_{:}$ it reduces to $(x_{\text{i}} + y_{\text{i}})\,(\text{ı :})$; and since the coefficient of each power of ı: up to the eleventh must be ı, in order to give the product known, we have two possible cases, which we may number Case i and Case ı :, calling the former that in which $x_{\text{i}} = 0$ and $y_{\text{i}} = \text{i}$ and Case ı : that in which $x_{\text{i}} = \text{i}$ and $y_{\text{i}} = 0$. Passing next to that term of the product of the polynomials which involves $(\text{ı :})^{\text{ı:}}$ we find it to be $(x_{\text{ı:}}y_0 + x_{\text{i}}y_{\text{i}} + x_0y_{\text{ı:}})(\text{ı :})^{\text{ı:}}$. Here, as before, the numerical coefficient, $x_{\text{ı:}}y_0 + x_{\text{i}}y_{\text{i}} + x_0y_{\text{ı:}}$ must equal unity, and since we have seen that either x_{i} or y_{i} vanishes, while x_0 and y_0 each equal unity, we have $(x_{\text{ı:}} + y_{\text{ı:}}) = \text{ı}$ each of our two cases splits into two. We had better not number any case as the third, but call Case ı . : that in which $x_{\text{i}} = 0$, $y_{\text{i}} = \text{i}$, $x_{\text{ı:}} = 0$, $y_{\text{ı:}} = \text{i}$.

Case ı . i $x_{\text{i}} = 0$, $y_{\text{i}} = \text{i}$, $x_{\text{ı:}} = \text{i}$, $y_{\text{ı:}} = 0$
Case ı ı : $x_{\text{i}} = \text{i}$, $y_{\text{i}} = \text{:}$, $x_{\text{ı:}} = 0$, $y_{\text{ı:}} = \text{i}$
Case ı ı i $x_{\text{i}} = \text{i}$, $y_{\text{i}} = \text{:}$, $x_{\text{ı:}} = \text{i}$, $y_{\text{ı:}} = \text{:}$

(1912 June 16.) This table ought to be instructive to you. The prime factors shown, with few exceptions (if any) are either $F(n)$s, or divisors of $F(n)$s, or are similarly related to another series of numbers of which ı . i and ı . . i and ı . . . i, are examples which instead of being, each less by 1 than a power of 2, are each greater by one than such a power. We might write $En = 2^n + 1$; but the $E(n)$s haven't the same importance as the $F(n)$s. ... The factors of each Fn were found by supposing that most of them would be each [larger] than a multiple of n. For instance in order to find the factors of F ı ı . . ı (that is $F25$), which gave me more trouble than any other, I assumed that after I had divided by 31, the next smallest factor would be 1 more than a prime number that was a multiple of 25; so that I actually tried dividing $\frac{33554431}{31}$, or

1082401 by 101 getting 85 as remainder
by 151 getting 33 as remainder
by 251 getting 89 as remainder
by 401 getting 102 as remainder
and by 501 getting 241 as remainder

before I finally tried 601 and found that went in exactly. The only factors I did not find either as divisors or as quotients in such trials were those of F ı . ı i ($F11$); and those I might have got so for $23 = 2 \times 11 + 1$ and $89 = 8 \cdot 11 + 1$. I was aided by a small table of prime factors in Hülsse's "Sammlung." I have carried the table a good deal further than the copy I give in this letter.

Before giving proofs of my four truths, I must tell you that the branch of mathematics to which those truths belong is unparalleled in mathematics for being the field in which not ignorant pretenders but the very greatest of mathematicians, such as Fermat, Leibniz, Euler, LeGendre, Gauss, Cayley, Sylvester, — perhaps even the impeccable Dirichlet have fallen into misstatements. Such faults in mathematics can have no other origin than looseness of definition: so, as definition is in my line, I will préface my proofs of the four truths by some strict definitions of arithmetical terms *in the senses in which I shall use them* quite regardless of any care for propriety of language.

To begin with, in order to express that a number is of the kind that ought to be used in counting (or that would be if we were to conduct that process with rigid attention to every theoretical condition of accuracy), and which may consequently be used to reply to the question, "How many?" — such a number I shall term a *cardinal*, using this word as a substantive.

Before this definition can be regarded as satisfactory, I must say what I mean by "*counting*"; and to make this definite I must define a *collection*; and that will involve defining the meaning of a single object, or say, a *singular*.

Beginning with this, I define a *singular* as an object which is neither *indefinite*, on the one hand, nor *indeterminate*, on the other. So I am called upon to explain what I mean by "indefinite" and "indeterminate."

By *indefinite* I mean having the character of not being subject to the (verbal, if you will) principle of contradiction, so that the same predicate *may be* truly affirmed and truly denied of it. Thus, "some man" is indefinite," since "*some man*" may be wise and "some man" may be not wise. So mere possibilities are "indefinite." Thus, I *can* open my mouth, and I *can* not open it; though one or the other must be false of what I *actually* do.

By *indeterminate*, I mean not (verbally, if you like) subject to the "principle of excluded middle," that either the affirmation or the denial of a statement is the fact. Necessities and would-bes are "indeterminate" in this sense. It is not true that I *must* work or *must* play because I can take my choice and pass my time differently on different days; and this is what I mean by saying that would-bes, must-bes, and universal rules are "indeterminate," while actual facts must have a character or not have it. Whatever is existent or actual is a "singular" in my use of words. Thus, the population of the United States at the instant I put my period at the end of this sentence will be either odd or even, and not both, and is so a "singular."

One singular of one category or nature may be the same existent object, in itself, as a plural of another object. To *count* is a certain operation performed upon a singular that might be, or be supposed, identical with a plural of another category, the singulars of this supposable plural, being called the "singulars counted"; and it is essential to what is meant that some one or more of a possible endless series of vocables shall, orally or otherwise, be attached to the singulars counted, and that in a certain succession. Since no actual series is endless, and yet, no matter how far the counting has to go, there will always be a definite vocable to be attached to the next object, it is plain that a perfect general rule for extending the series of vocables without ever coming to a stop, is essential to the idea of counting; and although that rule may not have been definitely decided upon, — as, for example, just at what point we ought to attach the word "billion," — yet we must have the notion of such a rule, or we haven't got even the *notion* of counting. Therefore, those writers who talk of savages who can "count" up to three (or seven) and no higher, talk nonsense. Savages in that condition of mind do not, properly speaking, "count," any more than birds do. They recognize the characters of small groups; but "counting" essentially involves a more or less definite conception of how to count endlessly, that is, without a definite highest cardinal, beyond which one would be without any notion what vocable to use. This is the only respect in which such jingles as "Eeny, meeny, mony, mi" fall short of *counting*.

Now let us not fall into the common blunder of saying that the cardinal numbers mean how many. For "how many" is a general term. Now the only sort of real meaning that can attach to a general is the conception of a general resolve to behave according to some principle. When I gave the doctrine of "pragmatism" the name it bears, — and a doctrine of vital significance it is, — I derived the name by which I christened it from πρᾶγμα, — "behaviour," — in order that it should be understood that the doctrine is that the only real significance of a general term lies in the general behaviour which it implies. Those who share with me a belief in pragmatism will, therefore, say, not that cardinals mean how many, but that "how many" means at what point in the series of cardinals the count of any given collection will come to an end, so making the general idea refer to a habit or manner of conduct.

But it is time to come down to a more detailed description of counting. The essence of it is that at the instant of each application of a new cardinal, an object shall receive a mark that withdraws it from the collection of objects not yet "counted," or connected with fresh vocables, and puts it into the collection of single members that have been counted. It is therefore a theoretical requisite of correct counting that the operation should begin by making sure that no object has already attached to it the mark that is destined to be attached to the single members as they are counted. It is theoretically requisite that this first step should be proclaimed like the others. The vocable "none" or "zero" may, for instance, very properly be pronounced as soon as the step is taken. It is certainly at first a theoretical possibility that it should be found that there are *no* objects of the kind proposed to be counted. At any rate, we frequently find occasion to say "There are *no* objects of such and such a kind;" and when we say this, the vocable "no" has the character of a numeral as fully as "ten" or "two" have. The vocable "zero", or "none," thus answers the question "how many," and moreover there is a stage of the process of counting at which it would be appropriate to call it out. These are sufficient reasons for including "zero" or "none" among the cardinals; and a person who thinks otherwise ought in consistency to use the ordinals in counting. At any rate, the series of numbers, :., ¡, ı :., ı ¡ etc. has certain characters common and peculiar to it, and having the greatest importance in arithmetic, should be distinguished by an appellation; and I am going in this letter to call it "the series of cardinals."

On the other hand all these cardinals except the first have the character of defining the possible multitudes of existing aggregates of singulars, as well as having a special importance in mathematics which *zero* by no means shares, since it alone cannot be a divisor that gives any existing quotient, even though fractional. These numbers, therefore, must likewise receive a name; and I shall call them "whole numbers," or "integers."

The integers not 1 form a less important class, yet ought to have a designation. I call them *plural integers*. ...

J. JAMES M. PEIRCE (L 339)

Milford 1893 Nov. 11

My dear Jem:

I bore you with my long letters, I fear. But I mean to confine this one to mathematics. I want you to be duly impressed with the force of my suggestion about a logical introduction to the theory of functions.

I have reviewed those two books for the Nation. Of course, I could not say 1/10 of what I wanted to say; and I doubt if they will ever print what I did say, though it involved the labor of reading Harkness and Morley *through* except parts of the last chapter, and reading about half of Forsyth, including his last chapters. (The last is a most enjoyable book. I thought the first was capital 'til I read that.) I remarked in that Notice that Prof. B. Peirce, as long ago as 1841, had published a text-book with the significant title of *Curves & Functions*, implying that Analytic Geometry and the whole Calculus were to be taught as parts of the Theory of Functions! In fact, this branch is the entire theory of continuous quantities and their relations. It embraces all pure mathematics except the Theory of Numbers, which it uses. Moreover, taking the young man who does not expect to pursue a mathematical profession, but who wishes to study mathematics for a year or two for its disciplinary value, beyond the Theory of Probabilities which every man wants to use, and Mechanics which everybody wants a general conception of, the only things he wants to study are the Theory of Numbers as far as Quadratic Forms, for the sake of the logic of finite numbers, and the general outlines of the Theory of Functions, for the sake of the conception of continuity. Of course, I include the outlines of the Calculus. But I would not bother with a good many details insisted on, about curvature etc. The general intellectual purpose of the study should rule the course.

Harkness & Morley's §54 has about as many logical faults as any demonstration I ever saw. Among others that of trying to reach a conclusion without having any premises. I mean, that they lay down, or definitely appeal to, no property of the points on a line. (They say a *straight* line, as if that had anything to do with it.) (I have not seen Fine's Number-System. If you can get Ginn & Co. to send it to me I will write a notice of it. But I cannot promise the notice will appear; for (*enim*. It is a pity English confuses *enim* and *nam*.) the Nation has to suppress part of my stuff, for fear of losing all their subscribers.) It is perfectly obvious that the proposition of §54 cannot be proved. This illustrates the necessity of putting this whole business of the conception of Continuity upon a very different footing.

Now, will you laugh? If so look at the footnote on p. 63 of Harkness & Morley. Here they have been talking about this ε for a whole chapter. There can be no clear thought about it, and no accurate reasoning, until it is fixed in the mind of the reader just at what point the selection of this quantity is to come in. Yet here is this essential explanation stuck in as an afterthought this way. The authors well remark that it was a long time before the importance of the distinction between uniform and nonuniform convergence was apprehended (p. 72 small print). This distinction is, however, entirely a question of the order of precedence of assigning values to two quantities. The logic of relatives shows that the necessity of that sort of distinction is precisely the prime point of difference between mathematical and simple syllogistic reasoning; and it shows when and how the order of selection can be changed and when not. A little practice of it shows further that such a subject as the theory of functions is simply sown with logical pitfalls for whoever does not attend to this matter. Thus, if Cauchy and others had just devoted a few days to the study of the logic of relatives as I would expound it in my book their minds would have been far clearer and they would have advanced more rapidly.

The same thing is the key to perspicuous enunciation. Observe, for instance, how clearly the logic of relatives obliges you to think such a proposition as "To every X corresponds a Y that is not Z." In the language of the logic of relatives that has to be expressed thus: — A relation can be found, such that, no matter what X you take, after it is chosen a Y can be found, such that it is not a Z, and is in that relation to that X. If we use ! for the sign of a *suitable selection*, and ? as a sign that it makes no difference what instance is taken; if, further we write $(A; B, C)$ to mean that B is in the relation A to C, and $(P; Q)$ to mean that Q has the character P; then we write algebraically (using F to mean a relation), and a line over a letter for negation

$$!F \ ?X \ !Y \ (F; X, Y) \ (\bar{Z}; Y)$$

or, using 1 to mean the relation of identity, and $\bar{1}$ for non-identity,

$$!F\ ?X\ !Y\ ?Z\ (F; X, Y)\ (\bar{1}; Y, Z)$$

The part of this formula, or writing, which has a row of letters preceded by ! or ? I call the *Indicator*. It corresponds to the subject of a sentence, because it tells what we are going to speak about. The function of the letters is like that of a demonstrative, personal, or relative pronoun. The remaining part of the script I call the *Boolean*, because it follows the simple rules of the Boolean calculus. It corresponds to the predicate of a sentence.

This method of expressing ideas has a considerable analogy with the old Egyptian language. Most words conveyed to them a pictorial idea. The arrangement of these makes a complex of pictures; and then the remaining words are either of the nature of pronouns or of prepositions (and these are essentially pronominal because they refer to the parts of the body and the like. Behind, before, etc. are words quite similar to this, that, I, you.)

A little training in the use of this system, will quickly develop mathematical distinctness of thought. Let us suppose we are considering values of x and of y which satisfy certain conditions, — say which satisfy the equations $\varphi(x, z) =$; $\psi(y, z) = 0$ for real values of z. As we are speaking of no other values, x and y may be used to mean such values only. Let us write fxy to mean $F(x, y) = 0$, rx to mean that x is real, and my to mean that the modulus of y is less than 1. Now, your pupils will learn very quickly to apprehend instantly such distinctions as those between

$$!x \cdot ?y \quad rx \cdot fxy + my$$

which means that if any value of y satisfying $\psi(y, z) = 0$ with a real value of z, has its modulus equal to or greater than 1, then there is a real value of x satisfying $\varphi(x, z) = 0$ with a real z, which satisfies the equation $F(x, y) = 0$ independently of the value of y.

$$!x\ ?y \quad rx(fxy + my)$$

means the same except that it positively asserts that there is a real value of x satisfying $\varphi(x, z) = 0$ with a real z.

$$?y\ !x \quad rx(fxy + my)$$

means that such a real value of x exists and that it satisfies $F(x, y) = 0$ for every value of y there may be (if any) that with a modulus greater than or equal to 1 satisfies $\varphi(y, z) = 0$ for a real z.

Instead of ! and ? we can use

ʾ to be read "some," meaning that a suitable individual is to be chosen.

ʿ to be read "any," meaning that any individual that comes can be taken.

The rules are about as follows. (I say *rules*. But I had better call them the *Charter* of the calculus, since they proclaim what you are at liberty to do, not what you must or must not do.)

1. Given different propositions all asserted. You have only to diversify the letters used in the different propositions, so that the same letter shall not occur in two propositions, and then you can multiply them together, and having done so, can bring all the indicators to the left. And the order of precedence of two letters which came from two different propositions is indifferent. It will be best to carry the "somes" as far to the left as possible. Skill must be exercised in arranging so as to effect the purpose in hand. Thus, suppose we have a surface which in *some* plane $x = a$, where a is unknown, has a real intersection with *every* plane $y = b$. Call this surface S and write

ʾx ʿc ʾz $(Sxyz)$

Suppose we have another surface S' whose intersection with every plane $x = a'$ has for one of its branches a line parallel to the axis of z. We write this

ʿx' ʾy' ʿz' $(S'x'y'z')$

We can either multiply these together in the order

ʾx ʿx' ʾy ʿy' ʾz ʿz' $(Sxyz)$ $(S'x'y'z')$

or in the order

ʾx ʿy ʾz ʿx' ʾy' ʿz' $(Sxyz)$ $(S'x'y'z')$

From each of these we can deduce propositions not deducible from the other by the following rule.

2. *Rule of Identification and Diversification.* Part I. *Identification.* Throughout the Boolian of any proposition we can substitute for any letter marked *all* any one of those to its left in the indicator, dropping the former from the indicator.

Thus in the proposition at the head of this page, identifying x' with x', y' with y, z' with z, we get

ʾx ʾy ʾz $(Sxyz)$ $(S'xyz)$

that is, the two surfaces have a point in common.

So in the second proposition, identifying x' with z, and z' with y we get

$$’x\ ‘y\ ’z\ ’y'\ (Sxyz)\ (S'zy'y)$$

That is, if the second surface be rotated about an axis equally inclined to those of x, y, and z, through 120°, so as to bring its axis into the position of z, its z axis into the position of y, and its y axis into the position of x, there will exist a plane perpendicular to the axis of x, which cuts the first surface at one or more real points for each real value of y; and at some one of these (for each real value of y) a line parallel to the axis of x will have a real intersection with the second surface.

Milford Pa 1904 Feb. 13

Dear Brother:

In studying the logic of quantity lately, a familiar group appeared in (to me) a new light. I will write the matter out so that you may judge whether it be worth printing.

The smallest universe in which there is a relation that is a 4th root of unity is one of four individuals; and in such universe there are three such relations, together with their converses. The universe may be imagined as consisting of the 4 diagonals (undirected rays) of a cube; say A, B, C, D. The four fourth roots will be quadrantal displacements of the cube about axes each through centres of opposite faces. I will denote these quadrantal relations by x, y, z, respectively converse to ξ, η, ζ.

$$x = D{:}A + C{:}B + A{:}C + B{:}D, \quad y = B{:}A + C{:}B + D{:}C + A{:}D,$$
$$z = C{:}A + A{:}B + D{:}C + B{:}D$$
$$\xi = C{:}A + D{:}B + B{:}C + A{:}D, \quad \eta = D{:}A + A{:}B + B{:}C + C{:}D,$$
$$\zeta = B{:}A + D{:}B + A{:}C + C{:}D$$

The squares of these, which will be reversals may be denoted by B, Γ, Δ.

$$B = x^2 = \xi^2 = B{:}A + A{:}B + C{:}D + D{:}A$$
$$\Gamma = y^2 = \eta^2 = C{:}A + D{:}B + A{:}C + B{:}D$$
$$\Delta = z^2 = \zeta^2 = D{:}A + C{:}B + B{:}C + A{:}D$$

Products of two different quadrantals will give four cube roots of unity together with their converses, which are represented by displacements of the cube about the diagonals through 120°. I will denote them by heavy letters **a**, **b**, **c**, **d**, converse to α, β, γ, δ. They are

$$\mathbf{a} = \xi\eta = \eta\zeta = \zeta\xi = A{:}A + C{:}B + D{:}C + B{:}D$$
$$\mathbf{b} = \xi y = yz = z\xi = D{:}A + B{:}B + A{:}C + C{:}D$$
$$\mathbf{c} = x\eta = \eta z = zx = B{:}A + D{:}B + C{:}C + A{:}D$$
$$\mathbf{d} = xy = y\zeta = \zeta x = C{:}A + A{:}B + B{:}C + D{:}D$$
$$\boldsymbol{\alpha} = zy = yx = xz = A{:}A + D{:}B + B{:}C + C{:}D$$
$$\boldsymbol{\beta} = \zeta\eta = \eta x = x\zeta = C{:}A + B{:}B + D{:}C + A{:}D$$
$$\boldsymbol{\gamma} = \zeta y = y\xi = \xi\zeta = D{:}A + A{:}B + C{:}C + B{:}D$$
$$\boldsymbol{\delta} = z\eta = \eta\xi = \xi\zeta = B{:}A + C{:}B + A{:}C + D{:}D$$

There are besides six products of quadrantals with B, Γ, Δ, represented by reversals about axes through the middle points of opposite edges. I denote them by X, Y, Z, Ξ, H, Σ

$$X = x\Gamma = \Gamma\xi = \Delta x = \xi\Delta = A{:}A + B{:}B + D{:}C + C{:}D$$
$$Y = y\Delta = \Delta\eta = By = \eta B = A{:}A + D{:}B + C{:}C + B{:}D$$
$$Z = zB = B\zeta = \Gamma z = \zeta\Gamma = A{:}A + C{:}B + B{:}C + D{:}D$$
$$\Xi = x\Delta = \Delta\xi = \Gamma x = \xi\Gamma = B{:}A + A{:}B + C{:}C + D{:}D$$
$$H = yB = B\eta = \Delta y = \eta\Delta = C{:}A + B{:}B + A{:}C + D{:}D$$
$$\Sigma = z\Gamma = \Gamma\zeta = Bz = \zeta B = D{:}A + B{:}B + C{:}C + A{:}D$$

These together with unity are the 24 substitutions of the four elements, A, B, C, D.

They may be divided into two equal classes; the *artiads*, or those of even order in the quadrantals, and the *perissids*, or those of odd order. Each class may be divided into three sets of four relations, such that any individual of the universe stands to any individual of the universe in one and only one relation of any one set.

In the multiplication-table I now give, the sets are separated; and the relations are differently ordered as *multipliers* and as *multiplicands*, in that the Greek and small Roman minuscules, not capitals, are interchanged [Fig. 1].

1 B Γ Δ	**a c d b**	**α δ β γ**	X Ξ ξ x	Y y H η	Z ζ z Σ
B 1 Δ Γ	**c a b d**	**δ α γ β**	Ξ X x ξ	y Y η H	ζ Z Σ z
Γ Δ 1 B	**d b a c**	**β γ α δ**	ξ x X Ξ	H η Y y	z Σ Z ζ
Δ Γ B 1	**b d c a**	**γ β δ α**	x ξ Ξ X	η H y Y	Σ z ζ Z
α γ δ β	1 Δ B Γ	**a b c d**	Y η y H	Z Σ ζ z	X x Ξ ξ
γ α β δ	Δ 1 Γ B	**b a d c**	η Y H y	Σ Z z ζ	x X ξ Ξ
δ β α γ	B Γ 1 Δ	**c d a b**	y H Y η	ζ z Z Σ	Ξ ξ X x
β δ γ α	Γ B Δ 1	**d c b a**	H y η Y	z ζ Σ Z	ξ Ξ x X
a d b c	**α β γ δ**	1 Γ Δ B	Z z Σ ζ	X ξ x Ξ	Y H η y
d a c b	**β α δ γ**	Γ 1 B Δ	z Z ζ Σ	ξ X Ξ x	H Y y η
b c a d	**γ δ α β**	Δ B 1 Γ	Σ ζ Z z	x Ξ X ξ	η y Y H
c b d a	**δ γ β α**	B Δ Γ 1	ζ Σ z Z	Ξ x ξ X	y η H Y
X Ξ x ξ	Y y η H	Z ζ Σ z	1 B Δ Γ	**a c b d**	**α δ γ β**
Ξ X ξ x	y Y H η	ζ Z z Σ	B 1 Γ Δ	**c a d b**	**δ α β γ**
x ξ X Ξ	η H Y y	Σ z Z ζ	Δ Γ 1 B	**b d a c**	**γ β α δ**
ξ x Ξ X	H η y Y	z Σ ζ Z	Γ Δ B 1	**d b c a**	**β γ δ α**
Y η H y	Z Σ z ζ	X x ξ Ξ	**α γ β δ**	1 Δ Γ B	**a b d c**
η Y y H	Σ Z ζ z	x X Ξ ξ	**γ α δ β**	Δ 1 B Γ	**b a c d**
H y Y η	z ζ Z Σ	ξ Ξ X x	**β δ α γ**	Γ B 1 Δ	**d c a b**
y H η Y	ζ z Σ Z	Ξ ξ x X	**δ β γ α**	B Γ Δ 1	**c d b a**
Z z ζ Σ	X ξ Ξ x	Y H y η	**a d c b**	**α β δ γ**	1 Γ B Δ
z Z Σ ζ	ξ X x Ξ	H Y η y	**d a b c**	**β α γ δ**	Γ 1 Δ B
ζ Σ Z z	Ξ x X ξ	y η Y H	**c b a d**	**δ γ α β**	B Δ 1 Γ
Σ ζ z Z	x Ξ ξ X	η y H Y	**b c d a**	**γ δ β α**	Δ B Γ 1

Fig. 1

This is the 24×24 square multiplication table of the group. The multiplicand is to be entered in the top row, the multiplier in the left hand column.

It will be noticed that a set of three reversals, as *Ξ*, *H*, *Σ* can put the cube into all its 24 positions, as well as quadrantal displacements, and just as directly, too.[1]

Your affectionate
C. S. Peirce

[1] A draft of this letter contains a diagram of a cube showing these displacements. (See Fig. 1, Vol. 4, 13 [MS. 328]. The entire paper is of interest here.) It also suggests

that if James thinks it "worth printing," he "might send it to the Mathematical Society." The figure given here is reproduced on a larger scale and looks less smudged.

Peirce owned a copy of the English translation by G. G. Morrico of Felix Klein's *Lectures on the Ikosehedron* (London: Trubner & Co., Ludgate Hill, 1888). It was presented by Mrs. Peirce to the Harvard College Library, 28 June 1915. Notations in Peirce's hand throughout are evidence of his attention to Klein's work in the theory of groups. On p. 30 at the end of the chapter on Regular Solids Peirce lists in the notation of relatives the quadratic group and the tetrahedral group and writes "see fly leaf at end." That page no longer exists since the book has been rebound. However, one can see the influence of Klein in the development of the contents of this letter as well as in MS. 328.

K. JOSIAH ROYCE (L 385)

[c. 28 May 1902]

My dear Professor Royce:

You seemed to want to know what I thought of your Supplementary Note. It seems to me an admirable thing, and a most useful thing. It is a very great step in philosophy to introduce these ideas and it is impressively put. I may say at once that I don't go along with you in all the uses you would put them to. I cannot admit that realistic metaphysics as you call it (mixing together, apparently, two positions that seem to me distinct) can be refuted so easily. I do not think that any criticism of perceptual facts can justify itself. *It* is open to criticism; but *they* are not. The realistic argument from those facts is solid, and your refutation does not seem to me so. Not that I object to your *conclusion*, any more than in the note where you praise Schröder's §9X, which to me seems nothing but a logical fallacy. In many places you remind me of Spinoza who uses arguments to my mind of no value but with an unexpressed undercurrent of weighty thought. However, these are not the things I intended writing about, I only meant to point out some small points where the thing might be improved. In what you say of the axiom "The whole is greater than its part," I think Cantor's definition of *greater* should be adduced.

I agree that it is better to found the theory of arithmetic on the ordinal numbers; but it is equally possible to found it on the cardinal numbers. There is also another interesting way; that of proceeding from the algebra of logic.

Veronese is not sound.

As for Hegel & Harris on infinity, they utterly mistake the idea, which has nothing to do with absence of limit, as Riemann said.

As to my own definition of finite, I am sorry you did not remark that it antedates Dedekind and that in that earlier paper I also put the Fermatian inference into strict logical form. My definition of the finite differs from Dedekind's hardly more than verbally. His form has the advantage of joining more naturally to Gauss's idea of *Abbild*, although I also used this idea in the same paper. In making *finity* a positive, not negative, conception, I distinctly said in what sense I did so; namely, that of the *finite* it is possible to reason in a peculiar way not applicable to the infinite, concerning which all ordinary modes of mathematical reasoning do apply. I do not think it *necessary* to introduce infinity into the calculus, but it seems to me quite unobjectionable and simpler. I am convinced it will be the method of the future.

There are other ways in which Cantor's ideas ought to be introduced into philosophy. For example, the principal argument for psycho-physical parallelism seems to be that it is the only way of escaping the contradiction of holding that matter can act immediately only on matter and mind only on mind. But that is not so. Matter may act on mind and mind on matter not *immediately* but by an infinite series. For example, if matter consists of vortex atoms in a fluid, itself consisting of vortex-atoms in another fluid, and so on *ad infinitum*, one of these fluids might act on the next and so on by material causation and then might come a series of souls, equally, infinite and acting on one another by mental causation. It is obvious that there are many places in philosophy where apparent contradictions could be refuted by similar hypotheses, going to support my general position that inductive reasoning is alone sound in metaphysics.

Important as is the idea of the endless series in philosophy, the proposition that there is no maximum multitude is at least equally so. (I regret you do not adhere to my practice of using "multitude" for Cantor's *Mächigkeit* and "collection" for his *Menge*.) *One and two* in my opinion in no way imply infinity. But *three* I grant you does. Although an infinite series and even a continuum may be *real*, yet the *actuality* which is perfect only with the reacting pair, diminishes as the multitude increases, until the true (not analytical) *continuum* is really no more than a concrete law.

The way I should put what seems to me the most important lesson of your first volume is that *purpose* and *opinion* are merely inseparable aspects of the same thing. Only in one case the purposive character is prominent, in another the cognitive character. That being fully understood, something like your theory of reality follows as a corollary.

You emphasize immensely the religious aspect. My religion rather fights shy of metaphysics, as my moral nature does of ethics. But from a logical point of view, your book has made me feel more strongly the importance of ethics. I suppose esthetics is also important for logic through ethics.

Very truly
C. S. Peirce

[c. 13 November 1903]

My dear Royce:

[If] you have really detected a fallacy in my proof that two collections cannot each be greater than the other, it naturally concerns me much to understand it. Couldn't you jot it down. For your convenience I will restate my argument.

1. A relation being a mere logical possibility, to assert that an indesignate relation of any general description *exists* and to assert that it is *possible* are the same. If sets of individuals such as a relation of the described form requires exist, there must be such a relation. Thus, there is no one-to-one relation of every individual of the collection, X, Y, Z to an individual of the collection I, J, because if r be the relation, and x be that one of the collection I, J to which X is r and y be that one of the same collection to which Y is r, x and y must be different, and any unit of the same collection to which Z should be r would necessarily be either x or y, contrary to the assumption that r is a one-to-one relation.

2. Hence, if there be no one-to-one relation in which every B stands to an A, it must be logically impossible that there should be such a relation.

3. But a one-to-one relation is not in itself absurd, nor can there be any contradiction in supposing that there is a one-to-one relation in which every B stands to *something*, since in fact every B does stand in the one-to-one relation of identity to something.

4. Hence the impossibility must consist in some existential limitation of the As. To show what I mean by an "existential limitation," take any other form of relation as an example. Suppose the relative ϱ to be such that, whatever individuals X and I may be, if X is ϱ to I then there is *just* one individual other than I to which X is ϱ and there are *just* two individuals other than X or each other that are ϱ to I. Then if the Bs, *by themselves*, are all ϱ to anything, the multitude of Bs is a multiple of 3. If, this being the case, nevertheless the Bs by themselves, are in *no such*

all to *A*s, it can only be that *A*s enough do not exist. Hence, the only way in which two sams could be each greater than the other would be by both being sufficiently *small*; and in fact if both are 0 they are each greater than the other, if you choose so to define.

5. However, in order to ascertain more clearly the nature of the existential limitation of the *A*s. Begin by supposing ϱ to be [as in Fig. 1] and also [as in Fig. 2]. That is to say if X is ϱ to I and X is ϱ to J then I and J are identical, and also every B is ϱ to an A. There are, if [*b*] is the multitude of the *B*s, $a^{[b]}$ such relations, that is there is a multitude of such relations, that is there is a multitude of such relations greater than the multitude of *B*s. (Supposing there are more than one *A*s)

6. But now suppose that in every such relation some A is not ϱ'd by any B; that is [as in Fig. 3].

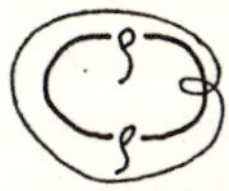

Fig. 1

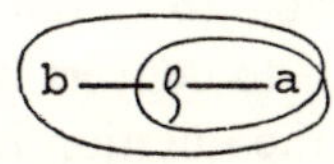

Fig. 2

or more fully

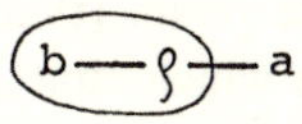

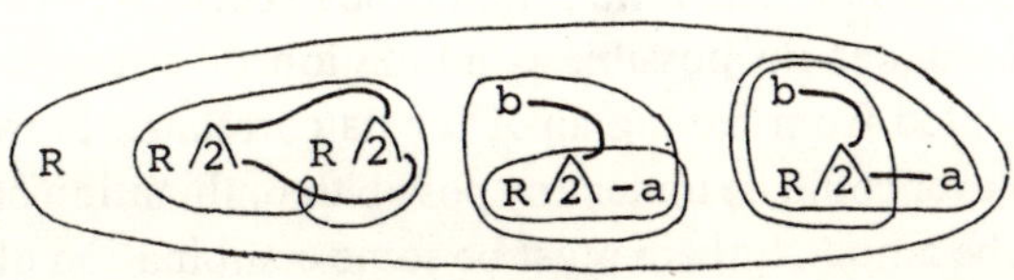

Fig. 3

Now perhaps you think that at the next step I reason in this way. There is room in the *coupé* for Mrs. Royce, and there is room for Mr. Royce, and there is room for Peirce; and therefore there is room for all three. But my reasoning is really this:

There is *under every combination of possibilities* room for Mrs. Royce, and under every combination of possibilities room for Mr. Royce, and under every possible combination room for C.S.P. *Therefore* there must be room for all three.

Possibly you have discovered that in some way which has escaped my notice I am reasoning that because, no matter how many may have got in, there is always room for one more, therefore there is room for all.

But I cannot see where either of these fallacies lurks in my reasoning which turns always upon *room*, or possibility. I do not *assume* that the correlates can be arranged in linear order, Cantorian or otherwise; but it may appear as if I did because it at once follows from my *conclusion* that they could.

But to come to my reasoning:

There are at least as many ways in which every *B* might be related to a single *A* as there are of different *sams* of *B*s.

Let us suppose, however, that there is not one among all these relations in which every *A* is ϱ'd by a *B*.

From this point of the argument, which we may call *P*, there seem to be several ways of proceeding, all sound.

In my lecture, I continued as follows: Let us add to the *A*s a collection of non-*A*s say equal to all the *B*s in multitude, and let the correlates of ϱ be changed so as to convert it, in every possible way, into a one-to-one relation. Then the question is whether every one of the vast collection of relations so obtained will have one or more non-*A*s as correlates of the *B*s. If this be the case, there must be a logical necessity that it should be so. But since it is expressly assumed that in all possible cases some *A* will be un-ϱ'd by any *B*, and since the only way in which any given *B* will be compelled to take a non-*A* for its correlate (that is to say, in which there will be no possible variation in which that *B* takes an *A* for its correlate) is that there is no *A* that is not a correlate of some other *B*, it follows that under no possible combination of circumstances is any single *B* prevented from having an *A* for its correlate. Now if no single correlate is or ever could be under any possible combination of possibilities compelled to be a non-*A*, there must be some combination of possibilities under which all *B*s should have *A*s for their correlates.

Can you put your finger on any fallacy there?

But going back to the point *P*, a second way of proceeding would be as follows:

If among the relations ϱ hitherto obtained there is any which is a one-to-one relation, that at once proves my point. But if there is no one-to-one relation, change the correlates of each ϱ in a manner to be described, so as to render the relation a one-to-one relation. First, for the sake of formal completeness, I will say that, if possible, the correlates are all to be changed to *A*s that are not correlates. But it will never be possible to convert the ϱ into a one-to-one relation in this way. For were this possible, the resulting one-to-one relation would have been one of the ϱs.

Secondly, (another formal division), we will suppose that correlates are to be changed from *A*s that are correlates to *A*s that are not correlates so far as possible. What can prevent this being done so that no *A*s remain that are not correlates? It is purely an affair of necessary logic. It must be possible or it must involve some contradiction. But the only possible ways in which it could involve contradiction are, 1st, that some ϱ should have no *A*s that were not correlates, and 2nd, that there should be no *A*s which could cease to be correlates of some *B*s without leaving them non-correlates of all *B*s. In the former case there would be a one-to-one relation in which every *A* stood to a *B*. In the latter case there would be a one-to-one relation in which every *B* stood to some *A*. What then? Shall we say that no change of correlates can reduce any one of the ϱ's to a one-to-one relation? This seems manifestly absurd. But if the change can be made it can be divided into two stages, the first of which shall change correlates into other *A*s in all ways that are not self-contradictory, the second part changing correlates into non *A*s. But this second part will never be reached, because no contradiction ever can arise until our point has been proved in one or other of the two ways just mentioned.

To me, as at present advised, this seems an absolutely necessary argument. There is no assumption that the conversion is to be performed in indefinitely many steps; but only that *two* steps are to be taken.

However, going back to point, *P*, there seems to be a third argument equally conclusive and even more direct. Almost anybody but you would find it convenient that I should repeat that the situation is that we have all the ϱ's which result from first supposing all the *B*s to be in a relation ϱ to one *A* and taking all the variations resulting from changing the correlate of each *B* to some other *A*; and it is supposed that, for all these ϱ's, some *A* is not ϱ'd by any *B*. That is, in view of the existential constitutions of the two collections, there would be a contradiction in supposing that every *A* was ϱ'd by a *B*. But the only way in which there could be a contradiction in supposing that some of the variations had produced such cases is this. No matter what ϱ you take, in changing this ϱ, so as to make some (or all) of the *A*s that it leaves not ϱ'd, to be ϱ'd, every such change of it would cause some *A*s that had been ϱ'd to be left not ϱ'd. (Of course, in general, even if all the *A*s that had been ϱ'd were left wholly un-ϱ'd, still the *A*s that had not been ϱ'd would not all become ϱ'd; but I only mention this to point out that I have said nothing that conflicts with it.) But this being thus, there can be no contradiction in supposing that some of the ϱ's so produced will be one-to-one relations; and therefore some of them certainly will be so. Consequently, there will be a one-to-one relation, in which every *B* stands to an *A*.

Now I am anxious to know what fallacies can be found in these three arguments.

Very faithfully
C. S. Peirce

L. FRANCIS RUSSELL (L 387)

New York 1896 April 26 (Address will be forwarded.)
Stewart's Hotel. Broadway and 41st St.

My dear Judge:

Yours of the 22nd reached me today. I shall be happy to write the article in question, leaving it to you to decide whether it shall appear under your name, or anonymously, so long as it does not so appear as to be taken for the work of any other person. I shall keep quiet, simply saying that the article represents my views, but telling Schröder about it, confidentially, perhaps. I am under a promise to write two articles for the *New York Herald.* I have also promised a friend, in case he pays my expenses, to go on to Washington and do certain work for him there. The weather is very cool here, at present; and I have no money to pay for a lodging, so that I am forced to walk the streets all night. This fatigue makes my work in the daytime, which has to be confined to the hours during which the libraries are open, slow. Then, I get very little food. I am strongly urged for an excellent situation salary 25000 fr. in Paris, which, were it to go to the most efficient man, I should be sure to get. I have an invitation to pass the summer at a château in France. If I could do so, I should be able to finish my Arithmetics during the summer, and should find assistants there for another profitable work, which could be commenced. If I could get an appointment as newspaper correspondent in Paris, with an advance of money to take me there, I should be able to go, and make a proper appearance. And making a proper appearance, I should be able to get my Arithmetics published to advantage. I have succeeded in getting in an application for a patent for an indispensable adjunct to a new machine for domestic lighting by acetylene. I have great confidence that that machine is going to revolutionize domestic lighting, and it has to be put in every house where the new mode of lighting is used. They will be forced to use my invention; and were I to be able to make a decent appearance, no doubt, I could make them pay me a dollar or more royalty; which would come to hundreds of thousands.

But unless I can get some start, what is to prevent their stealing it? I have other inventions whose real value is immense; but this is the only one I have taken steps to patent. So far, I am sole proprietor of it. You see, therefore, that if you were to get me a position as Paris correspondent of a newspaper, or any other good start, such that it would enable me at once to make a good appearance, it would come to very much more than itself, and would put me in a situation in which I could turn money your way, in return. And you may be sure my gratitude would be genuine and loyal. Of course, I do not describe the thing; because to *prove* to you what its value is would require many pages. But I would part with ¼ interest in the patent to a good business man who would exert himself actively to do the necessary business with either of the two parties who might buy it, namely, the owners of the machine for which mine is indispensable, and the great company with ten millions capital who would like to gobble that machine, but can't without me. For that ¼ I would require $5000 cash and would give you another ¼ to find me such a party. I should require you not to part with *all* your fourth; so that, we could protect ourselves against dishonest deals. Now if you want a description of the matter with an argument as to its value, let me know. I say that, supposing the patent passes, it is worth in the neighborhood of a million.

I dare say that after you have assumed certain properties of a circle not true in non-Euclidean geometry you can prove the 3 angles of a triangle are equal, by your definition of a straight line which makes the straight line the limiting case of a circle or of part of a circle. But that would not suffice to show that you can prove geometry without postulates. This would be as much as to say that two different constitutions of space are not possible. I think that erroneous; and would ask how you prove that space has three dimensions in any such way; how you prove, that space has no points, lines, or surfaces from which there are more or fewer ways of motion than from ordinary points; how you prove that space is all one piece, that every ring in space can shrink to nothing in space without breaking (if you hold this to be true, or if not, how you prove whatever you do hold true). I call your attention too, to the fact that all projective geometry rests on the assumption that every plane returns into itself, but so that if an object having three dimensions rests on the plane and sliding upon it passes through infinity and so back to its original position, it is now on the *other side* of the plane; so that the plane of projective geometry is what I call a *perissid* surface; and some writers call a *unifacial* surface. But the theory of functions deals with a plane which is a simple *artiad* surface. All the pairs of straight lines upon it meet twice unless they are parallel, and in that case they have two coincident points at infinity, and when an object passes along such a line through infinity, it returns to its original place just as it was. There then are two conflicting hypotheses about a plane with which mathematicians are working, without inconvenience. If therefore you can prove the propositions of geometry without postulates, — that is, without any initial hypothesis about the constitution of space, you ought to be able to prove that one or other of those two conceptions is absurd. And since both branches of mathematics make continual use of these very properties, you will utterly overthrow one or other of the two greatest branches of mathematical doctrine.

I have just completed a memoir I intended reading to the National Academy of Sciences in Washington this last week. But I was unable to get there. In the introductory part of this memoir, I undertake to state in general terms what is logically possible, and what not. This assumes that some things are impossible although they do not involve any contradiction, such for example as that there should exist only two or only three things, that there should be a relation which could not exist between a certain set of things although there were no contradiction involved, etc. etc. Having thus described the logically possible, I go on to consider the multitudes of collections. I succeed in that way in proving that of two collections not equal one must be greater than the other; although there is no contradiction in supposing that in every possible way of setting them off into pairs, one object of each pair belonging to the one collection and the other to the other, there should always remain unpaired objects among both collections. I also show that greater than the collection equal to all finite whole numbers there are a series of possible collections each next greater than the last, but infinitely greater than that last; and these collections are equal, in the multitude of them, to the finite whole numbers; and greater than them all is a possible collection, than which no collection can be greater. Having thus disposed of collectional quantity, I propose in subsequent papers to consider other kinds of quantity.

Very faithfully
C. S. Peirce

P.S. None of the collections mentioned at the end of my letter has the properties of an infinity of the first order. For in all cases, beyond the finite collections, the square of the multitude equals the multitude. The maximum collection equals its own exponential. The lowest infinite multitude has no logarithm. The rule of possibility would be worth telling. It is very simple. But I have no space for it, this time.

Milford Pa 1905 July 14

My dear Russell:

Decidedly I must send you my article of Jan. 1901. Your *summum bonum*, "life", is probably at bottom about the same as mine, though I view it more concretely. I look upon creation as going on and I believe that such vague idea as we can have of the power of creation is best identified with the idea of theism. So then the ideal would be to be fulfilling our appropriate offices in the work of creation. Or to come down to the practical, every man sees some task cut out for him. Let him do it, and feel that he is doing what God made him in order that he should do.

It appears to me that given any man there are certain propositions as to the truth of which he has no doubt whatever, or no discernable doubt. He would act upon them with no misgiving, not the slightest. And roughly speaking these are in the main the same for all men. At any rate there are propositions we all believe, every grown man. But all these propositions are *vague*, and as soon as we attempt any precise definition we fall into conflict with one another and oftentimes into self contradiction. Now there is, I think, no department of logic in which new work [is] more desirable than in working out the whole theory of vagueness. Take a regular tetrahedron *per se*. Any two of its vertices make up an object of which there are six. They are all just alike in every respect and yet each is different from the other five. Of these five there is one to which it is peculiarly related. Some pair of indistinguishable pairs of vertices is formed of three vertices only. Some pair is not so formed. The indefiniteness of the *some* makes the principle of contradiction inapplicable. How far is this, or something corresponding to this true of all that is vague?

A *law* fails to confer existence on its subject. That is, nothing perhaps fulfils the condition of the law. If it does, it does so independently of the law. A general then lacks existence.

Now take the principle that in every contest some contestant must fail. This principle *does* confer *existence*. What is it, then, which it fails to confer. What element of being is it that the vague lacks. You cannot say it lacks *determinacy*; for this is equally lacking to the general. In fact, determinacy is lacking to whatever is not a complete reality. The vague seems to lack what we may call *which-ness*, self-identity, that which makes the principle of contradiction pertinent. Is that so?

C. S. Peirce

1908 Sep 18

Dear Russell:

I wish you would acknowledge receipt of my letters. I wrote you a long letter which I *infer* you got (I haven't yet finished reading yours of the 16th inst.). But as you didn't acknowledge it, I sent you a reply postal to ask if you got it. It would have been easy to write

Yes

Russell

But it seems to have been too much trouble. You observe that I attend to your letters, though they make some draughts on my supplies of time and energy.

As to the book you say you want to persuade Carus to get out, the idea *has merits*. That does not say that it is on the whole good as it stands. To judge of it, it is necessary to consider what mathematics is. Now to get an idea of what any science is *in its present condition* — which is all that a man of sense will attempt, for any science but his own, — it is necessary to study the men at work on it; for what it is, is what they make it. The mathematicians are men with great powers of embracing in one thought great complications of very simple elements, and of thinking about them with great exactness. They also have extraordinary ingenuity in finding points of view from which the properties of their complexes may be discerned. Having these exceptional powers, of course they devote themselves to the exercise of them. The result is that mathematics is nothing but a collection of deductive reasonings. Of course these group themselves in certain ways. They have *external* groupings of very little value, but commonly noticed; and they have *essential* groupings according to the nature of the leading concepts. It is the rarest thing in the world to find a mathematician who is capable of logical analysis, so that he really understands what it is that he has been doing; the literature and history of mathematics are filled with proofs of the incompetence of the greatest masters in that, and except Leibniz, I should be puzzled to name another. Take Fermat who discovered the *substance* of the calculus, though it took Leibniz to invent the proper notation for it. Well Fermat made a discovery of that method of reasoning about integers whose essence is that whatever is true of zero and is true of the number next higher than any number of which it is true is true of all integers; because any integer whatever can be reached by successive additions of 1 starting from 0. But Fermat though he showed extraordinary skill in using his method, never seems to have been able to formulate it or give a reason for it. He calls it a kind of *reductio ad absurdum.* This simply shows that he didn't know that the *reductio ad absurdum* differs from direct reasoning in no essential particular but only in the order in which the ideas that have to be put together are taken.

The whole Weierstrassian mathematics, — that is to say the way mathematicians reason since Weierstrass showed how loose most of the geometrical reasoning is, showing for example that it does not follow because a real function is continuous that it must have a differential coëfficient, which he did by simply giving an obvious instance of its falsity, — well, the whole Weierstrassian mathematics is characterized by a *distrust of intuition.* Therein it betrays ignorance of a principle of logic of the utmost practical importance; namely that every deductive inference is performed, and can only be performed, by imagining an instance in which the premisses are true and *observing* by contemplation of the image that the conclusion is true. The image, as *singular*, must of course have determinations that the premisses, as *general*, have nothing to do with. But we satisfy ourselves that the particular determinations of the image chosen, so far as they go beyond the premisses, could make no difference. There is no other way of making a deductive inference than that, however simple it may be. Take for example the [conversion] of *E* the universal negative proposition. To ascertain whether or not this follows, I take the example "No bird is a phoenix." To be sure this is not the only universal negative proposition but it does not differ from any other in any respect which could make any difference as to what might logically follow from it or what might be consistent with it. Now I imagine a cage in which are contained all the birds but no phoenices. To be sure all the birds could not in fact be contained in the cage; but we can imagine that they were so compressible that they could, and that would not prevent phoenices from being birds, because there is plenty of room for the phoenices, if there are any outside the cage, and there are none in it. Now then if there are any phoenices they are either in the cage or outside of it. But they are not in it so if there are any they are outside. So then none of them is a bird since every bird is inside and not outside the cage. So then if no bird is a phoenix, no phoenix is a bird. It is true that whatever phoenix there is is a bird. But that does not prevent its being true that whatever phoenix there is is not a bird.

Let us take a different example. Given the two equations

$$ax + by + c = 0$$
$$\alpha x + \beta y + \gamma = 0$$

Required the solution. Add to them the identical equation

$$mx + ny - (mx + ny) = 0$$

where we may give any values we please to m and n without falsifying the equations. We know from the nature of a determinant that

$$\begin{vmatrix} ax, & by, & c \\ \alpha x, & \beta y, & \gamma \\ mx, & ny, & -(mx+ny) \end{vmatrix} = 0$$

whence $$\begin{vmatrix} a, & b, & c \\ \alpha, & \beta, & \gamma \\ m, & n, & -(mx+ny) \end{vmatrix} = 0$$

or $$-(a\beta - \alpha b)(mx+ny) + (\alpha c - a\gamma)n + (b\gamma - c\beta)m = 0$$

or $$mx + ny = \frac{(\alpha c - a\gamma)n + (b\gamma - c\beta)m}{a\beta - \alpha b}$$

when by putting $m = 1$ [and] $n = 0$, $x = \dfrac{b\gamma - c\beta}{a\beta - \alpha b}$ and putting $n = 1$ [and] $m = 0$, $y = \dfrac{\alpha c - a\gamma}{a\beta - \alpha b}$. The point is that we have to *observe* the equations to see that they do permit the first equation on this page.

Now as to the book you have in mind, its design as I understand it is, by specimens of mathematical reasoning to improve the reasoning of non-mathematicians. For that purpose, it would not often be advisable to give the whole mathematical memoir, but only the logically instructive parts; and those need not be given in the words of the original writer, though his mathematical *forms* should be conserved. The annotations would require a very strong logician, not the kind of man that the average editor or publisher takes to. Small men like small men: they have an instinctive antipathy to great minds; and it would be a pity the thing should be so done that its weakness should be shown up at once. The poor annotator would not be in an enviable situation, nor anybody else concerned.

If the ground were fully convered, the book could not be a very small one. It would have to be at least as large as Russell's Phil. of Math., — a weak performance, by the way. Existential graphs should begin the work, because there is no organ of definition and logical analysis that is at all equal to that. I have not yet brought out this feature of the system, though I have much in MS. When I do bring it out, those few who go through the exposition will be amazed. Then should be taken up integers. First the Cantorian *ordinals* though he himself puts multitudes first. But it is wrong. For ordinals are the general, multitudes the special. That is, the different multitudes of plurals (which Cantor badly denominates *Cardinal-Zahlen*, and *Mächtigkeiten* von *Mengen*) are simply the successive grades of plurals; and there are (if you like so to modify the meaning of the term) "cardinal numbers," or as I have called them, 'arithms' (which is better), to designate those grades. But *ordinals*, in general, are designations of *grades*, each grade being possibly occupied by what I will call *numerates*, or subjects of numeration. In the special cases in which the numerates are *plurals*, the grades are *multitudes*, and their ordinals become cardinals, or *arithms*. I shall show that all real rational numbers are ordinals, and that irrational reals are ordinals likewise; while imaginaries and quaternions are a special kind of complexes of ordinals. I may mention here that I fully recognize that integer arithms have a special interest owing to the fact that each plural is composed of singular members, while rational arithms denote grades of ratios of two multitudes, and irrational arithms are the arithms of ratios between first-abnumerable multitudes. I have completed or at any rate extended the system of fundamental concepts employed by Cantor, in a very important respect, which I will, if I have time, explain further along in this letter.

Between Grades, there are two highly important dyadic relations, whose corresponding relative terms I call "after," i.e. *counted later*, and "next," i.e. *immediately after*. It is obvious that "next" can be expressed in terms of after. In fact the graph *N*—next—*M* is equivalent to 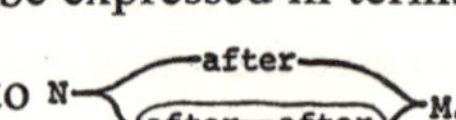

But "after" can also be expressed in terms of "next," (in a certain sense of "in terms of"); and this I am inclined to think is the truer analysis. At any rate, it lets us into the secret connexions between the numerates.

If we confine ourselves to the finite ordinals it is easy to define 'after' in terms of next. It is a part of the definition of ordinals that every Grade has a Grade next to it and that every grade but one is 'next' to a Grade; and no Grade is 'next' to a Grade other than a Grade to which it is 'next' nor has a Grade 'next' to it which is other than another 'next' to it. Moreover every Grade lacks a certain character which is possessed by a Grade next to it, and which is further possessed by whatever Grade is next to a Grade that possesses it. Now to say that a Grade is 'after' another is to say that it possesses every character of that kind that is not possessed by the Grade that it is 'after' but is possessed by the Grade that is 'next' to the latter. But this is not every kind of afterness, but is only that kind of afterness that alone occurs between finite Grades.

The higher kinds of afterness were introduced by G. Cantor. They are described by him in Vol XLIX of the Mathematische Annalen and in a form supposed to be adapted to philosophical readers in a little book Zum Lehre vom Transfiniten. Halle: Verlag von C. E. M. Pfeffer, 1890. There is also a very important critical report on the whole of Cantor's work on numbers by Arthur Schönfliess which forms the whole of Heft 2 of Vol. VIII of the Jahresbericht der deutschen Mathematiker-Vereinigung. It is called "Die Entwickelung der Lehre von den Punktmannigfältigkeiten." There has also been some fundamental work by Borel. I am not altogether satisfied with Cantor's work, greatly as I admire it.

I ought to have said above that it is essential to the idea of Ordinals that it is a linear system i.e. in *some* sense of any two Grades one is 'after' the other. In the sense just defined the finite system is *teres totus atque rotundus*. Nothing can be added to it. But there is another kind of *nextness* which can be introduced without modifying any of the above definitions. Namely a Grade may be "next," — not to any one Grade, but to an endless series of Grades; and Cantor postulates that every endless series of Grades has one, and but one, Grade 'next' to it which is not 'next' to any other Series nor to any Grade. For after Achilles has run through his geometrical series after the Tortoise, time still runs on and he with it, and the very next instant he is supposed to catch up. So this new grade the infinitesimal (i.e. the infinity-eth) is in a new sense 'after' all the finite grades. I will scribe the finitely after [as] $\underset{\overset{|}{0}}{\text{—after—}}$, and this new after [as] $\underset{\overset{|}{|}}{\text{—after—}}$ [Fig. 1]. Then Cantor denotes that which is 'next' to the whole series of finites by the ordinal ω. He notes that the *count* now depends on the order of counting; so that $1+\omega$ is

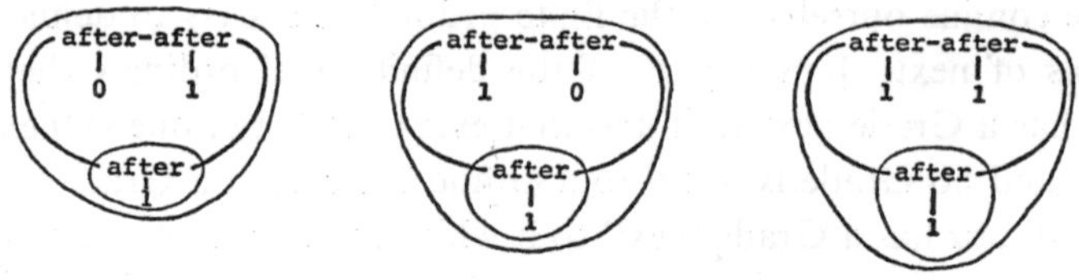

Fig. 1

no longer the same as $\omega+1$. I consider the *addend* as operating on the augend; and since, in my algebra of dyadic relatives, as well as in all my earlier algebras of logic I always wrote the operator to the left of the operand I write $\omega+1=\omega$ and $1+\omega$ as *next* to ω. But Cantor (and mathematicians generally) write the *addend* after the *augend*, which, in my view, is wrong. Of course it still remains true that every Grade has a Grade 'next' to it and therefore we run on to $\omega+\omega$, which Cantor writes 2ω. Then comes $2\omega+1 \ldots 2\omega+\omega = 3\omega$ and so on, until he has an endless series of endless series of ordinals. To say that there is a *next* Grade to this seems to me a new extension of the idea of nextness, since *this* next, is neither *next* to any Grade nor *next* to any endless series of Grades, but is only next to an endless series of endless grades. Here therefore it seems to me clearly that we have a new kind of afterness, which I write [as] $\underset{2}{\text{—after—}}$. But Cantor simply speaks of ω^2 without regarding himself as introducing any new conception; and his reason is that the *multitude* is unchanged. That is to say there is an order of counting according to which $\omega^2=\omega$. But it seems to me the better way of expressing this would be to say that the multitude of infinites, or transfinites, as he calls them, depends on the order of the count, and that the minimum count of ω^2 equals that of ω. It is of course easy to distribute the finite integers into any *finite* number of endless series.

Of course there will be a next to any endless series of numerates which is not next to any numerate, and there will be a next to an endless series of endless series of numerates which is neither next to an endless series of numerates each having next to another (nor is next to any numerate), and there will be a next to an endless series of endless series of endless series of numerates and so on indefinitely. And then (what we can only express indirectly by a sign describing a sign), the whole lot of numerates expressible by "series of series" etc. is supposed itself to have a next; though it is not evident that that does not land itself in an absurdity. This method Cantor supposes to be carried out endlessly; and he assumes (I can find no attempt to prove it) that in this way his Grades will attain every possible multitude. But that is an entirely illogical assumption. For I have proved that every multitude has a higher multitude, the plural of all plural[s] of *M*s being always greater than the plural of all *M*s. If Cantor has really proved, as he says he has that by his method, starting from ω he reaches 2^{ω} (or as he expresses it, ω^{ω}, which is the same multitude) and $2^{2^{\omega}}$ and $2^{2^{2^{\omega}}}$ and so on indefinitely, then he certainly has ordinals that count a series greater than any collection. But his multitudes being themselves a collection, and not skipping but always advancing to the next, it is impossible they should count a series more multitudinous than themselves; so that there must be an error somewhere. It may be the fallacy lies in not observing that the multitude of a collection is not the last ordinal in this or that count of it, but is the earliest ordinal that can count it, the count varying according to the order of counting. Or it may be owing to his not having proved that none of his descriptions of ordinals involve any absurdity. Schönfliess agrees with me that Cantor has not proved all he thinks he has.

Cantor does not start with the relation of "next" but with that of "after" and consequently fails to recognize any distinction between one kind of "after" and another, which is a great blemish.

In referring to his paper in Vol. XLIX of the Annalen, I omitted to say that it cannot be read without first mastering in minute details a previous paper in Vol. XLVI where he treats of multitudes. I should be much disposed to write a work myself on this subject if my logic were not far more important.

I don't know as I ever showed you how rational fractions are fundamentally ordinals. Start with the assumption that $\frac{N}{1}$ where *N* is an integer represents that integer, and with the assumption that $\frac{A+M}{B+N}$ where *A*, *B*, *M*, *N* are integers is *after* either $\frac{A}{B}$ or $\frac{M}{N}$ and that the one of these it is not after is *after* it. Of course, the same would be true of $\frac{A}{N}$ and $\frac{M}{B}$. Next assume that $\frac{1}{0}$ is after $\frac{0}{1}$. It follows that using Cantor's sign $P \prec Q$ for Q is after P, $\frac{1}{1} = 1$ is after $\frac{0}{1} = 0$ and that $\frac{1}{1} \prec \frac{1}{0}$. Then having $\frac{0}{1} \prec \frac{1}{1} \prec \frac{1}{0}$ we interpose $\frac{0}{1} \prec \frac{1}{2} \prec \frac{1}{1}$ and $\frac{1}{1} \prec \frac{2}{1}$ $(=2)$ $\prec \frac{1}{0}$. Next $\frac{0}{1} \prec \frac{1}{3} \prec \frac{1}{2} \prec \frac{2}{3} \prec \frac{1}{1}$ etc. Next $\frac{0}{1} \prec \frac{1}{4} \prec \frac{2}{5} \prec \frac{1}{2} \prec \frac{3}{5} \prec \frac{2}{3} \prec \frac{3}{4} \prec \frac{1}{1}$; and it is easy to prove that we shall then obtain every positive rational value *expressed in its lowest terms.* You can work out the proof for yourself, perhaps; though it is quite a theorem. Now we have supposed nothing at all about the values of these fractions nor have we supposed the intervals between them to be anything in particular, yet we have deduced all that is needed to establish the whole doctrine of fractions. Consequently, that doctrine has nothing to do with *measure* or exact equality, but only with ordinal sequence. You are at liberty to suppose that $\frac{1}{2}$ stands for two thirds or one tenth, and all the consequences will be true. The line between the numerator and denominator is not a sign of division. The fraction only expresses such a function that it has the properties stated. Of course you must recognize as a corollary that $\frac{P+P}{Q+Q}$ is intermediate between $\frac{P}{Q}$ and $\frac{P}{Q}$ that is, equals $\frac{P}{Q}$. So $\frac{7}{13}$ is between $\frac{1}{1}$ and $\frac{6}{12}$ i.e. is $>\frac{1}{2}$, between $\frac{1}{4}$ and $\frac{6}{9}=\frac{2}{3}$, between $\frac{2}{3}$ and $\frac{5}{10}=\frac{1}{2}$, between $\frac{2}{4}=\frac{1}{2}$ and $\frac{5}{9}$ etc. The assumption usually made (often with a puerile attempt at proof of it) that between all the values of a convergent series and the nearest rational value there is but a single irrational value, is perfectly gratuitous. There may be any multitude of such values of course "infinitely near" one another; and we have a perfect right to assume that there are (i.e. may be, are possibly) and so revert to Leibniz's differential calculus. For since rational quantities are mere ordinals, the assumption that because there is no "assignable difference" i.e. no rational difference between two values therefore there is no difference at all, is perfectly arbitrary.

Passing to the subject of true *continua*. The true definition of a continuum should be given; namely a continuum is an object all of whose parts are alike in their relations to *their* parts. So Kant. Then Listing's great memoir, I mean the one on the Census Theorem in the Göttinger Abhandlungen. But this should be annotated and brought into connection with projective geometry on the one hand and quaternions on the other. Besides which, the doctrine of Topical Singularities should be treated. And reference should be made to space of more than three dimensions. I haven't time to write more at present.

Very faithfully
C. S. Peirce

(Various things in the Theory of Functions should be brought in; ideas from the Galois Theory of Equations.)

Thanks for your surprisingly pretty card.

P.O. Milford Pa 1909 Jan. 1

My dear Russell:

A Happy and Prosperous New Year to you! I hope you don't read the reports from Sicily or Calabria. To me, to whom these places were so very familiar a generation ago, all this is enough to shake my reason.

You must think it very strange that, after my protestations, I should not have answered yours of Dec. 10th. But I will tell you how it is. My wife is very ill, and very energetic, conscientious, and the most particular housewife I ever knew by far. Even if I were not in the condition of penury in which I am, it would not (short of *wealth*) be possible for me to provide the servants that her state of nerves and of overwork require; and her devotion to me and acceptance of her share in my lot, when she might roll in luxury if she would leave me, tears at my heart strings. We have *no* servant. For it is better to have none than such as I could afford. Consequently a great part of my energy has to go to *chores*. Besides that my Hibbert article and some other things have caused me to be flooded with letters from people of consequence and which cannot be answered without much labour. In addition some matters of business consequent upon the death of one brother and the fact that the family take the attitude toward me that families generally do, — though I think *more so* than usual have thrown an additional mass of work upon me. Then there are my own logical studies which it is impossible for me to dismiss from my mind, and which become all the more urgent owing to my saying to myself that the time is brief in which I can be of service now. The result of all is that I have been so overworked that for more than a month I have not averaged 4 hours daily in bed and have worked all the time at the very top of my powers. Not only your letter but many other things of the greatest urgency have been utterly neglected.

As to the straight line, I think very differently from you. In the first place like the geometers of our time I think that projective geometry has nothing at all to do with measurement. Now the straight line is the chief thing in projective geometry and therefore I do not think it ought to be metrically defined. Your definition is metric. If I were to admit a metric I do no see any objection to resorting, as most geometers do, to the definition of the straight line (let me say *Ray* — meaning a limitless straight line) as the shortest [distance] between two points. (Your definition introduces the idea of a circle. But a *circle* is a metrical concept.) In particular, in 1873 Klein's paper on the *nicht-euclidische Geometrie* in the Annalen der Mathematik seemed to make it luminously clear to me, — even more so than it already was, — that measurement involves something clearly *additional* to the properties of space itself, something dependent entirely upon mechanics, and if there were no such thing as an approximately rigid body to suggest absolute rigidity and (as at present) no way of measuring time apart from measuring space, — what would become of the concept of measurement? Yet we could sight along a range and determine whether it were straight or not. But even *straightness* and *flatness* to my mind have nothing to do with space itself. If you read Listing's paper in the Göttinger *Abhandlungen* on the Census-Theorem, you will see that there is a geometry which knows no difference between a straight line and a curved line.

Imagine all space to be filled with an absolutely rigid glass in which three systems of equidistant parallel lines would be visible, the 3 being at right angles to one another and the lines to be either indefinitely close together or so that intermediate-visible lines would be produced at will. Then everything in metrics could be seen to be true if only that glass could be moved about interpenetrating other bodies without resistance. Every proposition in metrics could so be illustrated and if the glass were instead capable of moving like optical rays and shadows, all the propositions of projective geometry could be seen to be true in any special case. But suppose that the glass could be rendered plastic, though we will say incompressible, and distort it in any manner whatsoever. Then render it rigid again, and you will perceive that though the lines are no longer straight, there will be a perfect analogue to every proposition about straight lines.

It will be better however, to suppose that in place of the glass being movable like a rigid body, that it is capable of motion of a projective kind. You will thus see that

Any family of surfaces, each in one piece, of cyclosy one and of periphraxy one, which are such that any two of them have in common only one line in one piece and of cyclosy one, and any three of which have in common a single point, or else the three pairs have the same line in common and such that through any three points of space there is one such surface and but one unless a onefold infinity of them all having the same line in common, — any such family of surfaces are in all respects (not metric) related exactly as the planes of space are related to one another, and every non-metrical proposition of geometry has its exact analogue, concerning these surfaces.

Now if it be true that measurement is something quite additional to straightness, you will perceive that it follows that there *cannot be* any purely geometrical property of a straight line, unless of course you make assumptions concerning some other undefined line and define a straight line in terms of that.

All you can do, then, is to define a plane as one of a family of surfaces having the above properties. If you wish to render metric geometry true, you will simply add further that there is a certain one of these surfaces which every other of them cuts at right angles and there are some other conditions that I am not equal to working out tonight. I did once write a geometry in which the whole thing was worked out; but I could not get a publisher for the reason that I found the publishers all thought *they* knew something about geometry, whereas in fact they were in the deepest ignorance of it, knew nothing at all for instance of the 3 kinds of geometry and had notions in conflict with those of all the higher mathematicians' views. The MS was eventually lost, It was rather a pity as it was worked out very thoroughly and clearly. I defined a straight line as the path of a ray of light (though a ray of light is a fiction) and measurement by the properties of a rigid body and its looking glass image, though there is no rigid body known. However, there seems to be *room* for a ray of light were there such a thing, and for a rigid body.

As for Carus, he seems to me to have much ability in the way of language. His English is remarkable. But anything less like a scientific man, it would be difficult to [find]. In philosophy, I should say there were half a dozen men in this country who are his superiors. Take Lester Ward for example, Royce, and several men less known. What he is ambitious to achieve is *reputation*; but reputation in philosophy in this country without a professor's chair cannot be achieved without great real power; and that is not *born* in a man; it has to be worked out; and the first condition is that the man's soul should be filled with the desire to make out the truth, to the exclusion or almost to the exclusion of other aims; but Carus is full of himself; and it stands irremovably in the way of a thorough devotion to truth. He ought not to try to combine two aims so disparate and incompatible.

As to myself, I have had the misfortune to be interested in a science which nobody but me really cares for or even wants badly to know about. Now to pursue researches alone is so dangerous, one is in such constant peril of taking one's private opinions for the truth, that the greatest caution is necessary. I go over everything many times before I feel that I can venture to be confident of being right. Of course I go slow; and with the detestation of all exact thought requiring painful effort to apprehend it, which is so nearly universal, it naturally cannot happen that I should produce more than one book that is at once widely regarded as important and that really is so. Whether or not if I succeed in writing such a book it would be recognized before intellectual development has gone so far that it is no longer useful, I cannot know. I think it exceedingly dubious. At any rate, however, it cannot do me any good personally; and that is the brightest and most consoling thing about it. I can always say to myself that my motive was as white as the driven snow, and after all one's own good opinion is worth ten times that of all the rest of the world put together.

Why don't you get on with your logic?

With warm wishes

C. S. Peirce

P.S. Carus has offered me a hundred dollars for the copyright of my 6 articles on "Illustrations of the Logic of Science," I to revise as I please. I should print the first two with merely clerical corrections, but with an addition correcting the errors of doctrine which are implicit rather than explicit. The next on Probability I should not greatly alter but should *define* probability which has never been done correctly without a vicious circle. I should also state and take sides in the great controversy of Boole vs. Laplace. The one about the Order of Nature could stand substantially, but with an important addition. In place of the remaining two, I should give a brief detailed account of the method of scientific reasoning and the grounds of its validity, and also a long series of examples of historical scientific reasonings, very long. The whole would be polished and made as clear as possible; but readers would have to think. It would take me the better part of a year and a hundred dollars would be less than a starvation price. I shall see what another publisher will say. Anything from you in the way of faultfinding with definite passages of those articles will be valued by me.

Milford Pa 1909 Apr 15

My dear Mr. Russell:

I hope I am right in thinking your article in the April Monist utterly unworthy of you: to think otherwise I should be obliged to rank you with the circle-squarers.

There are two questions or aspects in regard to the non-Euclidean geometry. The purely mathematical question is whether or not the assumption that the sum of the three angles of a triangle is less or greater than two right angles does or does not lead to a perfectly self-consistent and mathematically interesting system of geometry. As to that *all* mathematicians worthy of the name are at one. You may not be aware of this. But you *do* know that in putting yourself in opposition to it, you are opposing the carefully formed conviction of Gauss, of Riemann, of Cayley and Sylvester and Clifford, of Helmholtz, of Klein, of W. E. Story, and many other profound mathematicians. Even my father, whose "Geometry" was his darling idea and was contrary to non-Euclidean idea, was forced to acknowledge that the non-Euclideans were correct.

The other question was whether experiential space may be in accordance with the non-Euclidean theory and contrary to the Euclidean postulate. In 1870, I put forth, perhaps before anybody else, the *suggestion* — without guaranteeing it, — that it might be possible rationally to consider space as Euclidean whatever discoveries might be made looking to the truth of one or other of the non-Euclidean theories, and the Nancy man whose name escapes me takes that ground strongly. As to how it may be permissible or reasonable to conceive of *real* experiential space is an open question. But it has nothing to do with your position.

For your sake, I must HOPE you have not put yourself into the mortifying position of having read all these great minds the lecture you have read them about over-haste and over-confidence without having read and duly studied at least the greatest of the works they have written in defence of their views; and therefore I am bound to assume that you are familiar with the following:

Cayley's Sixth Memoir on Quantics

Clifford's work on Metrics

My own few words about the division of geometry into Topics, Graphics, and Metrics, observing that the cross-ratio is not a question of metrics at all but only of Graphics

Riemann's celebrated contribution Ueber die Hypothetese etc.

Klein's two luminous papers in Vols IV and VI of the Mathematische

Annalen (though they must have slipped your mind when you said that nobody had *proved* that the non-Euclidean Geometry involves no contradiction; since otherwise you must have *read* those remarkably clear papers without fully comprehending them); and also Klein's *two volumes of Lectures on the Nicht Euclidische Geometrie*

W. E. Story's account of the non-Euclidean Geometry, its trigonometry, etc. which are particularly valuable, not only as showing its Graphical relations to the Absolute, but also as enumerating all the kinds of non-Euclidean geometry which admit the truth of the usual Graphics.

With these works before you, it was indeed a lawyer-like proceeding to confine yourself to a criticism of a translation of Lobatchewski's little book, which though a brilliant and elementary exposition and (barring his admitting the truth of the propositions that he numbers as 4, 5, 7, & 9) involves, I believe, no real error, unless a mere slip of the pen, which is rendered perfectly obvious by noticing what it is that his demonstration really demonstrates, be reckoned as an error,—which nobody but a lawyer presuming upon the gullibility of his jury would reckon as an error.

You seem to be that lawyer, for otherwise, why in the world should you have given that wordy proof that in the non-Euclidean geometry two lines may be parallel to one third without being parallel to each other, when Lobatchewski's original explanation of what he means by "parallel" shows that, in the geometry, every straight line has *two* lines parallel to it through *each point of space* that does not lie in the line itself? You cannot have failed to see that these two lines cutting each other are not parallel to the third was in obvious contradiction to the proposition that two straight lines both parallel to a third are necessarily parallel to each other.

I press the Question: Why did you not content yourself with this obvious proof of the incorrectness of his proposition No 25??

The answer seems to me obvious. If you had done that, your readers would have at once perceived that Lobatchewski merely made a slip of the pen and *meant* that two straight lines parallel to a third *toward the same side* are parallel to each other and then they would have seen that *this* is precisely what his demonstration of Proposition 25 correctly demonstrates and would have reduced your whole argument to a *quibble*.

Your calling Lobatchewski a "Zeno" because he acknowledges a line between all straight lines through any given point that cut a given straight line at a real finite distance and those that cut it at imaginary distances (I need not remind you of Klein's papers) is to my mind down-right inappropriate, — not to use an uncivil expression. All lines in the plane of a circle that pass through any point in that plane may be divided into those that cut the circle and those that don't, unless you deny the principle of Excluded Middle. But are there not two lines that are on the boundary between these two classes? Namely the two tangents to the circle? Do you call everybody who uses that very natural way of speaking, whether it be logically correct or not (and I think it is) a modern Zeno? If so, I should think it to be a compliment to be called *by you* a modern Zeno or by any other opprobrious and contemptuous epithet! For Zeno was not the man who held there was a limit between two classes of loci, but on the contrary was the man who did not rightly comprehend the idea of the *Limit*. To my mind, it is you yourself who do not see that two parallels in the elliptically non-Euclidean geometry of Lobatchewski (or may be it is called "hyperbolic": I have forgotten, at the moment) cut one another on the Absolute, or surface at infinity, at an angle equal to *zero*, just as they do in Euclidean geometry. The only difference is that in the latter they cut at two coincident points of the Absolute, since the latter consists in that case of two coincident planes placed, as it were, *back to back*; i.e. the two planes form a sphere of infinite radius having all real points in its interior; while in Lobatchewski's geometry the Absolute is a real quadric surface, an ellipsoid and may be called a sphere, since, without having any definite properties different from any other kind of ellipsoid so placed yet owing to the Absolute serving to define distance (apart from its unit) it must necessarily *be called* a sphere.

As to a straight line not having any *definition proper*, it is demonstrable that it cannot be, properly speaking, defined.

You define a straight line metrically. Every modern geometer will object that a straight line is a purely *graphic* conception and graphics knows no such conception as distance or length and does not distinguish a circle from any other ellipse nor from a parabola or hyperbola. Besides, if a *metric* definition were admissible unquestionably the proper definition is that a straight line is the shortest distance between two points.

And after all your talk about the error of the non-Euclidean geometry being due to the absence of a definition of the straight line (which is absurd, since the straight line is a graphical concept and the non-Euclidean geometry purely metrical, leaving graphics untouched), you produce a definition of the straight line from which it is demonstrably if not *manifestly impossible* to prove the sum of the angles of a triangle to be equal to two right angles!

And don't you think your remarks at the bottom of p. 289 are rather arrogantly applied to Gauss, Riemann, Klein, Helmholtz etc. rather than to yourself?

Faithful are the wounds of a friend.

Your friend
C. S. Peirce

Milford Pa 1909 Apr 24

My dear Russell:

. . . I began to write for you an account of a way in which, regarding geometry as a science of observation, you could establish the truth of Euclidean geometry, as nearly as anybody really knows it is true, without any doubtful postulate. But I give up; for I am confident you would not begin to understand it. You don't seem to me to understand the logical nature of the non-Euclidean geometry. For you say that ordinary intelligent men look upon the matter, as you do, and upon the mathematicians who assert its truth as apparently cracked. Now they certainly do not, and cannot, think so. For that would imply that they knew what the non-Euclidean position is. An intelligent man has no opinion at all about a doctrine the nature of which he does not know; further than that if the general body of well-informed and competent judges are known to be in accord about it, he presumes they are right, unless it be one of those subjects like metaphysics which has not yet been reduced to any universally acknowledged methods. Now mathematics is of all sciences the one about whose reasonings there is the least question among competent men, especially since Weierstrass.

The doctrine of non-Euclidean geometry is that if Euclid's 5th Postulate (erroneously called by some former writers the 11th Axiom, though it is not so in the good mss. nor according to the statements of ancient geometers, and certainly is not an *Axiom* and never has been generally thought to be so) may be supposed false either in excess or in defect, and the new postulate substituted for it will lead to a system of geometry which involves no self-contradiction.

Now everybody knows that geometry must rest upon postulates, and the logical nature of a postulate was correctly described by the ancient mathematicians as a proposition that the mathematician does not pretend to know is true, but that he must assume to be true in order to develop a mathematical theory, because he needs it as a premiss to support his reasonings. For it is not within the province of the mathematician to study the facts of nature and to pronounce upon such matters. An *axiom*, on the other hand, is a proposition supposed to have been already settled before the study of the branch of mathematics in which it figures as an axiom has been taken up. Such, for example, is Euclid's erroneous axiom that any part is less than its whole, which may, however, be freed from *error* by regarding it as meaning that geometry is restricted to *finite* spaces; although modern geometers have conclusively shown (and especially Cayley in his 6th Memoir on Quantics) that that view renders any comprehensive understanding of geometry as impossible as the exclusion of $\sqrt{-1}$ from algebra would maim and disfigure that science and the whole theory of functions.

Now ordinary people, far from thinking Euclid's 5th Postulate satisfactory, have almost universally and in all ages, protested that it was *not* so. Therefore, whether they are aware of it or not, they have really been on the side of the non-Euclideans, and if they think the latter hold a position contrary to common sense, it is because they do not know what their real position is, and imagine that *I*, for example, and people who occupy such a ground as I do are typical non-Euclideans. That is not so. I agree entirely with the non-Euclideans on the *mathematical* question so far as that can be called a "question"; but on the field of logic I hold a position that non-Euclideans generally would repudiate. Namely, while mathematicians generally hold Laplace's *Théorie des Probabilités*, as a sort of bible, on account of its great *mathematical* power, and I do not dispute this last merit, I maintain that the logical foundation of the work, the explanation of what probability is, is puerile in the extreme, and that no observation of facts can impart any *probability* whatever to any generalization so far as it goes beyond observation. The validity of induction is indeed closely *connected with* a correct theory of probabilities, but does not *consist in* any definite probability attaching to the inductive conclusion.

I can't give much more time to the matter; but if you purpose to write a book about logic, I should say that you ought by all means to begin by *thoroughly mastering* Cayley's Sixth *Memoir on Quantics*,—though if, when you say you have read *Story's papers*, you mean that you have thoroughly understood him, you must be mistaken, because if you had, you would understand not only Cayley but the whole ground of certainty of the non-Euclidean Geometry. Therefore, if you have any difficulty about that (as I presume you will) you have only to begin by reading the first two chapters of Lindemann's redaction of Clebsch's lectures on Geometry, a very luminous and beautiful treatise. I have not the original but only Benoist's French translation, so I can't give the German title nor say what the "Chapters" are called in the German Edition. They make about a quarter of the whole work. Any little difficulties you can't surmount, I can aid you in by referring you to other books or otherwise.

Some time I will endeavour to give you my own way of imagining the non-Euclidean geometry which enables me to answer any simple question about it instantly. After you have completely digested that immortal memoir of Cayley's,—perhaps the greatest luminary, it was, of all my mathematical life,—by all means get to the very bottom of Klein's papers on the non-Euclidean Geometry in Vols. IV and VI of the Mathematische Annalen which wonderfully clear up the whole matter. Don't satisfy yourself with half-understanding—*Master everything* as you go on.

C. S. Peirce

There are besides two volumes of Lectures by Klein on the non-Euclidean Geometry.

Milford Pa 1909 June 5

My dear Mr. Russell:

I can see no sense at all in trying to associate non-Euclidean geometry with the word "virtual." All I see is that Lobatchewsky's little book is perfectly elementary and if a person goes through it, so as to follow each step of the elementary reasoning, he ought to have sufficient acquaintance with the subject as to answer the question you put me; and you assured me that you had studied the book. But now it appears that you have not read it. Your eye may have passed over every line, but though the reasoning is as plain as a pike staff you have not taken it in. According to Lobatchewsky's kind of geometry the locus at infinity in a plane is an ellipse, which can be called a circle if you like. Therefore the straight line between two points both at infinity will itself be at infinity at those points only. A triangle inscribed in infinity will have each angle $= 0$ but the area of the triangle will be some finite quantity which may be different in different spaces [Fig. 1]. Call this area A. Cut the triangle by a line from one angle so as to divide the area into two equal parts, and plainly the area of each

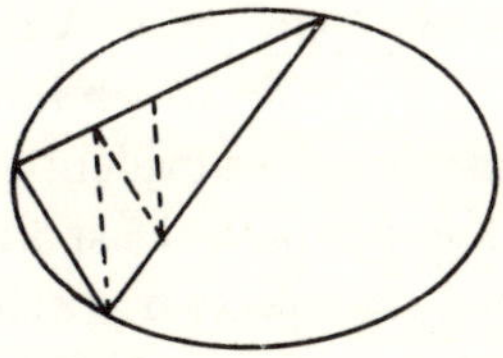

Fig. 1

will be $\frac{1}{2}A$ and the sum of the angles will be 90°. Bisect that again and the area will be $\frac{1}{4}A$ and the sum of the angles will be 135°. Bisect that again and the area will be $\frac{1}{8}A$ and the sum of the angles $157\frac{1}{2}$°. Of course A is a finite quantity.

Before you came out and contradicted all the real mathematicians in the world you should have read deeply the works of the great masters, in which number Lobatchewsky is not.

As for your friend's theological talk, I have skimmed it pretty well. The best parts seem to me questionable; and I do not think that men have made God. I hold God to be a reality. But anyway, it is theology, and I think that kind of loose theological speculation weakens religion, to which all that is utterly foreign. (See my paper in the Hibbert Journal for October 1908.) If it were put forward as a mere idle fancy it would do no harm and no church would persecute the man.

Very truly
C. S. Peirce

P.S. I scribbled on the back of my sheet,—having finished my letter,—without noticing that it was a letter. Instead of atoning for this incivility by copying it, I will do better, by adding as much more matter, which by the way costs time of which I am just now very short.

You speak in your last as if you supposed there were only two kinds of non-Euclidean geometry. This is not so. A non-Euclidean geometry consists in the hypothesis of a space of 3 dimensions with the same *topical* properties as Euclidean space; that is having its Chorisy = its Cyclosy = its Periphraxy = its Apeiry = 1 but differing in having a different locus at infinity (which locus Cayley named the "Absolute"). In Euclidean Geometry it is a plane or rather a pair of coincident planes with an *imaginary circular* line of nodes, so that any line that cuts this circumference has a simple contact with the Absolute and owing to this circular nodal line being imaginary any real curve that passes through it once must pass through it twice i.e. in the conjugate point. A circle is a conic which does this. A sphere is a quadric surface on which the Absolute lies.

Some of the non-Euclidean spaces evidently have properties different from our space. But that does not in the least diminish their mathematical Importance, since Mathematics makes no pretense that any of its hypotheses are true. If any of them are so, which I doubt as far as geometry goes, that is an extra-mathematical fact of not the smallest interest to the mathematician, as such.

One very interesting non-Euclidean Space (interesting particularly, however, to the non mathematician) is a space whose Absolute (or place that no movable body or part of a body can ever either reach or if there can leave) instead of having an imaginary circle for nodal line has a *real circle*, so that no plane can ever be rotated so far as to touch it and a plane that is touched by the Absolute can only be turned round this ring, either round from being nearly in its plane on one side to being nearly in the same plane on the other side, or else turned from one part of the circle to another. The interest of this sort of Space is that all its properties must be exactly the same as those of Euclidean space except that what is real in one will generally be imaginary in the other. Thus it enables us to solve a great number of intricate problems in a jiffy or less.

In the next place there is a Space in which the Absolute consists of two planes, with a line of intersection which may be real or imaginary. Then nothing can ever be turned through such an angle that if sufficiently prolonged it would collide with that ray of intersection.

In the next place the two planes may be imaginary in which case one could travel all round space in a straight path without going infinitely far. How great the circumference of space would be we can't say. To some animals it would seem a long journey and to others a mere step.

Next the Absolute might be a double cone in which case nothing could be so turned that any straight line in it should at one time lie on a tangent to the cone and at another time not. For such a turn would be through an *infinite angle*. And there are two varieties of this Space. In one of them it is the inside that is inhabited. In the other it is the outside.

Then there is a Space in which the Absolute has a Dice-box shape, or developable hyperboloid. The inside and outside would be *exactly alike* there being certain directions in which one could go all round space while in others you could not without passing through an infinite distance. Round those directions of axes there would be no infinite rotation while round the directions of axes along which the circuit of space is finite, you never could turn a wheel round so as to bring it back to its former position. Anything travelling round space in any of those spaces where such a thing is ever done is (I *suppose* so, for I am writing very hastily for lack of time) turned inside out. There is also a space in which the Absolute is a surface of the second order and the second class and people living outside of it find some planes in which geometry is Lobatchewski's while other planes have all triangles in them [with] the sum of the angles *greater* than two right angles.

In all *cases whatsoever* there is one and the same rule viz the distance between two points equals the logarithm of the cross ratio of those points, relative to the points where the ray through the first pair of points cuts the absolute.

While the angle between two planes equals the logarithm of the cross ratio of those planes, relative to the planes through the line of intersection of the first pair of planes, which (second pair of planes) are tangent to the absolute.

This was better than copying the first sheet.

C. S. P.

None of this is original with me. It is Cayley × Klein. KC which don't stand for Knot correct.
Of course you don't want to know the arithmetical value of the cross ratio. But you can take a perspective view of the four points or the four planes and then you can calculate the base of the system of logarithms in different spaces and maybe any values.

M. F.C.S. SCHILLER (L 390)[1]

P. O. Milford Pa 1906 Sep 10

My dear Mr. Schiller:

Let me thank you very particularly for sending me a copy of your last article which I have read with profit and entertainment, as I do all of your articles that I am so fortunate as to see.

Of course I agree entirely to most of what you say,—as well as I can understand it without having seen any writing of Taylor. For example, I agree that of the two implications of pragmatism that concepts are purposive, and that their meaning lies in their conceivable practical bearings, the former is the more fundamental. I think, however, that the doctrine would be quite *estropiée* without the latter point. By "practical" I mean apt to affect conduct; and by conduct, voluntary action that is self-controlled, i.e. controlled by adequate deliberation. But the neater definition you put into a footnote is worth fully all you claim for it.

However, it would be idle to write merely to note points of agreement. They are too many. Let me rather note that some of the ends which you mention as going to the meanings of concepts seem to me to form no part of those meanings. What the hundredth decimal figure of π means consists to my mind in just what any other figure means. For it is quite *conceivable* that it should be an important practical quantity. It is one of the beauties of pragmatism that it gives some symbols much more meaning than others, and the hundredth figure of π certainly has precious little. A much better question is, What on pragmatist principles is the difference between a rational and an irrational quantity or what it means to say that the diagonal of a square is incommensurable with its side? It is interpreted in the conduct of the arithmetician as such.

As for Cantor's "transfinites," there are very different kinds, the ordinal and the cardinal. Both have had practical applications. The first ordinal

[1] Professor Fisch writes that F. J. Down Scott is publishing the letters that Peirce mailed to Schiller in the *Journal of the History of Philosophy*.

transfinite is the "ordinal number" ω which denotes the first object which follows after an endless series, which series is supposed to be "wohlgeordnet" or, as I call it, a *Cantorian* series, for he deserves to have his name attached to so remarkable a conception. Now this is only an ordinal number. The objects numbered may be quantities themselves and they may either increase or decrease as the ordinal increases. A Leibnitzian infinitesimal of the first order is an assumed quantity smaller than any finite quantity. It is the first quantity we choose to notice after the series 0.1, 0.01, 0.001, etc. *ad inf*. It is impossible to prove that there is no such quantity. All you can prove is that there is no "assignable" quantity like that. This infinitesimal is not the quantity denoted by Cantor's ω, otherwise any greater quantity would be finite, while one infinitesimal of the first order may be infinitely greater than another of the first order. But it may be perfectly represented by a quantity whose ordinal place is Cantor's ω^n, where n is an indefinite number. Then an infinitesimal of the second order will be ω^{2n} etc. and thus one gets a form of the differential calculus far simpler (for those who are properly taught to grasp the concepts) than that founded on the doctrine of limits, and at least as logically valid.

As for Cantor's cardinal transfinites, though called numbers by him, they are not properly so called but are *multitudes*, or maninesses of infinite collections. The first is the multitude of the objects of an endless series of objects. I call it the *denumeral* multitude. The next is the multitude of all collections of objects involved in an endless series (by collection I mean simply a plural). The rest are each the multitude of all collections involved in a collection of the next preceding multitude. I call these the *abnumerable* multitudes. They run up from the first abnumerable through all the finite ordinal numbers endlessly.

These abnumerable multitudes are describable intelligibly and exactly, but only in general terms. No precise idea can be formed of the simplest of them; and they increase in difficulty at a frightful rate (that is in the characters that *would* make difficulty if it were surmountable at all). If anything violates the principle of pragmatism it is these. But I have no doubt whatever of the validity of the concepts. They are interpretable in the conduct of the logician or logico-mathematician in dealing with them. If they were not exact, so as lay definite logical obligations upon him they would be meaningless, or without definite meaning.

When you say that logical consequences cannot be separated from psychological effects, etc. in my opinion you are merely adopting a mode of expression highly inconvenient which cannot help, but can only confuse, any sound argumentation. It is a part of nominalism which is utterly antipragmatistic, as I think, and mere refusal to make use of valuable forms of thought.

19

MATHEMATICAL ITEMS FOR *THE NATION*

(Reviews Unless Otherwise Stated)

A. *AMERICAN JOURNAL OF MATHEMATICS* 1, 1878

Esposizione del Metodo dei Minimi Quadrati. Per Annibale Ferrero, *Tenente Colonello di Stato Maggiore, ec. Firenze,* 1876

Recent discussions in this country, of the literature of the method of Least Squares, have passed by without mention the views of the accomplished chief of the geodetical division of the Italian Survey, as set forth in the work above cited, which was first published, in part, in 1871. The subject is here, for the first time, in my opinion, set upon its true and simple basis; at all events the view here taken is far more worthy of attention than most of the proposed proofs of the method.

Lieut. Col. Ferrero begins by considering the principles of the arithmetical mean. A quantity having been directly observed, a number of times, independently, and under like circumstances, the value which might be inferred from the observations is, in the first place, a symmetrical function of the observed quantities; for, if the observations are independent, the order of their occurrences is of no consequence, and the circumstances under which they are taken, differ in no assignable respect, except that of being taken at different times. In the second place, the value inferred must be such a function of the values observed, that when the latter are all equal, the former reduces to this common value. The author calls functions having these two properties, (1st, that of being symmetrical with respect to all the variables, and 2nd, that of reducing to the common value of the variables when these are all equal,) *means.* There is a whole class of functions of this sort, such as the arithmetic mean, the geometrical mean, the arithmetic-geometrical mean of Gauss, the quadratic mean (this seems the appropriate name for $\sqrt{\frac{[x^2]}{n}}$), and many others instanced in the text. It is shown, without difficulty, that these means are continuous functions, and that their value is intermediate between the extreme values of the different variables, when the latter do not differ greatly.

Let o', o'', o''', etc. denote the values given by the observations. Let n denote the number of the observations; let p denote the arithmetical mean;

and let x', x'', x''', etc. denote the excess of the observed values over the arithmetical mean. Then write

$$V = f(o', o'', o''', \text{etc.})$$

for any mean of the observations. Develop this function according to powers of x', x'', x''', etc. We have

$$\begin{aligned} V &= f(p + x', p + x'', p + x''', \text{etc.}) \\ &= f(p, p, p, \text{etc.}) + \frac{dV}{dp}(x' + x'' + x''' + \text{etc.}) + \Delta; \end{aligned}$$

where Δ denotes the terms of higher orders.

Since $\qquad x' + x'' + x''' + \text{etc.} = 0,$
and $\qquad f(p, p, p, \text{etc.}) = p,$
this reduces to

$$V = p + \Delta.$$

In considering the value of Δ, we may limit ourselves to terms of the second order. As the partial differentials of any species and order, relatively to o', o'', o''', etc. all become equal when x', x'', x''', etc. vanish, we may write

$$\frac{d^2V}{do'^2} = \frac{d^2V}{do''^2} = \frac{d^2V}{do'''^2} = \text{etc.} = \beta$$

$$\frac{d^2V}{do' \cdot do''} = \frac{d^2V}{do'' \cdot do'''} = \text{etc.} = \gamma$$

then

$$\Delta = \tfrac{1}{2}\beta(x'^2 + x''^2 + x'''^2 + \text{etc.}) + \gamma(x'x'' + x'x''' + \text{etc.}).$$

But the square of $[x] = 0$, gives

$$\Sigma xx' = -\tfrac{1}{2}[x^2],$$

so that

$$\Delta = \frac{\beta - \gamma}{2}[x^2] = k\frac{[x^2]}{n},$$

where k is a quantity which does not increase indefinitely with n. Now, when the observations are good, $\frac{[x^2]}{n}$ is not large, and, therefore, in such a case no mean will differ very much from the arithmetical mean. The latter, being the simplest to deal with, may therefore be used without great disadvantage. Such is, according to Colonel Ferrero, the utmost defence of the principle which can be made to cover all the cases in which it is usual to employ the method; and all further defence of it is more or less limited in its application.

In very many cases, however, it is easy to see that either in regard to the quantity directly observed, or in regard to some function of it, the zero of the scale of measurement, and the unit of the same scale, are both arbitrary. For instance, in photometric observations, this is true of the logarithm of the light. In such cases, considering such function to be the observed quantity, we have there two principles, first proposed, in connection with a really superfluous third one, by Schiaparelli.

1st. The mean to be adopted must be such that if each observed value is multiplied by any constant, the result is increased in the same ratio.

2nd. The mean to be adopted must be one which is increased by a constant o, when each observed value is increased by the same constant.

$$V = p + A_2 + A_3 \ldots + A_n \ldots$$

where A_n is the sum of the terms of the order n in x', x'', x''', etc. The general term A_n is, therefore, of the form $A_n = a\Sigma x^n + \beta\Sigma x'^{n-1}x'' + \gamma\Sigma x'^{n-2}x''^2 \ldots + \zeta\Sigma x'^r x''^s x'''^t \ldots$ where Σ expresses the symmetrical sum of similar terms. In the general term $r + s + t + \text{etc.} = n$. Since ζ is evidently a function of p, we may put $\zeta = \phi(p)$, and it remains to find the form of this function. Multiplying every o by c, p is changed to cp, x to cx, and the general term $\zeta\Sigma x'^r x''^s x'''^t$ etc. $= \phi(p)\Sigma x'^r x''^s x'''^t$ etc. is changed to $\phi(cp)c^n\Sigma x'^r x''^s x'''^t$ etc. Since, therefore, V is changed to cV, we have $\phi(cp)c^n = \phi(p)c$. Putting $p = 1$, $\phi(c) = \dfrac{\phi(1)}{c^{n-1}}$. Denoting this numerator by ξ_1, the general term becomes

$$A_n = \frac{1}{p^{n-1}}[\alpha_1\Sigma x^n + \ldots + \xi_1\Sigma' x'^r x''^s x'''^t \ldots + \ldots],$$

where α, ξ, etc., are numerical coëfficients independent of p. From this circumstance it follows that the quantity in square brackets, which may be called A'_n, does not change when the same constant quantity k is added to all the observed quantities o', o'', o''', etc.; for such an addition only increases p by this same constant, and leaves x', x'', x''', etc., unchanged. Thus the mean in question, which may now be written

$$V = p + \frac{A'_2}{p} + \frac{A'_3}{p^2} + \text{etc.},$$

becomes, in consequence of such an addition,

$$V_k = p + k + \frac{A'_2}{p+k} + \frac{A'_3}{(p+k)^2} + \text{etc.}$$

But by principle No. 2, it becomes,

$$V_k = p + k + \frac{A'_2}{p} + \frac{A'_3}{p^2} = \text{etc.}$$

So that, $A'_2 = A'_3 = \text{etc.} = 0$, and we have

$$V = p,$$

or the arithmetical mean is the only one which conforms to the given conditions.

Another still more special case, is that contemplated by the demonstrations of Laplace, Poisson, Hagen, Crofton, etc. It is treated by our author, but need not be considered in this notice.

It may be of interest to see how Colonel Ferrero is able, without basing least squares expressly upon the theory of probabilities, to derive the formula for finding mean error. Using always the same notation, he terms

$$m = \sqrt{\frac{[x^2]}{n}}$$

the *mean residual* of the observations.

Suppose, then, that there be an indefinitely great *series of series* of observations of the same quantity, each lesser series consisting of n observations, and each having the same mean residual. Then, there being an infinite number of such series, the mean of their mean results may be taken as the true value, by definition. For the ultimate result of indefinitely continued observation is all that we aim at in sciences of observation. Then the number of the lesser series being q, the result will be

$$V = \frac{[p]}{q}.$$

Adopt the notation

$$\delta = p - V \quad \delta_1 = p_1 - V \quad \delta_2 = p_2 - V, \text{etc.},$$

then δ, δ_1, δ_2, etc., are the true errors of p, p_1, p_2, etc. Let y'_0, y''_0, y'''_0, etc. be the true errors of the first series of observations, y'_1, y''_1, y'''_1 etc. those of the second series, and so for the others. We have, then, $y = o - V = o - p + \delta = x + \delta$.

Squaring and summing for the nq values of y, we have

$$\Sigma y^2 = \Sigma x^2 + \Sigma \delta^2 + 2\Sigma x \Sigma \delta$$

or, since $\qquad \Sigma x = 0, \quad \text{and} \quad \Sigma \delta = 0,$

$$\Sigma y^2 = \Sigma x^2 + \Sigma \delta^2.$$

Now if η be the quadratic mean of the error of p, we have $\Sigma \delta^2 = nq\eta^2$, and

$$\Sigma y^2 = nqm^2 + nq\eta^2,$$

or the mean error μ of an observation is given by

$$\mu^2 = \frac{\Sigma y^2}{nq} = m^2 + \eta^2.$$

But it is easily shown (from the equality of positive and negative errors) that

$$\eta^2 = \frac{\mu^2}{n}$$

whence

$$\mu = \sqrt{\frac{[x^2]}{n-1}}$$

With regard to the mode of passing from the principle of the arithmetical mean to the general method of least squares, the best way seems to be first to prove that the solution of the equations

$$\begin{aligned} a_1 x &= n_1 \\ a_2 x &= n_2 \\ &\text{etc.,} \end{aligned}$$

is $x = \dfrac{[an]}{[a^2]}$. This is easy, after the rule for the error of a mean is established. Then, having given the equations

$$\begin{aligned} a_1 x + b_1 y + c_1 z + \text{etc.} &= n_1 \\ a_2 x + b_2 y + c_2 z + \text{etc.} &= n_2; \end{aligned}$$

first, consider these as similar to the equations just given; thus,

$$a_1x = n_1 - b_1y - c_1z - \text{etc.},$$
$$a_2x = n_2 - b_2y - c_2z - \text{etc.},$$
$$\text{etc.},$$

whence we obtain the first normal equation,

$$x = \frac{[an_1] - [ab]y - [ac]z - \text{etc.}}{[a^2]}$$

and the others in a similar way.

The treatise of Colonel Ferrero may be recommended to those desirous of having a thorough practical acquaintance with the method, as decidedly the best and clearest on the subject.

B. 10 NOVEMBER 1892

Logarithmic and Other Mathematical Tables. By William J. Hussey, Professor of Astronomy in the Leland Stanford Junior University, etc. Ann Arbor: Register Publishing Co. 1892.

For the semi-occasional user of logarithms, collections like Köhler's are best. But a person who is destined to use up several books of tables by the wearing-down of the paper under his fingers—which commonly happens to expert mathematicians—will prefer to be provided with four-place, five-place, six-place, and seven-place tables, since the expenditure of time in working with these is in the ratios of 1:2:3:4, respectively. Can the tables before us be recommended as being about as good as others? They are printed upon paper fairly opaque and quite free from sheen, substantial but rather cottony to the touch and too white. A small page is a recognized advantage in tables of logarithms. These pages are taller than those of any five-place tables we know except Hoüel's. The ink is not quite so black as we could wish, and some pages are a little gray. Very many figures look as if printed from worn types. The fourth figure of log 4092 comes from a wrong font. The alignment leaves much to be desired. The type is of the old pattern, which in our judgment is preferable to the Huttonian character (the pattern now common in ordinary printing, invented, it is guessed, by Dr. Charles Hutton in 1783), but inferior to the Egyptian, which are all of one height but without hair-lines.

We may examine the arrangement of the table of logarithms of numbers. Each tenth value of the argument is printed in Huttonian type. This gives it sufficient prominence; the large black round-numbers of Babbage are unnecessary. The table is arranged in a Newtonian block, which we deem more convenient than the columnar form, especially since it brings twice as many numbers on each page. The table everywhere opens to exactly 1,000 logarithms, not counting those on the last line, which are a sort of catch, or rehearsal of the first line of the next page.

This is a point of great superiority over Bowditch, Schlömilch, etc. The numbers in each tenth line are placed between horizontal rules, while the intervening nine lines are divided by leads into three sets of three. This is the plan of the highly approved tables of Bremiker; yet we prefer, with Schrön and others, the division by leads into sets of five. The first two figures of the five-figure logarithms are given only in the first column at the top of the page and where they change. Bremiker thus separates only one figure, while Bowditch gives all five in every column. The ten columns of the block are all separated by vertical rules, that after the fifth being extra heavy. This is the customary way, but we are fully persuaded that all these vertical lines are productive of error in following the horizontal lines with the eye. We consider the tables of Schlömilch, Oppolzer, J. M. Peirce, and others, which omit all but the line after the fifth column, as much the more comfortable.

The indication of a change of the figure in the last place of unrepeated decimals is by an asterisk prefixed to every logarithm affected. This is decidedly the best method. The proportional parts are exact to the sixth place. The practice of thus printing the proportional parts arose in consequence of Babbage, in his seven-place table, printing a dot under every terminal figure which had been increased. This he did on the ground that all information which could be given without disadvantage should be given — a good principle for seven-place tables, without doubt. Only, upon that principle, De Morgan's plan should be adopted of distinguishing the quarters of the last unit by means of the four ordinary punctuation marks, thus making the tables accurate to a fraction of the number entered equal to unity divided by a power of ten. However, Babbage's system was extensively adopted, and consequently it was necessary to give the proportional parts more accurately. Prof. Hussey prints a dash over every increased 5, whether it be terminal or not, and over no other increased numbers. It luckily happens that there is no case in the table of an increased 5 followed by three zeros, otherwise the system would break down. Now, we think a system illogical, and therefore inelegant, which can only be carried out by virtue of an accident. But what is the use of carrying the proportional parts to six places? Everybody must allow that it would be bad economy of time in computing to write down one's numbers alternately to five and to six places of decimals. Now, what difference does it make that we add the six-place numbers in our heads? A centimetre and a half at the bottom of each page of the table is devoted to giving the values of S and T, and that not unambiguously. This seems decidedly awkward.

There are trigonometrical tables, both logarithmic and natural, tables of addition and subtraction logarithms, etc. At the end of the book are given formulae and constants. The latter are pretty carelessly collected and copied. The velocity of light is made to be 296,944 kilometres per second! Clarke's value of the metre in inches, 39.370432, is given, although its error has been known for many years. First Prof. Rogers and then Gen. Comstock made fairly concordant determinations, very different from Clarke's. In fact, his was merely the result of measuring copies of Bessel's toise in inches, and then deducing the length of 443.296 lines of the toise, this being the number of lines of the *toise de Pérou* intended to make the metre at the time of the construction of the latter. But recently M. Benoît of the International Bureau has shown that the metre so deduced from Bessel's toise is too long by its 74,000th part. So, correcting Clarke's determination, and combining it, reduced to a weight of $\frac{1}{8}$, with the values obtained by Rogers and Comstock, we find:

	Inches.
Rogers	39.37027
Comstock	39.36985
Clarke, corr. by Benoit	39.36990
Weighted mean	39.37004

This makes 25.40003 millimetres in an inch. If we remember, then, that 39.37 and 25.4 should each be increased by one-millionth part of itself, we shall have the fact as accurately as it is known. We find this convenient rule used in the Yaryan Company's Tables. Prof. Hussey's book will do for easily contented computers.

C. 24 AUGUST 1893

An Elementary Treatise on Pure Geometry, with numerous examples. By John Wellesley Russell, M.A. Oxford: Clarendon Press; New York: Macmillan. 1893.

An Elementary Treatise on Modern Pure Geometry. By R. Lachlan, M.A. Macmillan. 1893.

Geometry in the Grammar School: An Essay. Together with illustrative class exercises, and an outline of the work for the last three years of the Grammar School. By Paul H. Hanus, Assistant Professor of the History and Art of Teaching, Harvard University. Boston: D. C. Heath & Co. 1893.

. . . The reason why analytical methods are more easily handled than the synthetical geometry is chiefly that the former arrange the whole subject in a perfectly definite and unmistakable manner. No wonder a pupil is puzzled to apply a theory consisting of some thirty fragments not connected by any intrinsic bonds. As long as this state of things exists, notwithstanding the infinitely greater elegance of the pure geometry, its great practical use will be to serve as a guide in the reformation of analysis. The older treatises upon modern geometry did not exhibit this loose articulation, for the reason that they dealt chiefly with projective properties, and introduced what little metrics they gave as corollaries to the projective theorems. This could no longer be thought of, yet it suggests the proper way of arranging the subject. No text-book of either synthetical or analytical geometry omits that grand proposition of Cayley, that every metrical fact is a projective fact about a certain fixed quadric, or in plane geometry about the section of this quadric by the plane; nevertheless, writers of text-books put them together as if they did not really believe this. If it be true, surely an eternal fitness requires that the projective geometry of rectilinear diagrams and conics should precede all metrical matter, and that the Euclidean geometry should be taught as a particular case of the non-Euclidean. . . .

D. 15 MARCH 1894

MATHEMATICAL FUNCTIONS

Theory of Functions of a Complex Variable. By A. R. Forsyth. Cambridge: University Press; New York: Macmillan. 1893.

A Treatise on the Theory of Functions. By James Harkness, Associate Professor of Mathematics in Bryn Mawr College, Pa., and Frank Morley, Professor of Pure Mathematics in Haverford College, Pa. Macmillan & Co. 1893.

Traité d'analyse. Par E. Picard, Membre de l'Institut. Paris: Gauthier-Villars. Tome I. 1891. Tome II. 1893.

Many good people fancy that the advances of mathematics, like those of jurisprudence, become manifest only when the state of things in one generation is compared with that in another; and that they are merely in the nature of extensions of old methods to new cases. In point of fact, there is probably no science in which the rate of *acceleration* of discovery, of the proportion of excess of the discoveries of one year over those of the year before, is so great as it is in mathematics, and no science but mathematics in which discovery seems to be becoming continually more and more fundamental. We are speaking of pure mathematics, not celestial mechanics.

Time was when geometry absorbed so overwhelming a proportion of the studies of mathematicians that "geometer" was understood to be synonymous with "mathematician"; and even to-day geometry is the most studied branch of mathematics. There are various reasons for this. One is that it is a comparatively easy subject. Besides that, the sensuous element of it seduces the mind, and carries it into excesses of study; and other causes there are. But next after geometry, in respect to the quantity of researches annually published about it, and far superior to geometry in its intellectual rank, is the subject of the theory of functions. For the last twenty years and more there has been a perfect freshet of original work in this line. Every year its tide is rising; every year increases the force and value of the new discoveries, which sweep on faster than they can be taken account of. In early days, enthusiasts would sacrifice hecatombs to celebrate the solutions of problems. Later, problems appeared less sublime; theorems were requisite to excite admiration. Now, theorems are as the sands of the sea; original methods alone can command mathematical dithyrambs.

At a not remote period in the history of mathematical thought, a Mystery (with a big M and in the darkest of black-letter) hung over the imaginary unit. It used to be written $\sqrt{-1}$; and what was the square root of a negative? But when it was found that the imaginary unit of algebra was only one of a class of units which, operating upon themselves, in a generalized sense of multiplication, produce -1, the mystery lost its capital; and after the philosophy of ordinary quantity had become better comprehended, the mysteriousness of the imaginary had vanished. The numbers, *one*, *two*, *three*, *four*, etc., are sounds which we have learned to pronounce in a certain order of succession, and which we do pronounce in telling over the individuals of collections. If such a collection is finite, we reach a last individual; and the number pronounced on coming to this last one affords the means of determining whether the individuals of two collections can be made to correspond, one to one, or, if not, in what manner they fail of that. Sometimes the things counted are really in succession like the numbers. Such are trees in a row, degrees of temperature, and years. In other cases, the succession of counting is purely artificial, as in enumerating the population, or the pounds of flour in a barrel. But the counting does not, on that account, cease to be useful, because, in whatever order the individuals are counted, the final number will, in counts of any one collection, always be the same. Even the separation into discrete units (as the gallons of water in a lake) may be artificial, provided that, if it were effected in various ways, it would always lead to the same resulting number. It will be noticed that this is not a nominalistic account of numbers—it does not make them *flatus vocis*, only—but it makes their existence *in re* consist in an experiential constancy; that is, it assigns to reality three elements, (1) sensuous quality, (2) compulsiveness, (3) generality. Besides the system of whole numbers, we often make use of a scheme of quantity connected together like the points on a line. This is useful even when there is no perfect continuity in the things to which it is applied. The scheme of imaginary quantity is simply one that is connected like the points of a plane. Certain natural phenomena, especially in hydrodynamics, correspond exactly, in theory at least, with such a scheme. But since any line upon such a plane is connected like ordinary real, continuous quantity, the usefulness of imaginary quantities extends to almost all cases in which real quantity is used.

The questions to which the theory of functions, so far as yet developed, chiefly addresses itself arise out of the supposition that a correspondence between the points of two different planes of quantity has been established by an equation. It considers the nature of the resulting continuity (so far as this is not resolved in the theory of plane curves), and, more especially, the modes of representing the relation, both geometrical and analytical. The main object of the whole study is to find out how to make use of differential equations, especially such as are the immediate dicta of mechanical laws.

The disciplinary value of the theory of functions is superior to that of any other branch of mathematics. For many minds elementary geometry serves, directly or indirectly, as their model of reasoning. But elementary geometry is so artificial, and is so permeated with fallacies and caprices, that it must be and ought to be difficult to a healthy and ingenuous young mind; and much of the perverse logic that is current in the world is to be laid at its door. Algebra has done much for every educated man; it has given him an exemplar of perfectly accurate abstraction. It would put a mighty weapon in his hands if the application of it, in the elementary books, were not pretty much restricted to two problems, elimination between linear equations and the solution of the quadratic. The theory of probabilities is most instructive and useful, but that is only applied algebra. The theory of numbers is an admirable school of reasoning, as far as it goes, and it goes so far that reflection upon it will counteract much of the poison that the text-books of logic inject into the current of thought. Projective geometry imparts the most precious secrets in generalization while making no fundamental analyses. As for analytical geometry and the calculus, all *that* ought to be taught (as in Prof. Benjamin Peirce's "Curves and Functions" it *was* taught nearly fifty years ago) as part of the general theory of functions. The theory of functions is, in the first place, intrinsically, quite easy—we mean to follow, not to invent. Of course, it is capable of being obscurely stated. Its logic is the most fundamental conceivable, and, at the same time, is the very subtlest that can anywhere be found; so that no man is too eminent never to have made a slip in it. The outlines of the theory ought to be known to every educated person.

There has hitherto been no treatise in our language on the modern development of the theory. At length the same year presents us with two. Though the first has 700 pages of royal octavo, and the other 500 of common octavo, yet the subject is so vast that a considerable part of the contents of either is excluded from the other, and much that we might desire to see is absent from both. Dr. Forsyth has been well known for some years less than half a generation as an indefatigable investigator of functions, and he has already produced two profound treatises on differential equations. The present work contains many not unimportant contributions of his own. Messrs. Harkness and Morley are younger men, but, as this volume shows, thoroughly versed in their subject. Dr. Forsyth keeps as much as he can to the general theory, treating such a special subject as elliptic integrals, for instance, with the greatest brevity possible, and at the same time in such a way as to afford a bird's-eye view of it. Messrs. Harkness and Morley, on the other hand, seem to have been of the opinion that it was better to go somewhat deeply into a smaller selection of topics. Many things are crowded out to make room for long chapters on Elliptic and Abelian Functions, while, at the same time, these very subjects are not treated with all the fulness which is requisite for the practical applications of them. This, at least, is certainly true of elliptics. Practical applications of Abelian functions ought, perhaps, in the present state of things, not to be thought of. Certain preliminary branches, absolutely indispensable to the comprehension of the theory of functions, such as the logic of infinity, continuity, etc., and the doctrine of the convergency of series, are entirely omitted by Forsyth, while he inserts matter about substitutions which the reader will be glad to find thus at his hand, but which really belongs in a treatise on algebra. The other writers have followed the opposite course in these respects, though we cannot quite content ourselves with their attempted reproduction of Cantor's logical ideas. Dr. Forsyth imitates, in a general way, the French lucid style of exposition, though the French accuracy of statement and neatness of demonstration are often wanting in his book. Messrs. Harkness and Morley express themselves in the German manner, which makes the exposition as easy as possible for the writer—and never mind the reader. For an illustration of what we mean it is sufficient to open the book at random. At the top of p. 352, we read (with a slight modification of notation for our printer's sake):

The symbol MN equals 1 or 0, according as M and N do or do not contain a common letter.

Now, it is inaccurate to speak of a *symbol* being equal to a number; and since M and N are single letters, there can be no question of their containing a common letter. But the authors mean that when they are replaced by the duadic symbols which form one of four or five different ways of expressing the same thing, *those symbols* have or have not a common letter, according as the corresponding quantities equal 1 or 0. The opposite page, 353, presents several singular instances of saying one thing while meaning another; and it is stated that a certain notation "will be" used, which notation is incontinently dropped without another word, and another one, not defined, is constantly used for many pages. So, on p. 355, a notation is defined, in no apparent connection with anything in the vicinity, and is never used for many pages, until it suddenly springs up after we have forgotten all about it. These examples are not culled.

It would be unfair to convey the idea that Forsyth is quite impeccable in his expressions. This is far from being true. Thus, at the beginning of chapter iii., in enunciating Cauchy's fundamental theorem on the expansion of a holomorphic function, the important words "unconditionally and uniformly," as describing the mode of convergence, are omitted, as they are overlooked in the proof given. The first page of chapter vii. has but sixteen lines of text, yet they contain no less than three faults of expression, if not of logic. Indeed, Forsyth is really too negligent in regard to terminology. Thus, that category of surfaces, curves, etc., which the Germans call *Geschlecht*, the French *genre*, and which we should do well to term *genus*, instead of the usual word "deficiency," Forsyth most confusingly designates as the *class*. Both books will be found serviceable to students, alike to those of higher and of lower grades. We may mention, by the way, that Forsyth is rich in illustrative examples, Harkness and Morley pretty poor. But we cannot sincerely pronounce either of them quite satisfactory, whether as a handbook or as a textbook, and both handbook and textbook are certainly needed. The latter ought to be so clear of all pedantical details as to be fit for the use of every young person who seeks a broad intellectual education.

Picard's very admirable work is not properly a treatise on the theory of functions. Indeed, when the first volume appeared, the author's purpose was to treat this subject rather slightly, and we were informed that volume II would deal with Differential Equations. Instead of that, only one of the seventeen chapters relates to that subject. It is evident the theory of functions has been growing in importance in the author's mind. Hence it is, probably, that instead of embarking frankly in the vehicle of imaginaries, the author thinks it more philosophical to deal chiefly with real functions, thus making many things difficult and crabbed which would in Cauchy's hands appear delightfully facile. The work is one of truly considerable power. It cannot justly be called a classic. Members of the French Institute are apt, for an obvious reason, to be over-praised. While we admit the great value of this work, we must say that some comments upon it we have seen from men to whose opinions we should naturally be inclined to defer, appears to touch the point of extravagant laudation.

E. 19 APRIL 1894

An Elementary Treatise on Fourier's Series and Spherical and Ellipsoidal Harmonics. By William Elwood Byerly. Boston: Ginn & Co. 1893.

Lectures on Mathematics, delivered in August and September, 1893, at Evanston, Ill. By Felix Klein. Reported by Alex. Ziwet. Macmillan & Co. 1894.

Notwithstanding its name, so redolent of Helicon, there is mighty little poetry in Spherical Harmonics. The blessed, after a thousand years' performance on harps, may possibly betake themselves to setting one another problems in modern geometry; but to spherical harmonics we may confidently assert they will not resort. This subject might be called the conveyancing of mathematics, since it teaches how to express facts in a form which, though it affords no insight into causes or essences, but on the contrary is blind and bewildering, is for all that quite indispensable for making the mathematician master of his data. The usual problem is this: A certain quantity has a value at every point of some surface — most usually, that of the earth. This value — it may be elevation above or depression below the sea level, or the distance of the sea level from the centre, or the force of gravity, or a magnetical constant, and so on — has been ascertained at many points, and is assumed to vary continuously. (Most experts will say no such assumption is made.) Then, spherical harmonics shows us what we may presume to be the approximate values at points where the quantity has not been observed. Moreover, it affords a general expression for the value; still further, it shows how to cut up the quantity into parts, each of which is susceptible of further mathematical treatment. It is, thus, a theory (for so mathematicians use the word theory) of great utility; and, like other utility-mathematics, is tedious, difficult, disagreeable, and unbeautiful. This is a circumstance which breeds many loathers of mathematics, because these disagreeable branches are taught first.

The present treatise is undoubtedly the best in our language upon this subject. Its only rival, that of Todhunter, always an unnecessarily dry book, is now pretty antiquated likewise. Mr. Byerly adheres to one point of view pretty consistently, exhibits the doctrine under its best aspect, and leads us into it by the easiest road. It is a branch which nobody but a practical mathematician will care for, and which every practical mathematician has to master.

When we turn from this book to Klein's lectures, we seem to be passing out from a tremendous, rattling factory, with its grimly earnest, unlovely economy, into the pure meadows with the really vastly greater, but infinitely calm, agencies of sunshine, breeze, and river. Here, in only a hundred pages, the moving impulses of modern mathematics are set forth in a way in the highest degree instructive and interesting to every mathematician, without any tax upon his energies. Felix Klein, we need hardly say, is generally considered as the most interesting, if not the greatest (certainly *not* in all respects), of living mathematicians. For such a hundred pages as these the mathematician may search in vain. The small compass renders the process of mathematical cogitation all the clearer, and strips it of details which in other books obscure it; and particularly of details of demonstration that are often wrongly taken to be the soul of mathematical thinking. Such a lesson as this book affords of the conduct of mathematical research the younger student (it is not for beginners) will not easily find. Those who know Klein need hardly be informed that the lectures range over a large part of recent mathematics. The following passage (in which we take the liberty twice to put *experience* in place of "conception") is interesting:

> We are forced to the opinion that our geometrical demonstrations have no absolute objective truth, but are true only for the present state of our knowledge. These demonstrations are always confined within the range of space experiences that are familiar to us; and we can never tell whether an enlarged experience may not lead to further possibilities that would have to be taken into account. From this point of view, we are led in geometry to a certain modesty, such as is always in place in the physical sciences.

Appended to the lectures are ten pages on the history of modern mathematics in Germany.

F. 4 JULY 1895

Riemann and his Significance for the Development of Modern Mathematics. By F. Klein.

Die Grundbegriffe der Ebenen Geometrie, Vol. 1. By V. Eberhard.

. . . One of the leading mathematicians and mathematical philosophers of our age, Klein, gives another of those instructive comments upon mathematical procedure of which we have enjoyed a number from his pen. He again dwells, as he had already done, upon the importance of attentive intuition — in other words, of the *observation* of diagrams and the like — as an essential element of mathematical reasoning. Considering that mathematicians have long held that mathematics covers all exact reasoning, quantitative or not, it will be seen that Klein is going over to a logical doctrine which has had defenders in this country and in England. According to this, our assurance that $(2.4.6.8.10) \div (1.2.3.4.5) = 2^5$ is of the same nature as our assurance that sulphuric acid is precipitated by baryta, we having satisfied ourselves in each case that a single experiment is sufficient; only in one case we observe Nature, in the other our own construction. Dr. Klein remarks that although pure mathematics deals with a purely imaginary world, yet the course of its development is not arbitrary. He seeks to explain its orderly growth by historical continuity, and by the fact that as older questions get solved new questions "naturally" arise. No doubt they do, but that is precisely what needs explanation. We should like to know in what this "natural" succession of ideas, without any external *nature* to guide it, consists. Is there, for example, some Hegelian dialectic, or is there a different general law for this growth of pure concepts, or is it lawless?

The paper contains so many points of interest that we can only recommend it to the reader. The logical importance of Riemann's memoir on Trigonometric Series is pointed out. Klein says (in the absence of the original, we have taken the liberty of rubbing down some of the angles of the English translation):

Riemann's collected works are neither numerous nor extensive. They are comprised in an octavo of 550 pages, and but half of that matter was published by him. Yet his sway over the minds of mathematicians was the most potent of his time; nor is it even yet come to an end. This is owing to the *originality* and the *penetration* of his mathematical thought.... Passing by the latter character, I desire to point out that Riemann's originality lies in a unifying central idea, the source of all his achievements.... He devoted much time and thought to physical theories.... These are preserved in fragmentary form in his posthumous papers. All have in common the hypothesis (since made prominent by confirmations of Maxwell's theory of light) that space is filled with a continuous fluid serving as a common medium for the propagation of optical, magnetic, and gravitational phenomena.... Those physical ideas are the mainsprings of his investigations in *pure mathematics*.... The mathematical work of Riemann is the counterpart of the physical researches of Faraday.

He remarks, too, that Riemann's growing influence is illustrated by two new French books, Picard's "Traité d'Analyse" and Appell and Goursat's "Théorie des Fonctions Algébriques."

Dr. Eberhard's work is a welcome mathematical treatise on plane point-systems. We mention it because some thirty pages of the long preface are occupied with the basis and purpose of geometry. As in some other recent German work in the same direction, while the need is for a purely logical analysis, we are furnished with semi-psychological and epistemological reflections which would not answer the purpose even if they were beyond criticism. It illustrates, however, the purely empirical ground upon which geometry is now placed by mathematicians, considered as a science of actual space, although the very same men may be high idealists when they come to geometry of *n* dimensions or other pure mathematics. From generalities which might almost have been taken from Bain's "Senses and Intellect," or some such psychological treatise, Eberhard passes at once to the intersections of planes and straight lines, in entire forgetfulness that in order to exhibit the foundations of geometry he ought first to have treated topology, or that branch of geometry which relates to lines and surfaces whose exact forms remain undefined—a branch which includes the theory of knots, [Euler's] formula for the number of summits, edges, and faces of a polyhedron (which need not have plane faces), etc. This plainly underlies the optical doctrine of straight lines and planes....

G. 21 MAY 1896

The Number Concept: Its Origin and Development. By Levi Leonard Conant. Macmillan. 1896.

This volume is made up of tables of the numerals of a great many (perhaps 500) different languages, with a slight connective commentary, drawing attention to the signification and composition of the words. The shortcomings of the work are numerous and regrettable, though by no means fatal; its merits are few and simple, but considerable.

The title is a misnomer, and the author shows that his own number concept is in a low stage of development. Numerals are not themselves concepts at all, nor do they signify concepts. They are simply a scale of vocables, which we use very much as we use a foot rule. We apply them to a multitude, and mark how far on the scale that multitude will go. In explaining this, we explain what the number concept really is: it is the intelligent conception of the purpose and method of the system of numerals. It is entirely unnecessary that this should, in the form of a concept, or intellectual product, be in the minds of those who use numerals. It is sufficient that they should know by experience that counting is somehow useful, that it aids bargaining, etc., and that they should be habituated to the use of a series of words in counting. The continual use of the word "concept," instead of speaking of "words" or "terms" and their "significations," is a German way of speaking, very inferior, both in logical accuracy and in perspicuity, to our English idiom. At any rate, the real subject of this book is numerals and their modes of formation.

Very little is said of the number-concept (which is really of very late development), nor of the idea which the tribes mentioned may entertain in regard to number in general; and what little is said is not worthy of criticism. Not only does the author fail to discriminate the number-concept from the use of numerals, but he also falls into a confusion of thought which must greatly embarrass his mathematical pedagogy, namely, a confusion between *number*, in the sense of the result of counting, and *multitude*. He tells us that all tribes "show some familiarity with the number-concept." Yet he mentions Bolivian tribes which are said to have no numerals whatever. Still, he says they show "a conception" of the difference between *one* and *many.* In another place, he says that the "number concept" of ordinary people is imperfect, in that they have little sense of the different degrees of multitudinousness of high numbers. On the contrary, this has nothing to do with the accuracy of their "number concept," or of their power of applying numerals to the purpose for which they were invented. It is true that to the mind trained in certain branches of applied mathematics the word "trillion" carries associations of rigid statistical uniformity which the word "million" lacks. Such a mind may be said to attach different conceptions to the two words; and the distinction is useful to such a mind. But this has nothing to do with the use of numbers as numbers. The person considered will put all that out of his mind when he has any definite numbers to deal with, and will perform his arithmetical calculations just like anybody else. A system of numerals is an apparatus for counting. Those numerous tribes which have names only for *one*, *two*, and *three* which express four by *two twos*, five by *two* and *three*, etc., evidently did not count at the time their language was formed, and probably do not count now. They, like all men, recognize pairs and triplets by their configurations, fours as pairs of pairs, etc. The so-called numerals of such tribes are, properly speaking, not numerals at all. When a tribe has a numeral system based upon *five*, *ten*, or *twenty*, the evidence is that they possess the art of counting. They are quite prepared to count indefinitely as soon as they can count at all, provided they have the power, possessed by most savages, of unconsciously coining a name as soon as they need it. The limits of their numeral words mark the limits of their need of such words.

From a philological point of view, the execution of the book is slovenly. The author copies the various transcriptions of the writings from which he has compiled the lists, without explanation, and omitting all diacritical marks. We do not know whether *c* is to be pronounced *k* or *sh* or *tsh* or *th* or *dh*, whether *g* represents the German guttural *ch* or the velar *k*, whether *x* stands for *ks*, for *h*, or for the Arabic *ghain*, whether *j* has the English, French, German or Spanish sound, etc. When we remember that the English word *fox*, pronounced by a Cherokee, and transliterated according to a recognized system, but with the diacritical marks removed, appears as *kwagisi*, we see that, for the purposes of comparison of languages, this book presents nothing but an imperfect list of references. There is little notice of Semitic numerals, none of the Egyptian, and scarcely any of the Babylonian. There is no mention of the so-called Chaldean names for the Arabic figures found in Latin twelfth-century works. There is no classification by races; but North American and African languages, the furthest remote from one another in their spirit of any of the tongues of men, are shovelled in together. Of many minor faults we take no notice.

The merits of the work are that it exhibits all the modes of formation of numerals, that it shows the universality of the bases 5, 10, 20, and the non-existence of any true binary scale or any use of 6 or 11 as a base, that it affords evidence that many tribes do not count, and consequently have no proper numerical system, and that there are the greatest differences in the arithmetical capacity of races equally barbarous.

H. 25 MARCH 1897

NOTE ON J. J. SYLVESTER

Very few mathematicians of the strength and originality of the late James Joseph Sylvester, who died in London on March 15, have ever lived. There have been great analysts whose secret was a symbolical method — it was, so to speak, their little game, by which they made problems of a certain class easy; others have accomplished great things by turning problems into geometrical shape; others have carefully avoided problems that were not adapted to their peculiar powers; but Sylvester seemed ready to attack any problem, provided only it was difficult — even problems in geometry, for which he was wanting in the peculiar knack that some men have; he never employed symbolical methods, but seemed to create a method specially adapted to each problem he took up. Perhaps he was not, on the whole, a mathematician of the greatest kind; but for naked logical strength but two or three have ever equalled him. He was of Jewish extraction, and was born September 3, 1814, in London. In 1837 he graduated from Johns, Cambridge, being second senior wrangler. After an inappropriate appointment to the professorship of natural philosophy in the University of London, he accepted a professorship in the University of Virginia, where he stayed less than a year, however. Other experiments, in England, were equally uncongenial or unfruitful. In 1876, at the instance of Peirce, he was called to the Johns Hopkins University. He accepted with much diffidence, for he always said he had not much mathematical reading. Nevertheless, his occupancy of the chair proved a glorious success, and a school of enthusiastic and very able young mathematicians grew up under his guidance. His students were always introduced to the matter that was glowing upon the anvil of his own workshop, and so learned how to make researches. In this way he conferred upon this country an inestimable benefit. He established here the *American Journal of Mathematics*, which continues to occupy a more than respectable position among journals of discovery. In December, 1883, he was elected Savilian professor of geometry in the University of Oxford, and thereupon returned to his native land, where, at length occupying a position such as ought to have been imposed upon him forty-five years earlier, he immediately began to stimulate the development of mathematics as he had done here.

I. 13 OCTOBER 1898

The Story of the Mind. By James Mark Baldwin.

. . . The only other chapter in which we discover any defect of lucidity is the last, concerning genius. If the word "genius" bears any unambiguous meaning, it surely signifies a very extraordinary native departure from the ordinary proportions of the faculties in a man, such as goes far toward fitting him for very extraordinary achievements. In this sense, there is no other department of science in which there is half the opportunity for genius that there is in mathematics. Hence, it is not a bad plan, when a generalization about scientific genius is made, to test it by comparison with the great mathematicians. Now, if we understand Prof. Baldwin, which seems hardly possible, in proposing as the criterion for distinguishing between the genius and the crank, that the true genius "and society must agree in regard to the fitness" of his ideas, and "for the most part his judgment is at once also the social judgment," he means that if the society of the day regard a man's ideas as unsound, he is no genius. But the history of mathematics simply swarms with instances in which work utterly neglected and despised in its day was found by a later generation to be fundamentally important, so that mathematicians proceeded to build upon it. The world had to grow up to it.

There was one Girard Desargues, who in 1639 printed a volume entitled 'Brouillon Project d'une atteinte aux évènements des rencontres d'un cône avec un plan.' Although Descartes praised Desargues in a private letter, as did the young Pascal in a work never yet printed, and Fermat somewhere (very likely on the margin of a book he was reading), yet he was so generally regarded as a crank by his contemporaries that early in this century he would hardly have been remembered at all, except for some sparse contemporary allusions to his "faiblesse pitoyable," etc. In fact, Montucla's great history of mathematics, enlarged by Lalande and extended by Delambre — five bulky quartos, which certainly did not intend to overlook any French name of the least account — does not recognize the existence of Desargues. As far as we are aware, no printed copy of the original book has ever yet turned up. But it happened one day in 1845 that M. Michel Chasles, the great geometer, started so early for the Monday meeting of the Institute that he lingered on the quay and began turning over the books exposed for sale on the parapet. He came across an old MS. copy of the book mentioned, which he purchased for a trifle. He took it home, and, having done so, he violated all book-buyers' manners by sitting down and reading it. He found it to be what, had it been written the day before, would have to be considered a very able treatise on that projective geometry which was rightly reckoned as the great glory of the nineteenth century in pure mathematics (though of course with important lacunae), and he further found that Desargues had made an immense stride in advance of modern geometers in recognizing the fundamental importance of a relation which he called involution, and which, under the same name, is still regarded as a cornerstone of geometry! There are certain *minutiae* of the history into which we cannot enter. The above account is as correct as its brevity permits.

Space does not permit us to set forth other like cases, not even a poor half-dozen selected from those of our own century—say the cases of Galois, Listing, Lobatchewski, Plücker, Green, and Hesse. Of these great names we find but two in the body of Phillips's Index; and one of these has no reference, so that it was doubtless inserted by the searching editor of the second edition. The man in vogue cannot escape the influence of the psychological law which causes him to desire to deny such facts; but they have their own sullen way of mutely but stubbornly continuing to exist. Happening to open the works of Beaumarchais at that piece in which his genius first found its strength, one would be surprised to read as its title, "Le Barbier de Séville, comédie en quatre actes et en prose, représentée et tombée sur le théâtre de la Comédie-Française, le 23 février, 1775." A public may need a little education even to appreciate a farce. On the other hand, nothing is more amusing than to compare those whom the society of their day looked upon as immense with some contemporaries who passed unknown, beginning with Alexander of Ales and Albertus Magnus against Petrus Peregrinus and Roger Bacon. Prof. Baldwin speaks of the "supreme sanity" of Newton—a decidedly unfortunate instance from various points of view. But here we only note that since Newton considered his commentary on Daniel to be his greatest work, it follows that, according to the criterion seemingly proposed, he would have to be reckoned as no genius.

J. 22 DECEMBER 1898

DARWIN'S TIDES.

The Tides and Kindred Phenomena in the Solar System. By George Howard Darwin. Boston: Houghton, Mifflin & Co. 1898. 8vo, pp. 378.

. . . The only chapters of the book with which we cannot feel ourselves quite satisfied are those which deal with mathematical arguments. We are here sometimes at a loss to understand what it is that the author is aiming at. He seems to be *explaining* the reasoning of the mathematician. But mathematicians, especially when they are dealing with the most difficult applications of mathematics, have not been inventing abstruse and difficult ways of reasoning; on the contrary, they have been trying with all their might to find the simplest and easiest ways; and they are men of great genius and training in finding out simple methods of reasoning. By far the shortest way to understand the reasoning of the mathematician about the tides is to begin by buying a book on the calculus; and when that is mastered, to go through with the rest of the course required for a thorough understanding of hydrodynamics. Any pretended "explanation" of the reasoning shorter than this either is fallacious, or covers only a small and insufficient piece of the reasoning for even a vague conclusion, or it is open to both criticisms.

The author, in his preface, has this remark:

> A mathematical argument is, after all, only organized common sense, and it is well that men of science should not always expound their work to the few behind a veil of technical language, but should from time to time explain to a larger public the reasoning which lies behind their mathematical notation.

There is more than one fallacy here. In the first place, the term common sense, the middle term of the first half-expressed syllogism (whose strict conclusion Mr. Darwin would not have found it convenient to state), is ambiguous. Common sense, in the proper acceptation of the term, resides mainly in the subconscious department of the mind. It informs us how things go in this world of ours, and in this world exclusively. It has nothing whatever to say about the pure hypotheses of the mathematician, and therefore has no bearing whatever upon mathematical reasoning. But it appears that in the sentence quoted something quite different is meant, which is called "common sense" only because these words impart that popular tone and that *bonhomie* which the author is endeavoring to assume. What seems to be meant is, that each step of the mathematical argument is perfectly evident to the vulgar equally with the learned, to the mathematically dull as much as to the mathematically bright, *provided he clearly apprehends the premises.* But this proviso contains the whole difficulty. Every mathematical premise contains so many elements, so intricately related, that most minds are unable to apprehend the proposition distinctly. The ability to do so is precisely what makes the mathematical mind. It is impossible to dissect a mathematical reasoning so that the mathematically very dull can apprehend it, because, when the dissection reaches a certain point, the logical relations become different; and the result is a fallacy.

It is not true, then, in any acceptation of the terms, that a mathematical argument is only organized common sense. As for what is said about a "veil of technical language," it involves an egregious begging of the question. The technical language of science is composed of words intended to aid the search after truth by facilitating the work of the mind in dealing with complicated conceptions. If, instead of fulfilling that purpose, it is that Talleyrand dialect which veils the thought, it ought to be altogether spurned, and will be so treated by every real devotee of science. But the "language" which Mr. Darwin has in mind is not speech—it is the language of algebra and the calculus. To the disciple of Lagrange and Laplace, the analytical formula is simply the most perfect possible description of the hypothetical phenomena. It is something into which geometrical representations ought to be translated, being itself as near pure thought as it is in the nature of thought to be. When it comes to such a question as the phase of a forced oscillation, especially of an oscillation in two dimensions (and such is the problem of the tides), the frankest way is to leave the mathematical argument untouched in that utmost simplicity to which generations of the most skilful reasoners have been able to bring it. By all means illustrate any steps of it that you can, by parallels drawn from familiar experience, but do not attempt to "explain" that which, on the contrary, must explain your explanation. Is there one bicyclist in five hundred who thoroughly understands why his wheel behaves as it does? Is there one in fifty who does not imagine that it stands up because of its rectilinear velocity? How utterly visionary, then, it is to attempt to popularize the mathematics of any less familiar and still more intricate subject.

K. 22 MARCH 1900

The Teaching of Elementary Mathematics. By David Eugene Smith. [Teachers' Professional Library.] Macmillan.

This attractive-looking volume makes pleasant reading, too, for it contains many a curiosity. A further merit is that it directs its reader to many books well worth his examination, although others of the greatest importance are overlooked. Most of the recommendations of the writer are well enough; but they are, on the whole, trivial. Vulgar arithmetic is, after reading and writing, unquestionably the most practically important subject taught in the schools. For the immense majority of scholars, it would conduce far more to their success in life to be good arithmeticians and bad spellers than good spellers and bad arithmeticians. The effort of the schools should, therefore, be largely concentrated upon making practical cipherers, at any rate, in the teaching of arithmetic. Yet, having looked over forty or fifty of the arithmetics in vogue, we are in a condition to say that pupils never are made really skilful at figures, and Mr. Smith betrays the fact that he, like the other pedagogists, is not himself a master of this low but necessary art. He discourses about trifling matters, and neglects most of those that are really weighty from a practical point of view.

About teaching geometry, too, he seems to us to be quite at sea. All that is of direct practical importance in the geometry usually taught might be put into a nutshell. But it is an indispensable preliminary to mathematical reading, and has always been acknowledged to be of great value as a mental discipline. If geometry were properly taught, it would train and strengthen a number of faculties: first, an important species of imagination; second, ratiocinative invention; third, logical precision of statement; and, most important by far of all, the power of generalization. But, to those ends, it is requisite that other branches should be taught than merely metrical, or ordinary, geometry — branches not essentially less elementary, but rather more so.

That Mr. Smith should have any notion of the educational treasures of topical geometry is more than we could expect. But we should think that a teacher from whose mouth the names of Desargues and Steiner and Von Staudt drop quite glibly (though we notice that the more available book of Cremona passes unmentioned) might recognize that even projective geometry is more fundamental and nearer the beginning of the matter than metrics, and that at least so much of it as is involved in perspective could advantageously be taught before attacking the somewhat artificial, and therefore confusing, logic of Euclid or Legendre. But even within the old, traditional limits we find nothing very useful in these chapters.

In regard to algebra, it is not easy to go quite wrong, owing to the perfection of the science. But there is here no symptom of a power of inculcating a real comprehension of the methods of analysis. As to the author's objection to applied problems of algebra, we must emphatically dissent. He quotes the dicta of some English mathematicians; but these are entirely misunderstood unless we are aware to how great an extreme this sort of thing has been carried in England, and that in reference to higher parts of mathematics. In moderation, it is now generally recognized by Continental mathematicians that the English practice is right. Besides, we are at present speaking only of the instruction of school-boys; and, in our opinion, considered simply as logical exercises, those practical problems that Mr. Smith contemns are most useful in training the power of disentangling a state of things.

The volume has an introduction by the editor in three pages, of which two are devoted to the logic of mathematics. We shall not imitate his brevity by undertaking to say here what we think of his views, because to do so might smack of superficiality.

L. 19 JULY 1900 AND 27 NOVEMBER 1902

Theory of Differential Equations. By Andrew Russell Forsyth. Cambridge: At the University Press; New York: Macmillan. 8vo.Vol. I., 1890, pp. 340; Vol. II., 1900, pp. 344; Vol. III., 1900, pp. 391.

The previous 'Treatise on Differential Equations' (Macmillan, 1885) of Prof. Forsyth, now Cayley's successor in Cambridge, presented the subject in an elementary way, suitable for a beginner. The present work is addressed to such as have at least a thorough elementary knowledge of the subject, and invites the attention of all who have penetrated more deeply into it. It does not, perhaps, display as much high talent as Émile Picard's treatise (which, of course, is of a different nature); but the thoroughness with which the substance of the memoirs has been worked over, quietly improved in many places, and set before the reader in a compact and easily read form, can be known only to those who will put themselves to the trouble of making the comparisons. The reader is everywhere referred to the greater memoirs, as well as to any other presentations of the several branches of the subject that may have been found specially luminous by Prof. Forsyth. Lesser papers, even in cases where they add something material, are, for the most part, passed by. For example, at the end of chapter ii. of volume ii., which is devoted to Cauchy's existence-theorem, there are fifteen references to thirteen different memoirs and treatises. Among these, of course, figures the classical work of Madame Kovalevsky; but the connected investigations of Delassus are not mentioned, nor does his name appear in the index. It is true that these, like Madame Kovalevsky's, relate mainly to partial differential equations, and are thus, in a formal view, foreign to the subject in hand. But then, would not the whole question of the existence of integrals have been advantageously set forth, at least in its outlines, connectedly?

This brings us to remark that the arrangement of Prof. Forsyth's matter seems to have been determined by the circumstance that such and such a part was ready for publication. Certainly, it was most desirable that as soon as any division of the work was ready it should be set before the mathematical world at once; nor do we intend to imply that the selection of parts to be prepared was a random one. Volume i. treats of total equations and mainly of Pfaff's problcm; volumes ii. and iii., of ordinary equations not linear. Now, on the principle of treating the general before the special, it was certainly desirable to place Pfaffians thus early; and with them a great part of the subject of partial differential equations. But a considerable part of the latter subject remains, and unless Prof. Forsyth returns to it, will remain, unconsidered. There is also room for doubt as to whether it was expedient to postpone ordinary linear equations to equations not linear. However, no arrangement could have been adopted that would clean up the whole matter to be disposed of at one systematic sweep. There must inevitably be considerable odds and ends requiring a subsequent gleaning. The author proposes to treat ordinary linear equations in "an additional volume." A complete collection of all that has been contributed to that subject would make a library exceeding the entire impedimenta of many an able mathematician; and as to its being compressed into one volume, a doubt may be permitted. So, considering how much else remains over, we may hope that two, if not three, volumes are yet to be added to this admirable and beneficent work.

This is not the place to consider it more in detail; but seeing that the doctrine of differential equations is one of the most practical of the branches of mathematics, which every young man who aspires to apply exact principles to the affairs of life more than his predecessors have done, has to know at least as well as he knows how to spell, we will point out to beginners the advantage there now is in seeking their first introduction to this discipline at the hands of Forsyth's "Treatise," since his "Theory" is now at hand to supplement it in directions in which they may desire to push their studies further. A better practical mastery of differential equations can be attained by beginning with Forsyth's "Treatise," followed by his "Theory," than by the aid of Jordan, Königsberger, or any other extant guide. It is a lucid and delightful work, and has the advantage of offering a careful selection of those exercises for the student which are much more needed here than in other equally advanced branches of mathematics, for the reason that there is no perfect calculus enabling one to tilt at a differential equation like a knight in armor. Strategy and experience are nowhere more demanded.

Theory of Differential Equations. Part III (Vol. IV.): Ordinary Linear Equations. By Andrew Russell Forsyth. Cambridge (Eng.) University Press; New York: Macmillan. 1902. 8vo. pp. xvi, 534.

The successor of Cayley here truly presents us with the fifth volume of his useful work on differential equations; for his more practical "Treatise" on the same subject really forms an essential part of this. No further volume is promised; but the author does not declare that he will give no other, and we are inclined to think that one more will come. Certainly, enough topics have been passed over to furnish forth two, and that richly. Although it cannot be called a complete handbook, the work has a tolerably definite plan and a judicious one. Whatever would more particularly interest an algebraist or student of the theory of groups is omitted. Whatever stood right across the path of a writer on Differential Equations is treated in a general way. Works of several American mathematicians are necessarily expounded in this volume. Prof. G. W. Hill's method forms the subject of one of the ten chapters, and that is meagre allowance. An unfinished MS. treatise of the late Prof. Thomas Craig has been made use of. Theorems by Osgood, Bôcher, Van Vleck, and others appear. A peculiar sort of suggestiveness attaches to that of Bôcher.

Of course, such a book can have no interest to the generality of our readers except in this respect. The premises upon which pure mathematics rests are few and simple. So far as they are capable of definite numeration, there are about a score of them. The reasoning is wholly deductive. If, therefore, deductive reasoning were what the logic books represent it to be; if, as Kant says, it merely explicated what is confusedly thought in the premises; if, as Mill says, it merely registered what had already been accepted, then the total number of mathematical conclusions could not exceed the total number of possible combinations of premises—or, say, something like a million, including the most trivial. By this time, then, pure mathematics ought to be approaching exhaustion. Doubtless, the current impression among ageometrical persons is that such is the case. Yet the four volumes of Professor Forsyth's "Theory" present, in somewhat full outline, only about two-thirds of the discoveries made during the nineteenth century in a subject which has occupied about one-tenth of the total energy of mathematicians; and far from there being the slightest sign of exhaustion, the bulk of the new work is increasing in geometrical progression, while it is constantly growing more and more profound and broad. As many new *methods* of value now appear in a decade as there were born of new *theorems* in the same interval a hundred years ago. Here is a subject dealing with nothing but the abstract creations of the mind; a subject, too, in which comparatively few are able to make discoveries; and yet it may be doubted whether sixty volumes could give a very much fuller account of the mathematical discoveries of the nineteenth century than could be given in the same space of the discoveries in so rich and universally accessible a field as biology. At any rate, the fact that there is no utterly overwhelming discrepancy affords food for rumination.

M. 18 OCTOBER 1900

A Brief History of Mathematics: An Authorized Translation of Dr. Karl Fink's Geschichte der Elementar-Mathematik. By Wooster Woodruff Beman and David Eugene Smith. Chicago: Open Court Publishing Company. 1900. 8vo, pp. 333.

The original of this work enjoys the sort of reputation that the approval of students in German universities can confer. Whatever that may be worth, it may probably show that readers similarly situated in this country, wanting some information about the history of arithmetic, algebra, geometry, and trigonometry, provided it be compressed and generalized, may find the translation will answer their purpose; although there is a good deal about old German books of no importance which had better have been replaced by notices of English writings that really had something to do with the development of mathematics, such as those of Shirewood, Bradwardin, Tonstall, Sacrobosco, Dee, Recorde, Digges, Oughtred, Blundevill, etc. Those readers who do not approve of an historian's wasting time in trying to make out how one event led to another, will find less of that sort of thing in this volume than in anything called a history that can easily be brought to mind. Those who are curious to know in what the mathematical interest and value of any works of mathematicians of the past really consisted, must, of course, seek their information elsewhere than in this little manual.

Its admirers praise its "breadth," and in truth it carries this quality to a high pitch — so high that the reader may oftentimes gather quite a false notion, until he becomes accustomed to the way in which Dr. Fink, in common with many another German professor, uses language, after which the same sentences will convey no notion at all. For instance, we read: "The earliest writer giving us information on the arithmetic of the Arabs is Al-Khowarazmi. The borrowing from Hindu arithmetic stands out very clearly." Now, considering that that writer was no "Arab," nor even an Arabian, but a foreigner called from Chorasmia because, within two centuries from the Hejra, the wild tribe of Mohammed had become possessed of such treasures that some sort of accounts had to be kept, and Chorasmia was the country where the art of computation had been most perfected; and considering that what was set forth in his treatise was substantially all the arithmetic the Arabians ever had — to speak of that work as giving us information about "the arithmetic of the Arabs," if it conveys any definite idea, is likely to convey a wrong one. So it is with the sentence that follows. It is true that Brahmagupta represents an earlier stage of development of the same arithmetical art as Al-Khowarazmi; and so does Bhaskara, although he wrote long afterward. But instead of there being any clear borrowing from the Hindus, many facts lead to a strong suspicion that it was in Bactria or some country north of India that the art in question originated and developed. Again, speaking of the celebrated mathematical papyrus, which we now know goes back in substance, as it professes to do, to the Tenth Egyptian dynasty, Dr. Fink says, "as the measure of the area of the isosceles triangle with base a and side b, ... $\frac{1}{2}ab$ is found, and for the area of [etc.]. These approximate formulae are used throughout and are evidently considered perfectly correct. The area of the circle follows, with the exceptionally accurate value $\pi = \left(\frac{16}{9}\right)^2 = 3.1605$." Now, anybody not acquainted with the German professorial style of expression would suppose that the papyrus contained something about an isosceles triangle; but it is not so. There is a figure of a triangular field of which two sides are marked as 4 "che" and 10 "che," intended perhaps for 400 and 1,000 cubits. The figure is not badly drawn, for the shorter marked side is very near 0.4 the longer one, perhaps 0.408 or 0.409. Being so accurate in that respect, it can hardly have been intended to be isosceles, for the angles at the base differ by about 11°, being 34° and 73°. Multiplying half the base by the side inclined to it by 84° after all made an insignificant error, less than half of one per cent. What is that in the value of a piece of land, presumably agricultural? It is not enough to dig a grave on. Now there was nothing theoretical about the Egyptian of the Tenth Dynasty. A Cincinnati porkpacker could not have a greater contempt for small quantities. He measured his field in a common-sense way. So, likewise, the Egyptian says nothing about the ratio of the circumference of a circle to its diameter, but calculates the area of a circular field by squaring 8–9 of its diameter,which is not a very awkward way of getting a 1 per cent. approximation. What Dr. Fink means by calling 3.1605 an exceptionally accurate value for π must remain a secret, for few of the vast number of evaluations that have been published have been so far from the truth; and if a comparison with equally ancient evaluations is intended, there can be nothing exceptional where there are not instances enough to base a rule upon.

The same sort of "breadth" pervades the modern parts of the book. Thus, on p. 250, we read that "the results of Desargues were more important for theory than for practice. More valuable results were secured by Taylor with a 'linear perspective' (1715)." One would hardly guess from this that Desargues applied his method to stonecutting quite in the style of Monge, while Brook Taylor's mathematically admirable work (why is its title not printed with capital initials and in italics instead of so as to suggest some optical instrument rather than a book?) involved no application except to drawing. On p. 131, after several pages devoted to the writings of the great Grassmann and to others closely related to his, mention is made of "Grassmann's Formenlehre," with no hint that this book, in which a microscope could hardly detect any originality, was by quite another Grassmann. Remarkable English and American contributions to the same branch of mathematics are passed by in silence. The editors ought not to have permitted this. About tables, no notice is taken of the valuable 8-place French logarithmic tables, nor of Mansion's and other tables for finding 12-place logarithms and their anti-logarithms; and, if our memory does not play us false, the statement about Dase conveys a false impression, his name alone being on the title-page of an assigned part of the great factor table, which the book seems to say he did not calculate. The circumstance that he was insane all the time he was making the calculations is interesting enough to insure its not being mentioned by Fink.

There is, to be sure, a conclusive reply to all such criticisms. It is simply that the book is neither intended nor adapted for the use of persons who care particularly to have their information minutely accurate. Readers for whom "breadth" will cover pretty much all other sins, will find it to their taste. There is a good deal of human interest in the history of mathematics; but all such unscientific stuff has been ruthlessly excised by Dr. Fink, no matter what its significance might be for the development of mathematics. So, those who found fault with Cajori's book for being too entertaining ought to find this one perfectly unexceptionable in that respect.

The translation is excessively literal, in many places too much so to convey the precise meaning of the author. We do not doubt that the accomplished mathematicians who have executed it have in many places improved upon the original, in point of accuracy. They seem to us, however, to have committed a grave mistake in changing the title, albeit they give notice of having done so on their title-page. They justify their step by saying that the author in many places strays so far beyond the bounds of the elements that "the original title is misleading." That may be, but what of it? Is this the logic of modern mathematicians? Because the book is very badly laid out for a history of elementary mathematics, does it necessarily follow that it must be well describable as a history of mathematics in general? Perhaps there is nothing about which the sort of readers to whom the book will most appeal wish more to be informed than the origin and early history of the differential calculus, probably the most important of all those events concerning which the history of mathematics can afford us any satisfactory account. To call a book, however brief, a history of mathematics, without a fuller narrative of that revolution than this book gives, seems a terrible compromise. It was a good idea to append an alphabetical list of mathematicians with brief notices of them, had it only been longer; and as it is, to look it over and see who is in it and who is not in it, will supply a little of the amusement which the body of the work sternly refuses to bend to. There is a satisfactory index.

On the whole, the volume will probably prove one of the most useful of the meritorious series to which it belongs; although, had the translators been willing to take the trouble to produce a history of their own, they would have earned a larger measure of gratitude from American readers.

N. 1 OCTOBER 1903

THE DECLINE OF MATHEMATICS IN ENGLAND

To the Editor of The Nation:

Sir: The statement made by you in the last issue of the *Nation*, page 229, may be easily answered by quoting such important branches of mathematics as theory of numbers and modern theory of functions, which are almost entirely due to benighted Continentals. Some time ago, when the English had three eminent mathematicians left yet — a Hebrew, a half-Russian and an Irishman—they used to say that they had generals, but no soldiers, in mathematics. Now the generals are dead and no soldier has risen to the rank of general. What a well-minded Englishman thinks of the present state of mathematics in England may be inferred from Professor Greenhill's review of the German translation of Professor Perry's book, published in *Nature* about a year ago. Without the slightest intention of composing a "Sovvenire di una gran nazione" to the English people, I cannot help quoting the momentous words of Henry John Stephen Smith: "A decline in the mathematical productivity of a nation amounts to a retreat on the whole line."

Yours very respectfully, H. T.

September 17, 1903.

(Our comment will be found on another page. — Ed. Nation.) [Commentary by C.S.P. follows.]

BRITISH AND AMERICAN SCIENCE

A fortnight ago, in speaking of Lockyer's appeal to his countrymen for the support of British science, we showed how to apply the old saw that the way to get the best performance from a human being is to encourage him. To-day we will try applying the same maxim in speaking of the future of

American science. We wanted to allow the Anglophobists (who never allow the sacred fire to die on their hearth) ample time to dispute our proposition if they could—the proposition, we mean, that for three hundred years not a single conception has taken sovereign pre-eminence in science that has not been largely—in most cases, even without contest—of British parentage. But our proposition remains undisputed. Its substantial truth seems to be tacitly acknowledged.

Whatever could be said to blunt its point has, no doubt, been indicated in a letter in another column from a rarely accomplished and ingenious scholar, signing himself "H. T." Whether this subtle writer is serious, or whether he is only making believe that he opposes us, our readers can guess as well as we, if they note that he does not explicitly deny our proposition, and that he throws in our way two or three very pertinent suggestions in support of it. Certainly, two more striking examples of what are *not* conceptions of sovereign preëminence in science than those he furnishes would be sought in vain—the theory of numbers and the theory of functions. A famous mathematician, being asked why he should be so in love with the theory of numbers, replied that he loved it because it was a pure virgin that never had been, and that never could be, prostituted to any practical application whatsoever. Not only no *practical* application, but (so far as one can look into the future) no *scientific* application, either, is likely ever to be made of one or other of those two theories, outside of pure mathematics itself. In short, they are as narrowly technical as anything can be. They are Leibnitzian monads whose activity, intensely interesting in itself, is imperviously secluded from the business of life and from the main business of science.

Only compare this isolation with the loud resonance awaked in every harp and organ of science by those discoveries which truly have been sovereignly preëminent through science—the inductive philosophy; the corpuscular philosophy with the atomic theory and its progeny; universal attraction; the differential calculus (certainly discovered by Newton, not certainly also by Leibnitz after enjoying Newton's conversation); the theories of elasticity (Boyle and Young), of heat as *vis viva* of molecules (Bacon and others), of electricity (Gilbert, Faraday, etc.), of light as transverse vibrations (Young) and transverse vibrations of an electromagnetical kind (Maxwell); natural selection (Darwin and Wallace); universal evolution (Spencer). We forgot to mention one of the greatest discoveries of all, made by an humble clergyman of the Chruch of England, Gay—the discovery that the association of ideas is the autocrat (or, at least, the first of two consuls) that governs all the activities of the human mind, so far as they are subject to any mental law. For the theory of numbers and the theory of functions it can be said that they far surpass chess both in beauty and in their broadly intellectual character; but to call them ideas of sovereign pre-eminence in science would be to fall into one of those extravagant statements in which mathematicians are only too prone, as when Henry John Stephen Smith—one of the protagonists of the theory of numbers, for all his being an Englishman—spoke of a decline in a people's mathematical activity as if it differed from all other historical developments in having but a single possible cause.

We have to thank "H. T." for another suggestion contained in his remark that, of the pure mathematicians of England's last half-century, the two most prominent were the Hebrew Sylvester and the half (or quarter?) Russian, Cayley. To be sure, there is a simpler and more conclusive proof that the singular relation which Great Britain has sustained to science has not been due solely to Teutonic blood, namely, the density of the entirely uncultivated German (if this species be not quite extinct). The days are very, very distant when it will be possible to disentangle the causes of national character; but as to the matter in hand, there is one cause that strikes any good American observer of intellectual English society. Montesquieu, who possessed an intimate knowledge of so many countries (he was, by the way, a Foreign Member of the Royal Society), put on the title-page of his immortal book this motto:

"Prolem sine matre creatam"
(offspring produced without a mother).

A close friend asked him in what sense this had any particular application to the "Esprit des Lois." He replied, after some reluctance: "A truly great work must owe its birth not only to a man of genius as its father, but also to a society of intellectual freedom as its mother." Probably in no country is thought of almost all kinds so completely untrammelled as in England. The chief external hindrance, everywhere, to supremely original scientific speculation lies in a certain spirit to which, for want of a better word, we may give the name of *pedantry* (one could hardly call it *obscurantism*). We mean that spirit which caused Poggendorff to refuse for his journal the now far-famed paper of Mayer about the thermodynamics of gases; that spirit which for a whole generation silenced, through the German universities, every contribution to philosophy that was not Hegelian, and which to-day as completely silences there everything that *is* Hegelian.

And this brings us round to the brief word we proposed to say to the young scientists and philosophers of America. Good and sufficient reasons have in the past acted to conceal the scientific genius of the American people in money-getting and in settling the order of things in this country. But now that those reasons are losing their force, and that you are turning to pure science, above all trust to your own wings. Beware of excessive subservience to the opinions that happen to be in vogue in the German universities. Imitate the Germans in those things that deserve imitation. Emulate them, for example, in that which has contributed not a little to German scientific leadership, their national self-confidence; in their persistency as well. Those two qualities have made that people the world's leader in all *Fach-Forschung* (if we may be allowed to coin the word)—a preëminence all the prouder that it is founded on two moral virtues. You have infinitely more reason to believe in your own scientific powers than they had, one brief century ago. In the nature of things, you will soon outgrow your school-boy deference to your master's *dicta*, and, trusting to your own genius, will surely develop a new and more philosophical type of scientific man.

O. 22 OCTOBER 1903

[REPLY TO A LETTER ON THE PRACTICAL APPLICATION OF THE THEORY OF FUNCTIONS]

An apposite instance! Had we known of it, we should have softened our remark. Nor, in making it, did we forget that several applications have been made of propositions worked out by Cauchy and earlier mathematicians before the "theory of functions" was christened, and for which (though they are now incorporated in it) it can take no more credit than the theory of numbers could to the carpenter's rule of three-four-five, which was known to Pythagoras, but the principle of which may to-day figure in a treatise on that theory. But now, for the first time, we meet with one genuine case of an application of the theory of functions, upon which we may ground some hope for further such triumphs. It is a convincing and striking proof that a line of thought which seems to relate exclusively to impossible states of things may, if resolutely pursued, eventually bring great light upon familiar experiences. Let us give this instance a permanent place in our memory alongside of the fact that Pascal, after his wonderful discovery about conic sections, abandoned that study as an idle pastime having no application to any matter of importance.

All this, however, does not in the least touch the point that our remark was designed to make; for at the time when it could be said that the British were neglecting the theory of functions (which is no longer true), there was no glimmer of reasonable hope that it could ever be of any use. As for the theory of numbers, the first application of it has, we believe, yet to be made, unless, perhaps, Cayley used it for his theory of chemical "trees." But that concerns the *partition* of numbers, a separate branch of mathematics which the English have perfected, we believe.

We are heartily of opinion — but it is no longer a matter of opinion — that the younger generation of physicists are going to reap a rich harvest from their studies of the higher mathematics. In this they are only following a time-honored custom, for almost all the great physicists, from Galileo down, have been strong mathematicians. At the same time, there are instances enough — like the beautiful researches of Le Bon on phosphorescence and peculiar radiations — to show that, even in these days, the consciousness of a decidedly deficient capacity for mathematics need discourage no young man, nor young woman, from devoting himself or herself to physical investigations. — Ed. Nation.[1]

[1] The previous item on the decline of mathematics in England (p. 1033) elicited a letter from James McMahon of Ithaca, New York, on October 10, printed in the October 22 issue of *The Nation.* The reply to that letter, given above, is Peirce's.
The first of three volumes entitled *Charles Sanders Peirce*: *Contributions to The Nation*, compiled and annotated by Kenneth L. Ketner and James E. Cook, is now available (Texas Tech Press, Lubbock, Texas, 1975).

P. 14 APRIL 1904

Lectures on the Logic of Arithmetic. By M. E. Boole. Oxford: Clarendon Press; New York: Henry Frowde. 1903. 12mo. pp. 144.

Elements of the Theory of Integers. By Joseph Bowden. The Macmillan Company. 1903. 12mo, pp. 258.

Mrs. Boole's is not a work like Dedekind's "Was sind und was sollen die Zahlen." The lectures are supposed to be addressed to children. "The earlier chapters are suited to little children, the later ones for children of fourteen or fifteen." There are twenty-three lectures or chapters, and "not more than one chapter is intended for use in any one term." The book can hardly be said to relate to the teaching of arithmetic, for, as the authoress says, it "is not intended to interfere with ordinary methods of teaching arithmetic." It aims to take advantage of a class having been formed in arithmetic, to teach the children a little logic. The logic, not too definite for the infant mind, is wholesome — occasionally quite refreshingly so, as when the writer says; "The sentimental people who assert that everything in arithmetic can be 'proved' to children have, usually, no idea of what rigid proof means." If she would delete "usually," every exact logician would agree with her. But we wish Mrs. Boole would treat the grown people for whom she writes, in this volume and in others, a little more as she says children should be taught. "If a teacher has anything to say to children as a statement (presumably she means as an assertion or as a proposition or in the indicative mood), he should say it, not exactly as a dogma which they are bound to believe, but as a working hypothesis which they are to assume as a basis for the present." This may be going a little too far, even in reference to assertions made to grown folk; but the tendency of it is good. If we are to believe everything in Père Gratry's "Logique," let it be when we come to see the truth of it.

George Boole was one of the great vehicles of truth of all time. Like all the great philosophers, he possessed the power of working his way to the truth with ideas that could not, at his time, be rendered entirely distinct, for he lived before the true logic of mathematics had been called into existence. He not only had this power, common to Aristotle, Descartes, Kant, Leibniz, and the rest, but he did what only one or two of the great philosophers was able to accomplish — he positively proved the truth of his idea. Gratry is no more to be compared with Boole than the lustre of Vega or Capella is to be compared with that of Jupiter. With some *aperçus* of high truth, his mind was too much in the attitude of prayer and of preaching to see the necessity of proving what he advanced. The infusion of Gratryism into Mrs. Boole's mind does the children no harm. We only fear that it may obstruct the reception of her wholesome ideas by teachers; as when she says, for example, "Arithmetic seems to some people dry and unbeautiful; but that is because they have not soaked it in that solvent which is called sympathy."

She hardly discusses the fundamental question whether or not it be desirable to cultivate the minds of young children in the direction of deductive logic. That is a matter to be carefully considered. The reviewer believes that she is right in thinking that it is desirable. This kind of logic — the logic of "Mamma says it is wrong" — we believe to be the very kind of thought for children. But it is too momentous a question in education to be hastily decided. Once grant, however, that logical conceptions are to be developed so early, and Mrs. Boole's methods of developing them are certainly exceedingly skilful and quite admirably adapted to the minds of children.

Prof. Bowden gives an independent development of his subject. That is a merit. Any person unacquainted with the logic of arithmetic could gain enough from the book to pay for the trouble of going through it carefully and critically. Further than that we cannot praise it. Its first paragraph is as follows:

> The concept of number, in its simplest and original sense, is a fundamental concept. It is incapable of definition — that is, it cannot b^ expresst in terms of ideas simpler than itself.

This is not so; and if it were so, there would be no use in such a book. There is no possible account of the logic of number that is not based on the logic of relations, whether consciously or not; and number does not express a simple relation, whether the ordinal or the cardinal numbers are regarded as primitive. This has been made perfectly clear in more than one of the books with the titles of which the footnotes of this volume are ornamented.

On page 3 we meet with this: "*Axiom.* Any number is equal to itself." A poor sense of logic must a man have to entitle this an *axiom*, when on the page before he had said, "To (the idea of sameness between two numbers) we give the special name *equality*."

Whoever wishes to understand the logic of Integers should begin with Dedekind's little book, of which a translation is published by the Open Court Co. There is a good deal more to be read besides if one's appetite holds out.

Q. 19 MAY 1904

Notes on Analytical Geometry. By A. Clement Jones.

There are great advantages in learning any mathematical theory which is subsequently often to be applied, from a logical syllabus expounded in lectures. The propositions should be numbered in the syllabus in one series, from beginning to the end; and propositions very frequently employed in the proofs should have, besides, brief names. Then, in the case of a corollarial proof—that is, one not requiring the introduction of any subsidiary lines or quantities—mere references to the numbers of the premises, sometimes with the number of times each is to be applied, will generally suffice. In case this is not enough, brief indications of how those premises whose application is not quite obvious are to be applied may be added. Theorematic proofs—that is, such as depend upon some ingenious addition to the conditions of the proposition to be proved—will require this addition to be stated; after which the proof becomes corollarial, and should be treated like any other corollarial proof. The student having become perfectly familiar with the arrangement and general contents of such a syllabus, by working through it, will ever after find it an invaluable work of reference, in which any result may be found directly, together with the logic of it in a nutshell. Dr. A. Clement Jones's "Notes on Analytical Geometry: An Appendix" (Oxford: Clarendon Press; New York: Henry Frowde) approaches to being such a syllabus in a duodecimo of 172 pages, of which 30 are occupied with examples and hints for their solution, with a perspicuous one-page index. The work is confined to plane curves of the first two orders, together with unicursal cubics, the whole treated from the metrical point of view, with Cartesian coordinates and occasional references to polar coördinates. The most serious omission is the problem of two conics without contact, with single contact of 2, 3, or 4 points, and with double contact, which is often wanted and is not very readily worked out. A better book of the same sort might be made; but, as it is, Dr. Jones's "Appendix" will prove a lifelong blessing to many a student.

R. 8 SEPTEMBER 1904

The Collected Mathematical Papers of James Joseph Sylvester. Volume I. Cambridge (Eng.) University Press; New York: Macmillan.

We receive with delight this first instalment, a beautiful and comfortable volume closely matching in outward appearance Forsyth's "Theory of Functions." It contains Sylvester's work from 1837 to 1853. At this moment, when the chill of senility begins to be perceptible over the very formalistic mathematics that has been and still is in vogue, the virile genius of Sylvester needs to be more fully appreciated. Doubtless there are many memoirs more significant than his and of broader conceptions; but we doubt if there be any whose thought has the peculiar mathematical quality in a higher degree. There are more flawless gems of mathematical workmanship, there are papers of more perfect polish in their execution; but we are strongly inclined to think that there are none quite so instructive in the heuretic art, partly for the very reason that these have not been so finished as to conceal the brush-marks, partly because of the personal originality and singularity with which they are stamped, and partly because Sylvester's garrulity led him almost constantly to tell how he came by his ideas. It would be well worth the while of a student of methodeutical logic to take up the theory of invariants, just for the sake of comparing the ways of thinking of Cayley and Sylvester, as exhibited nowhere so well as in this volume, with those of Clebsch, Gordan, etc.

Sylvester's habit of throwing his whole being—or only sparing to poetry and sentiment their strictly necessary aliment—for long year after year into the development of a single system of ideas, while recording his progress every two or three months (every month, in his most active years), renders this collection instructive beyond measure. Nor is the interest exclusively mathematical. Logical remarks of value are constantly occurring, and other philosophical suggestions are not rare. In one place we read, "Universal geometry brings home to the mind with an irresistible conviction the truth of the Kantian doctrine of locality." Verily, metaphysics is the Paris of the intellect: no sooner do the most scrupulously severe reasoners find their feet on this ground than they give the loosest reins of license to their logic. Universal geometry can testify concerning no other Space than its own, which is a space, not of three, but of an indefinite number of dimensions; and nothing is more striking in this generalized geometry than that it is decidedly easier for the human mind to comprehend a space of four dimensions than one of three. Give a higher geometer sixty days to accustom himself to a four-dimensional space, and he would be ever so much more at home there than he ever can be in this perverse world. Meantime, the dynamics of rotations asserts downright that the rotational part of motion, at least, is not relative; and as for the body alpha, the epistemological difficulties of this disguise of the reality of space are too serious; and if the fixed stars, or the whole universe, be identified with the body alpha, the difficulties become downright absurdities. The only body alpha that epistemology can admit is the body of space itself. Meantime, even if one were to prove that three-dimensional Euclidean space is native to the mind, that would be no argument in favor of Kant's position that it is *merely* an affair of the mind. On the contrary, the proper presumption would be that, in view of the unity of the universe, if such space is native to the mind, probably it is native to the outer world of reality also.

The volume is not without glimpses of human nature. Lagrange's so-called demonstration of the principle of virtual velocities "is contrary alike to sense and honesty"—yes, that is the color of it, "albeit sanctioned by the powerful oral authority of an ex-Cambridge professor." One wonders how its case would have gone without that orally powerful sanction. As for Lagrange's proof, it appears to us that cavillers mistake the purpose of it. At any rate, it does convince all reasonable living doubters, for the reason that every man's experience has given him an instinctive and virtual knowledge that work always has to be paid for, which the proof of Lagrange either tacitly takes for granted, or, as we interpret it, supposes to have been expressly admitted for the case of one pair of blocks and tackle. But we must confess to not having looked into the immortal book for many a year.

In every way this publication is a precious benefaction to mathematical students, especially to those who have been treading too exclusively the boulevard of dominant ideas. The four earliest papers relate to the mathematics of physics. Then, for five years, from 1839 to 1844, Sylvester was occupied with elimination and multiple roots. Between 1844 and 1847 the editor, Mr. H. F. Baker, has found nothing. Has he searched

Adrain's Diary? Three papers of 1847 relate to the integer equation

$$x^3 + y^3 + z^3 = Dxyz$$

The two following years are blank; but from 1850 to the end of the volume (1853) papers follow one another at an average rate of one every month. Here we find the Essay on Canonical Forms, well-known discoveries in Determinants, the "law of inertia," and the great memoir on syzygetic relations and Sturm's theorem, which last was the first of Sylvester's papers to be ushered in with a poetical motto. It is those lines, "How charming is divine philosophy," in which the Lord Chancellor so successfully disguised his detestation and disgust of all the philosophy of his day.

We shall endeavor to keep our readers apprised of the appearance of future volumes, and we hope to find fewer clerical errors in the next. A chronology of Sylvester's changes of place, travels, and, if possible, of his forming of mathematical acquaintances, would be of advantage.

S. 30 MARCH 1905

The Phase Rule and its Application. By Alex. Findlay. With an Introduction on the Study of Physical Chemistry, by Sir William Ramsay. Longmans, Green & Co. 1904.

There are numbers of highly successful men who have never been able to master elementary geometry or algebra. In the case of many of them, the defect is undoubtedly merely due to bad teaching. For others, there may be some peculiarity of metrical ideas which prevents these men from mastering them. But, with a third class, the truth probably is that there is a weakness of the understanding which causes all conceptions of the slightest intricacy to become confused. Now every man ought to endeavor to understand his own capacities; and the subject of the Phase Rule may serve as a touchstone to show him whether he belongs to the third class or not.

But what is this phase rule that has for so long set all the chemists agog? Take a number of materials, each either a pure chemical substance or a homogeneous mixture, but none of them such that any chemical reaction that could take place could produce it from the others, and (calling these the *components* of your system) put them into a tight cylinder in which a piston works, so that the aggregate can be subjected to varying pressure and temperature. Further, let there be some means of varying the proportions of the components present in the cylinder. Then, since the total quantity is a matter of indifference, there will be, as independently variable conditions, one for each of the components *minus* one, *plus* the pressure and the temperature; that is, on the whole, one more than the number of components. Now give your system time to come to a settled state of equilibrium while there is no change of the above conditions. This equilibrium must be *stable*; that is, if you vary those conditions a little and restore them, the state of the system must be restored. Now count the number of *phases*, that is, the number of substances or homogeneous mixtures separated from one another by surfaces like the surface between water and air. Then it will be found that the degrees of freedom — that is, the number of conditions (pressure, temperature, and the proportions of components) that can be independently varied *without altering the number of phases or destroying the equilibrium*—are equal to two more than the number of components less the number of phases. For example, can water, ice, and aqueous vapor exist together? Here there is only one component, and we suppose three phases separated by surfaces. By the rule, the degrees of freedom amount to zero. That is, if the temperature and pressure are both fixed right, such a state of things is possible. But when it exists, neither pressure nor temperature can possibly be varied, the equilibrium being preserved, for the least change will cause either the vapor, the water, or the ice to pass away entirely. Suppose, however, that air be added, so that now there are two components, and, for phases, first, air mixed with a little vapor; second, water containing a little air in solution; third, pure ice. There will now, according to the rule, be one degree of freedom, and the pressure may be changed so long as the temperature is correspondingly changed. Suppose we let the ice melt. There will then be two components and two phases, and consequently by the rule two degrees of freedom. That is, both pressure and temperature may be varied, but only on condition that the proportions of water to air in the vapor and in the solution are correspondingly changed at the same time.

Such is the phase rule. Its mathematical form is identical with that of Euler's topical theorem that the combined number of faces and summits of any polyhedron is two more than the number of its edges; and, like topical geometry, the phase rule has nothing to do with *measurement*, but only with *counting* three sets of things—components, phases, and degrees of freedom. Nor do the experiments for verifying it require any measures to be made, but only that more and less should be discriminated. Beyond that, there is no mathematics in the subject; but the chemical applications of the rule have a certain moderate degree of intricacy, not much less than that of elementary geometry and algebra. They will be beyond the powers of those who are unable to master those mathematical subjects owing to general weakness of the understanding, but will presumably (not certainly) be found to be just complicated enough to afford entertainment for other men.

The phase rule was discovered not by a chemist, but by Willard Gibbs, a mathematician, the ablest, in his particular line, of his day; the most unaffectedly modest man (for his strength) that search could find; the sort of man who is sure to be depreciated until, by the action of a sort of phase rule, the ice breaks up and men suddenly discover that it is best to laud him. For ten years and more the phase rule slumbered peacefully in the Proceedings of the Connecticut Academy of Sciences until the chemists had grown up to it, till Roosebaum and others brushed the dust off the volumes and carried them into their laboratories. But Maxwell fully appreciated Gibbs long before the phase rule was discovered.

Dr. Findlay's illustrations from chemistry are as replete with interest as they are abundant, and each one detailed. Upon all that side the work is simply splendid. Its weakness is that the abstract definitions, and the possibilities which the phase rule leaves open, but which have not yet been met with in nature, are not forced upon the reader's attention in all their definiteness sufficiently to enable him to comprehend exactly what the phase rule does for him in each case. In short; the work presents a picture admirably rich in its coloring, but whose outlines are too soft for a scientific purpose—on rare occasions, quite indefinite or even wavering. The practicable remedy for this in a new edition would be an appendix to which the reader could always turn to find there the abstract parts of the subject defined with abstract precision. The reviewer must confess that, though the subject is not new to him, he has here and there in the perusal of the work found his ideas losing their definiteness; and this state of things was not remedied until he had drawn up for himself the sort of appendix that was needed. However, taking the book as it is, anybody who adds to a turn for the mode of thought a fondness for chemistry, will find the volume replete with matters of singular interest.

Sir William Ramsay's introduction contains a very good sketch of the history of physical chemistry without any such high merit as one would have expected. The language is that of a man who, when he thinks, imagines apparatus and experiments, and who has little training in the use of words. Thus, on page lxi, he finds fault with Julius Thomsen because, in 1854, he believed it possible to measure the force of chemical affinity by measuring the heat evolved in reactions; and his objection is simply that heat is not force, but energy, a quantity of different dimensionality. Now this is hypercritical in the extreme; for though the word *energy* had been proposed by Young, it was not at all in use, even in England, before Rankine's paper of 1853, and in Denmark was quite unheard of; the word "force" being much more loosely used than it now is.

T. 19 OCTOBER 1905

[Note on the Contributions of George William Hill to Celestial Mechanics and Volume I of his *Collected Mathematical Works.*]

The severest of all touchstones of mathematical skill is universally acknowledged to be the working out of an exact numerical account of the way our satellite performs its intricate motions—the "theory" of the moon, as the mathematicians still call it, after Ptolemy. Well, inquire, say in Berlin, or in Pulkowa, or Paramatta, or Tacubaya, or in any corner of the earth where high mathematics is cultivated, who in our time has shown the most surpassing mastery of the theory of the moon, and the answer of any competent authority will come unhesitatingly, "It is Mr. G. W. Hill of Nyack Falls, N. Y." Had that village been aware of its renown, it might not have changed its name, alluring as the melody of "West Nyack" no doubt is. But Mr. Hill is the reverse of the kind of man to whom the Sunday *Herald* devotes a page, and it is probable that the villagers know him only as the genial but retiring gentleman who so loves the paternal farm on which he was born and where he still lives. The next most rebarbative problem of celestial mechanics, after the moon's, is perhaps the theory of Jupiter and Saturn (which have to be treated together), and in this Mr. Hill has outdone all other astronomers. But this is as nothing to his achievements in the theory of the moon. For here the method he pursued launched him on an unknown sea, requiring an entirely new chapter to be added to the calculus; and here, by means of the staggering conception of an infinite determinant, he succeeded in the hardy enterprise of virtually solving a differential equation of an infinite order. The boldness of the undertaking consisted in this: that Hill introduced into mathematics a kind of reasoning unrecognized by the mathematicians (albeit they had often unconsciously employed it), namely, the experimental reasoning of physics. For, an infinite determinant being a complete novelty, it was as yet unknown whether the particular type of such a complex series required for Hill's method of solution was convergent or not, or, if it were, whether it possessed the particular kind of convergency that would adapt it to the operations of the calculus.

Hill accordingly treated its satisfying this requirement as he would have treated a physical hypothesis, and proceeded to put it to the test of experiment, by calculating, on that theory, the rate of revolution of the axis of the moon's elliptical orbit, which, of all the elements of the solar system, is observationally the one by far the most sensitive to any erroneous assumption about the perturbations. He relied upon the knowledge that if his mathematics were wrong, there was every reason to expect that his calculated motion of the perigee would be sensibly—would be enormously—at variance with observation. It turned out, however, to agree with observation as closely as the results of observation were known. Yet it must be confessed that it is not as clear as the noonday sun that Mr. Hill himself, any more than previous mathematicians, perceived that he was applying Baconian reasoning to mathematics. In any case the brilliant demonstration of Poincaré was needed to enable future astronomers to apply Hill's method with entire confidence to all problems of three bodies. Nevertheless, when we consider that it would, after all, only be to physical questions that such complicated differential equations would ever be applied, Hill's procedure is seen to be of a piece with all the other reasoning that would go along with it, and therefore logically to be beyond criticism.

By such means our countryman abridged the labor of certain numerical calculations from months to hours, while vastly increasing their exactitude. Mr. Hill's work upon the Moon, originally published over twenty years ago, has since been perfected in some parts and improved in others by an Englishman, Brown. Still, excellent as Brown's work is said to be, the chief merit of the new method confessedly belongs to our neighbor across the river, eighteen miles above High Bridge. Therefore, with a stately quarto, Volume I. of "The Collected Mathematical Works of George William Hill," published by the Carnegie Institution, there came to us a visitor too infrequent of late years—we mean the oldtime glow of exultant American feeling. The volume is prefaced with a long and most interesting account of Mr. Hill and his work from the pen (on the whole the most competent and suitable that could have been selected) of M. Henri Poincaré. It is in French, of course; and we find M. Poincaré writing *collège* with an acute accent, a practice which is redolent of Nancy as it was before the war, the Nancy of old Dr. Poincaré. The volume falls but little short of being a handsome one; paper and type are good. There is a pretty good portrait of the man; but in the pose of the head, though it is not foreign to Mr. Hill, we see more of the photographer than of his subject.

20

APPENDICES

A. [DYADIC VALUE SYSTEM] (6)

Of course, the simplest such cyclical value-system is one of two numbers. Properly speaking, it cannot be called cyclical; for to speak of the order of sequence of fewer things than four as cyclical means nothing, since no way of arranging 0, 1, 2 is more cyclical than another. Indeed, in the case of two things there can hardly be said to be any sequence. When you have taken one, if you intend to take all, you have no choice but to take the other. Consequently, in the dyad, considered as a value system, the ordinary significance of numbers evaporates. There is simply one character and another. *Black* and *white* would answer just as well as 0, 1 in that system of arithmetic. In fact, the whole relationship between the values in that system may be summed up in two propositions,

first, that there are different values,
and second, that there is no third value.

With this simplest of all value-systems mathematics must begin. Nay, all reasoning must and does begin with it. For to reason is to consider whether ideas are *true* or *false*. Now, the relationship between truth and falsity is precisely that of the dyad value-system; namely it is stated in the following propositions:

First, the principle of contradiction, nothing can be true and false, at once;

Second, the principle of excluded middle, a fact must be true or false, there is no third character for it to possess.

The orderly and philosophical proceeding will be, then, not to apply arithmetic to this system, but, on the contrary, to develop arithmetic out of this system as a germ. This is obvious enough; but, nevertheless, it will be found convenient to employ some of the arithmetical signs, because of their familiarity and the consequent facility of working with them, and also in order that when the general number-system is developed, the conceptions of the dyad system may fall into their proper places.

Since the study of this dyad value-system must precede the strictly logical study of number, we ought not to call it arithmetic. We may rather call it *logistic*, a word which, when it was in use, meant the art of numerical calculation, which we now call arithmetic, but which, now that that meaning is almost forgotten, may, on account of its sounding like *logic*, be readily made to take the meaning of the art of logical calculation.

B. THE THEORY OF MULTITUDE (24 and 114)

Multitude is that character of a collection by virtue of which it is more than another collection and less than a third. The main purpose of the theory of multitude is to ascertain what the different possible *multitudes* (i.e. grades of maniness) are, and how they are interrelated. This purpose can only be attained by exact reasoning; and in order to have the requisite premisses, we must first form a pragmatistic definition of a *collection*. That is to say, we must get a distinct notion of those formal propositions concerning collections which must be true, since we should not call anything a collection of which they were not true.

Let us take an example. Take the collection of blue horses. But we do not know that there are any blue horses. Is there, then, any such collection? I reply that while the collection of blue horses will not have the same mode of being in case there should be no such horses as it would have if there were, nevertheless the very fact that we know what a blue horse would be assures us that there is a totality of whatever blue horses there may be, although it be nothing. There is a great principle here. The world must be intelligible. It has to be intelligible or it could not become world. It is true that it is a mere hope of ours that there will prove to be any real world; but it would be the weakest insanity to stultify reason by abandoning that hope. We must, therefore, assume that there is some intelligible whole of whatever we can suppose; that is, *could* suppose, were there any basis of fact for the supposition. This is to admit a sort of being, a being as possible, to that which may not actually occur in a given universe of experience. The reality of a thing consists in its retaining its own characters quite independently of whatever opinion or fancy you or I or any man or generation of men may entertain about it. Reality has its grades. Any object which maintains its characters with sufficient steadiness to make one proposition false and another true has sufficient reality for the purposes of the mathematician.

The first noticeable character of a collection, as such, is that it is an *ens rationale*, a creature of thought. Not that all mankind, for example, forms a collection by virtue of being actually thought as such; but it forms a collection by virtue of the possibility of thinking it. Its being lies in being thought.

Secondly, like most creatures of reason, it is an individual. In this mankind differs from man. Man is neither white nor non-white. Some man is white and some man is not white. But to mankind the principle of excluded middle applies. There is but one mankind, and it is particolored.

A *collection* is an individual object concerning which no facts whatever are true except such the truth of which consists in the truth of facts concerning other individual objects independent of the collection and of each other; and these objects are called the members of the collection.

Although such seems to be the essential idea of a collection, yet the conception admits of enlargement in several ways, to advantage. In the first place, all the independent individuals of the universe that are not members may be considered as contributing negatively to the determination of the collection, almost as much as the members. The members of the collection may be called its *ineunts*, the other coordinate independent individuals, which are non-members, may be termed its *exeunts*. In the next place, although the definition contemplates a plurality of members, there are strong reasons for admitting collections of one. Such a collection is quite a different thing from its sole member, consisting almost as much in excluding all other things as in [including] that, while the independent object does not refer to anything but is whatever it is by its own energy. The practical necessity of this view will appear in almost every instance in which collections in general are considered. Thus, suppose we have occasion to consider the persons whom different individuals love. For each individual these form a collection. It will be most inconvenient to put into a separate category the totality of the loved of a person who happens to love only one person. In the third place, and for the same reason, we have to conceive that as there is one collection which has every individual as ineunt and none as exeunt, so there is a collection which has every individual as exeunt and none as ineunt. Suppose for instance that we are distinguishing different groups of persons according to the different ways in which those groups bring together persons loving different lots of people. Shall we not consider two such groups as different if one includes and the other excludes a person who does not love anybody? This illustrates the practical utility of admitting the zero collection. Of N individuals there are 2^N possible collections; but if the zero and unit collections are excluded, to many other inconveniences we add that of this view giving the complicated expression of $2^N - N - 1$, which is analytically troublesome. There are 2^{2^N} collections of collections in the former case; $2^{2^N - N - 1} - 2^N + N$ in the latter.

A *relation* is a fact real or fictive, generally any one of a collection of facts, relating to the members of a collection. Thus, the fact that they belong to a collection is a relation. If anybody says that such and such phenomena are "utterly unrelated," the fact that he says this constitutes a relation between them. It affords a cheap way of making sure of telling an untruth. If the fact involves more than two individuals, then if these remain unspecified, it is called a *polyad*; if just two, a *dyad*; if just one, a *monad*; if none, a *medad*. In the last case, nothing is left unspecified.

Isaac blesses Jacob in place of Esau is a medad;
Isaac blesses —— in place of Esau is a monad;
—— blesses Jacob in place of —— is a dyad;
—— blesses —— in place of —— is a polyad.

The commonest relations to be considered are dyadic relations. A dyadic relation is any one of a collection of facts about pairs of individuals. We distinguish between the two individuals, calling one the *relate*, the other the *correlate*. We often denote a relation by a letter, as *r*; and use the expression *A* is *r* to *B*, or in the same sense, *B* is *r*'d by *A*. We mean by that that there is a collection of collections of two objects each, which two objects are distinguished grammatically as the active and the passive subjects. We denote this collection of collections by the letter *r* to show that we do not refer to any particular individual collection of collections, but to any one of a collection of collections of collections.

Two collections, say the *A*s and the *B*s, may be such that there is a possible relation, *r*, such that every *A* is *r* to a *B* to which no other *A* is *r*, while there is a possible relation, *l*, such that every *B* is *l* to an *A* to which no other *B* is *l*. When that is the case, the two collections are said to be *equal*, or to have the *same multitude*.

Again, two collections, the *A*s and the *B*s, may be such that there is a possible relation, *r*, such that every *A* is *r* to a *B* to which no other *A* is *r*, while there is no possible relation, *l*, such that every *B* is *l* to an *A* to which no other *B* is *l*. In that case the *B*s are said to be *more* than the *A*s, or the collection of *B*s to be greater than that of the *A*s, or to have a *higher* *multitude* than that of the *A*s. The *A*s are said to be *fewer* than the *B*s, the collection of *A*s to be *less* than that of the *B*s, and to have a *lower multitude.*

This last definition is, however, redundant. For if there is no relation, *l*, whatsoever, such that every *B* is *l* to an *A* to which no other *B* is *l*, it can be shown that that very fact constitutes a relation, *r*, such that every *A* is *r* to a *B* to which no other *A* is *r*. This proposition ought to be called the *fundamental theorem of multitude.* Like other fundamental theorems, it has justly been deemed excessively difficult of proof. Cantor after studying multitude profoundly and with great power for many years, only produced a demonstration of this theorem in one of his last papers. It is a very roundabout way, anything but clear. To my mind it is clearly fallacious. This is not said to depreciate Cantor, whom I greatly admire, but because I wish to show that the occasion is a good one for putting the power of the logic of relatives to a test.

C. MULTITUDE (25)

Art. 34. The theorem that there are indefinitely divident relations between the units of any denumerable collection is so important that, although what I have already said makes it almost evident, I will give a formal demonstration of it. It is a somewhat difficult proposition to demonstrate because it is an assertion of logical possibility, which as I have already remarked, is always difficult to prove unless by showing how it is possible. Now the relation between the one-to-one relation of the primal arrangement of the denumerable collection and any indefinitely divident relation is so extremely complicated that a rigid demonstration by showing how the latter relation is possible is extremely complicated. It is best to take a middle course by showing in a general way how the divident relation can be constructed, without showing precisely how. This method often affords convenient demonstrations of such theorems.

We often speak of the *permutations* of enumerable collections. Each such permutation represents a distinct relation between the units of the collection which has all the properties of the indefinitely divident collection except one. For a permutation presents the collection arranged in a horizontal row. Now if *p* is the relation of "being in the permutation further to the right than," then calling the units of the collection the *L*s, no *L* is *p* of itself (that is no unit is in the permutation further to the right than itself); but of any two different *L*s one is *p* of the other (that is one is in the permutation further to the right than the other); and if one *L* is *p* to another, the former is *p* of everything *p*'d by the other (that is, if one is in the permutation further to the right than the other it is further to the right than anything that is further to the left than the other).

The first step in my demonstration consists in a formal proof that between the units of any enumerable collection whatever, say the *L*s, there are such permutational relations. The only object of giving a formal proof of a proposition like this which everybody knows perfectly well to be true is that it enables us to see just how one proposition depends upon another. If the multitude of the *L*s is λ, I say that the multitude of distinct permutational relations among the *L*s,—distinct in the sense that taking any two of them p_1 and p_2 some pair of *L*s, L_1 and L_2, can be found such that L_1 is p_1 to L_2 but is not p_2 to L_2 or *vice versa.* In order to show that $\lambda!$ distinct permutational relations exist among the *L*s, I will first describe the definitions of certain relations. I will, then, show that $\lambda!$ distinct definitions of the given description can be framed; and finally I will show that each of these has those properties which I have just stated as the essential properties of a permutational relation. I proceed, then, to describe the definitions of certain relations among the *L*s which I will call the $\overline{\varpi}$-relations of the *L*s. Everyone of these definitions refers to a one-to-one relation, *r*, which is generative of the *L*s. I have first, then, to demonstrate that there is a one-to-one relation generative of the *L*s. By this I mean that starting with a certain *L*, which we may call L_0, and passing from every *L* reached to the *r* of it, we necessarily pass to every *L*. In other words whatever character belongs to L_0 and belongs to the *r* of every *L* to which it belongs, belongs to every *L*. Such a relation will exist in the case of every enumerable collection; for if it exists for every collection of a given enumerable multitude, it exists for every collection which exceeds that by unity. For if *r* be the relation for a given collection we have only to take any unit L_1 of that collection of which L_2 is *r*, and add to the collection any new unit, *X*, and define the relation *r′* as follows: Taking any object which is *r* of another, that object is *r′* of that other, unless the former be L_2 and the latter L_1; and *X* is *r′* of L_1, which L_2 is *r′* of *X*. Then let *c* be any character which belongs to L_0 and which belongs to the *r′* of every *L* to which it belongs. Then, if that character belongs to L_1 it belongs to *X*, and if it belongs to *X* it belongs to L_2. Hence if it belongs to L_1 it belongs to L_2 the *r* of L_1. Hence if it belongs to any *L* of the smaller collection it belongs to the *r* of that *L*; for the *r* of every *L* of the smaller collection except L_1 is the same as its *r′*. Hence, by the definition of *r* it necessarily belongs to every unit of the smaller collection and therefore to L_1. But we have just seen that if it belongs to L_1 it belongs to *X*. Hence it belongs to every unit of the larger collection. That is to say, *r′* is a one-to-one relation generative of the larger collection. Thus, we have proved that if every collection of a given enumerable multitude has a one-to-one generative relation so has any collection that exceeds that multitude by unity. But we have seen it is the most fundamental of all properties of enumerable multitudes that whatever character belongs to *zero* and if it belongs to any enumerable multitude belongs to the next greater belongs to all enumerable multitude. Now this character belongs to *zero* so far as it has any meaning to predicate it of *zero*. Hence, it belongs to any enumerable collection whatever so far as it has any meaning as applied to that collection.

I may add that by the syllogism of transposed quantity if there is in any case any L at all that is not r'd by any L, there is, at most, but one. For since nothing belongs to the collection of Ls which does not belong by a logical necessity owing to the facts that L_0 belongs to the collection and that every r of anything that belongs to the collection likewise belongs to the collection, it follows that every L is r to an L, unless L_0 be an exception. If L_0 is not an exception, then every L is r to an L; but (r being a one-to-one relation) no two different Ls are r of the same L. Hence, by the syllogism of transposed quantity, every L is r'd by an L. But if L_0 is an exception, we have only to consider the relation r' defined as follows: 1st if one L is r to another L, the former is also r' to the latter, 2nd, a certain one of the Ls not r'd by any L, if there be any L not r'd by any thing is r''d by L_0. If every L is r'd by an L, then since r is a one-to-one relation, by the syllogism of transposed quantity every L is r to an L without exception. Hence if L_0 is by exception not r to any L, it follows that there must be some L which is not r'd by any L. Consequently, every L is r' of an L. For if it is not L_0 it is r to an L and the relation r is one mode of the relation r'; while if it is L_0 then by the definition of r' it is r' to some single one of the Ls that are not r'd by any L if there be a single such L. But we have just seen that such there is. Every L_0 then is r' to an L. But no two different Ls are r' to the same L; for if neither is L_0 r' is the same as r, and r is a one-to-one relation, and if either is L_0, it is r' only to an L which is not r'd by any L. Hence, both premises of the syllogism of transposed quantity hold. Every L is r' to an L and no two different Ls are r' to the same L. Hence, the conclusion follows that every L is r''d by an L. But were there more than one L that was not r'd by any L, according to the definition of r' only one of these would be r''d by any L. Hence, it is absurd to suppose that there are two different Ls that are not r'd by any Ls. I have thus proved that if there is any L at all that is not r'd by an L, there is, at most, but one of which that is true.

I now proceed to describe the definitions of the $\overline{\varpi}$-relations. The first clause of the definition of any $\overline{\varpi}$-relation, which we may call $\overline{\varpi}$, asserts that nothing whatever is $\overline{\varpi}$ to itself. The second clause asserts that a certain designate unit of the collection of Ls is not $\overline{\varpi}$ to any other L; and this L which we may call the Π_0 of the relation $\overline{\varpi}$, will be in different

$\overline{\omega}$-relation to each one of the *L*s. Since, then, the multitude of *L*s is λ, the $\overline{\omega}$-relations will consist of a collection of λ different classes of $\overline{\omega}$-relations, the $\overline{\omega}$-relations of any one class having the same *L* for their Π_0s. The following clauses of the definition of $\overline{\omega}$ assert each that a certain designate unit of the collection of *L*s, which in different $\overline{\omega}$-relations will be each one of those units which it can without contradiction be, is $\overline{\omega}$ to that unit which was designated in the clause last preceding and is also $\overline{\omega}$ to everything that is $\overline{\omega}$'d by that unit. Thus, each of these clauses subdivides every one of the classes of $\overline{\omega}$-relations already obtained into classes each of which is distinguished from the other sub-classes of the same class by the definitions of all the $\overline{\omega}$-relations in it designating the same individual in this clause. In none of these clauses of the definition of $\overline{\omega}$ can the individual *L* that is designated be the same as the individual designated in any preceding clause. For any collection of clauses is an enumerable collection of clauses since it is formed by the successive addition of clause to clause. Now [a] clause asserts that the individual *L* that it designates is $\overline{\omega}$ to the *L* designated in the preceding clause and to every *L* $\overline{\omega}$'d by that *L*. Hence, it follows that every *L* designated in any clause is $\overline{\omega}$ of every *L* designated in a preceding clause. But by the first clause nothing is $\overline{\omega}$ to itself. Consequently, the same individual *L* cannot be designated in two different clauses without contradiction. But I have expressly stipulated that no *L* shall be designated which cannot be designated without contradiction; and therefore the same *L* is not designated in two different clauses. These clauses succeed one another without cessation as there is any individual which a new clause of the same description could designate without contradiction. Finally, a concluding clause asserts that no *L* is $\overline{\omega}$ to any *L* to which it is not necessitated to be $\overline{\omega}$ by the clauses of the definition that have gone before.

It is very easy to see that $\lambda!$ is the multitude of the $\overline{\omega}$-relations referring to an enumerable collection of units of multitude λ. For $\lambda!$ is that multitude which is the product of a collection of multitudes which is defined as follows: 1st, One of this collection of multitudes is λ, 2nd, Given any multitude of the collection, the multitude which is less than that by unity is a multitude of the collection, provided it exceeds zero, 3rd, No mul titude belongs to the collection which is not logically necessitated to belong to it by the first two clauses of the definition. But this is precisely the multitude of the $\overline{\omega}$-relations. For the multitude of sub-classes into which each clause of the different definitions subdivides the different classes that result from previous clauses is the same for each class. For it is equal to the multitude of the units that can be designated without contradiction, and that multitude is one less than the multitude of those classes which the next preceding clause produced. Since then the multitude of clauses is enumerable and the first one is the multitude of one collection they are for any one clause equal, and are equal to the multitude of units in an entirely different equal collection. Therefore, the whole multitude of the $\overline{\omega}$-relations equals the multitude of different sets of units which could be formed by taking one unit out of each one of a collection of collections of units of which collections of units each was less by unity than the collection immediately preceding it, and of which the largest collection was λ and the whole collection of collections was as great as it could be consistently with the condition that every collection of units must contain some unit. Now this, by the definition of multiplication is the same as $\lambda!$.

It now remains to show that each $\overline{\omega}$-relation has the properties of a permutational relation. That is to say, I have to prove, 1st, that no L is $\overline{\omega}$ of itself; 2nd, that of any two Ls one is $\overline{\omega}$ of the other; and 3rd that if one L is $\overline{\omega}$ to another, the former is $\overline{\omega}$ of everything that is $\overline{\omega}$'d by that other. The first of these propositions is identical with the first clause of the definition of $\overline{\omega}$. The succession of clauses of the definition of $\overline{\omega}$ does not cease until every individual L has been designated; and since the collection of Ls is enumerable the succession of clauses will be so completed. Hence every L is individually designated in the definition of $\overline{\omega}$. Hence, every two Ls are individually designated. But no one clause designates more than one L. Hence, the two are designated in different clauses. The different clauses follow one another one by one. Hence, of any two Ls one is designated after the other in the definition of $\overline{\omega}$. But I have already proved that of two Ls one designated after the other in the definition of $\overline{\omega}$ the former is $\overline{\omega}$ of the latter. Hence, of any two different Ls one is $\overline{\omega}$ of the other, which is the second essential property of a permutational relation. Finally wherever in the definition of $\overline{\omega}$ one L is said to be $\overline{\omega}$ of another (unless this latter be Π_0 by which according to the definition nothing is $\overline{\omega}$'d) it is also stated that that L which is $\overline{\omega}$ of another is also $\overline{\omega}$ of everything that is $\overline{\omega}$'d by that other. Consequently if one L is $\overline{\omega}$ of another L the former is also $\overline{\omega}$ of whatever there may be which is $\overline{\omega}$'d by the latter. This is the third and last essential property of a permutational relation.

I have thus proved that between the units of any enumerable collection of multitude λ there is a multitude $\lambda!$ of permutational relations. This is the first step toward proving that between the units of any denumerable collection whatever there is a vast multitude of indefinitely divident relations.

It will be convenient here to give two theorems relating to the permutational relations of enumerable collections. Of course, nothing prevents the existence of permutational relations among the units of a denumerable collection; but of them these theorems are not true.

One of these theorems is that if the *L*s form an enumerable collection and *p* is any permutational relation among the *L*s, then there is some *L* which is not *p*'d by any *L* The definition of a permutational relation is as follows: 1st, No *L* is *p* of itself; 2nd, Of two *L*s one is *p* of the other; and 3rd, If one *L* is *p* of another, it is also *p* of everything *p*'d by that other. Now were every *L p*'d by an *L* you might choose any *L* and one of its *p*s might be chosen from among the rest; and of any *L* that was chosen one of its *p*s might be chosen. Then, the relation of being "the immediately chosen *p* of" would be a one-to-one relation; for only one *L* would be chosen as the immediate successor of any given *L* and no *L* would be chosen as the immediate successor of two given *L*s. Thus, by the definition of a denumerable collection, a denumerable collection would be chosen from among the units of an enumerable collection. But it is impossible for a denumerable collection to be a part of an enumerable collection; so that it is absurd to suppose that every *L* is *p*'d by an *L*. As a corollary from this theorem, there is also an *L* which is not *p* to any *L*. For the definition of a permutational relation does not change its meaning if "*p*'d by " is interchanged with "*p* to" throughout.

The other theorem is that if the *L*s form an enumerable collection, and *p* is any permutational relation among the *p*s, then taking any *L*, say L_1, which is *p*'d by an *L*, another *L* say L_2 can be found which is *p* of L_1 but is not *p* of any third *L* that is *p* of L_1. This can be proved by the same method as the former theorem. For were it not true, there would be an *L*, say L_1, such that although it is *p*'d by another *L*, yet taking any *L*, say L_3, which is *p* of L_1, no matter what L_3 may be so long as it is of this description, L_3 is also *p* of some *L*, say L_2, which is *p* of L_1. Then the relation of being the "immediately chosen *L* that is *p* of L_1 and is *p*'d by" would be a one-to-one relation, say *p*′, such that L_2 would be an *L*, and there would be a *p*′ to L_2 that would be an *L*, and there would be an *L*, *p*′ to any given *L* that was *p*′ to an *L*. Consequently, there would, by definition, be a denumerable collection existing as a part of an enumerable collection. It is, therefore, absurd to suppose that any *L*, say L_1 which has a

p at all has not an *immediate* p, that is an L that is p of L_1 without being p of any p of L_1, so long as the Ls form an enumerable collection.

We have already seen that the aggregate of any enumerable collection with an enumerable collection is an enumerable collection. From this, it immediately follows that if of any denumerable collection, A, any enumerable collection, B, be a part, the complementary part, C, of A is denumerable.

Now consider any denumerable collection, the As. I propose to define though not very precisely a certain relation, d, among the As. For this purpose, let the Ls be taken from among As, and let them form an enumerable collection among the As. Except that the Ls must form an enumerable part of the collection of As, they may be anything whatever. The relation, d, is to be such that it is permutational with reference to the Ls. That is, no L is to be d to itself, of any two Ls one is to be d to the other, and if any L is d to an L it is to be d to every L d'd by that L. Since the remainder of the As which are not Ls form a denumerable collection, we may suppose Ms to form a collection a part of the collection of As that are not Ls, and to be equal to the collection of Ls. And let q be any permutational relation among the Ms. Then, since the Ls and the Ms form two equal collections, we may define the relation c as follows: 1st, that M which is not q of any M shall be c of that L which is not d of any L; 2nd, if any M, as M_i, which is q'd by an M is c to any L as L_j which is d'd by an L, then the immediate q of M_i shall be c to the immediate d of L_j; 3rd, no M which is not necessitated by the foregoing statements to be c to an L shall be c to that L. Since, then, the two collections of the Ls and the Ms are equal and enumerable, it follows that every M will be c of an L and every L will be c'd by an M. For if not every M is c to an L, the only circumstance which, according to the definition of c can prevent it is that some M which is q'd by an M is c to that L which is not d'd by any L. For if every M which is q'd by an M is C to an L that is d'd by an L, then that M which is not q of any M is c to an L, and if any M is c to an L (since it is in every case c to an L that is d'd by an L), then that M that is the immediate q of that M is c to an L, and so, by the Fermatian inference, every M is c to an L. But if some M is c to that L which is not d'd by any L, then every L is c'd by an M. For by the third clause of the definition of c, no L is c'd by an M unless it is necessitated to be so by the foregoing conditions. Now the only circumstance which can necessitate any L except the one which is d to no L to be c'd by an M is that the L of which it is d is c'd by an M that is q'd by an M.

Hence, if that L which is d'd by no L is c'd by an M, then taking any L whatever except that which is d to no L, if it is c'd by an M, then the L which is d'd by it is c'd by an L; and consequently, by the Fermatian inference, every L is c'd by an M. Thus, if there were an M that was not c to any L, every L would be c'd by an M. Now since the two collections, the Ls and the Ms, are equal, there is some one-to-one relation, which we may call c', in which every M stands to an L. Let us, then, define a new relation, r, as follows: 1st, if any M is c' to an L, and that L is c'd by an M, then the former M is r to the latter; 2nd, if any M is not necessitated by this condition to be an r to an M it is not r to that M. Then, it is evident that every M is r to an M; for every M is c' to an L, and every L is c'd by an M, so that the former M (which may be any M) is r to the latter. It is also evident that no two Ms are r to the same M; for both c' and c are one-to-one relations. Hence, since every M is r to an M, and no two Ms are r to the same M, and the Ms form an enumerable collection, it follows that every M is r'd by an M. Thence, by the definition of r, every M is c to an L. And in precisely the same manner it could be shown that every L is c'd by an M.

So far we have only described the effect of the definition of d insofar as it is a relation between two Ls. I will now describe the effect [in] a new article of the definition of d. That is to say, supposing that the effect of the definition is that insofar as d is a relation between the units of any enumerable collection, the Ls, it is a permutational relation, then the effect of this new article is that insofar as it concerns the units of another enumerable collection, the L's which is the aggregate 1st of the Ls, 2nd, of the units of a wholly distinct collection, the Ms, which is a part of the denumerable collection of the As, and is equal to the collection of the Ls, and 3rd, of one other unit of the As, which is neither an L nor an M, d shall be subject to the following conditions, in the statement of which I introduce a relation, c, which will be defined below. 1st, every M which is c to an L is d'd by that L; 2nd, every L, say L_1, which is d'd by an L, say L_2, but is not d'd by any third L which is d'd by L_2, is d'd by that M which is c to L_2; 3rd, that L which is not d'd by any L is d'd by some one A which is neither an L nor an M; 4th, if any L' (that is, an L, an M, or the A just mentioned which is neither an L nor an M) is d to an L', it is d to every L' that is d'd by that L'; 5th, if an L' is not necessitated by the foregoing conditions to be d to an L', then it is not d to that L'. The relation c here mentioned is such that q being any definite permutational relation among the Ms, 1st, that M which is not q of any M is c of that L which is not d

of any L; 2nd, if any M, say M_i, that is q'd by an M, is c to an L, say L_j, that is d'd by an L, then the immediate q of M_i is c to that L which is d of L_j but which is not d of any third L that is d of L_j; and 3rd, no M not so necessitated to be c to an L is c to that L.

I am now going to show that d is not only, as we assume that it is, a permutational relation among the Ls, but also that it is a permutational relation among all the L's. That it has the third essential property of a permutational relation among the L's is stated in the fourth clause of the new article of the definition, namely, that if any L' is d to an L' it is also d to every L' that is d'd by that L'. I will next show that it has the second property, namely that of any two different L's one is d to the other. If both are Ls, this is already assumed. If one of the two L's is an M, then it is d'd by an L, by the first clause of the new article of the definition. If one of the two L's is that L' which is neither an L nor an M, then by the third clause of the new article, it is d of an L. If one of the two L's is any M except that M which is not q to any M, it is d of an L. Namely, it is c of an L which is d of a second L without being d of any third L that is d of that second L; and the M is, by the second clause of the new article, d of that second L. Hence, of any two different L's neither of which is an L, one is d of an L, which we may call L_a, while the other is d'd by an L which we may call L_b; and I will show that in every case L_a is either identical with L_b or is d of it. If one of the L's which is not L is not M, it is the only one which is neither L nor M and by the third clause of the new article is d to that L which is d of all other Ls. Hence, by the fourth clause of the new article it is d of all those L's and is d to every L' except itself. If one of the L's which is not L is that M which is not q of any M, then by the first clause of the new article it is d'd by that L which is d'd by every L except itself, and consequently the M is d'd by all those Ls and by every L' except itself. If neither of the two L's which are not L is either not M or is not q to any M, then both of the two L's are at once d of an L and d'd by an L; and the L by which either of them is d'd is d of the L of which it is d. Hence, by the fourth clause, one of them is d of the other. Thus, if both or neither of the two L's is L, one is d of the other. If one of the two L's is L and the other not L, the latter is either d of every L' or is d'd by every L' or is d'd by one L and d of another L, the former being d of the latter but not being d of any third L which is d of that latter. Consequently, the former is d of every L of which the latter is d while the latter is d'd by every L by which the former is d'd. Hence, the Ls can in reference to any L' not an L (but which is both an M and is q of an M) be divided into two classes those which are *d* of that latter L' and those which are *d*'d by that L'. Thus, in every case, of any two L's one is *d* of the other. It only remains to show that no L' is *d* of itself.

It appears then that whatever denumerable collection the *A*s may be and whatever enumerable collection forming a part of it the *L*s may be, there is a relation, *d*, which is, in the first place, permutational with reference to the L's, and in the second place, is such that if it is permutational with reference to the units of any enumerable collection that forms a part of the *A*s is so also with reference to the units of any enumerable collection aggregated of that and any other collection forming part of the *A*s and one greater than that. From those premises it follows that *d* is permutational with reference to all the *A*s.

But it is easily shown that it is not only permutational with reference to the *A*s but also that it is indefinitely divident. That is to say, taking any two *A*s whatever, A_1 and A_2, if A_2 is *d* to A_1, there is an *A*, say A_3, such that A_2 is *d* to A_3, and A_3 is *d* to A_1. For whatever *A*s A_1 and A_2 may be, they will, as has been shown, form part of some enumerable collection with reference to the units of which *d* is permutational. Then, calling that collection the *L*s, there will be among the *A*s a collection [of] *M*s equal [to] that of the *L*s, such that whatever two *L*s be taken, one of which, say L_2 is *d* to the other, say L_1, there will be an *M*, such that L_2 is *d* to that *M* while that *M* is *d* to L_1.

This, then, completes the proof that there is a vast collection of indefinitely divident relations between the units of any denumerable collection.

D. CONSIDERATIONS CONCERNING THE DOCTRINE OF MULTITUDE (27)

The doctrine of multitude (*Mächtigkeiten*) is not a theory of pure mathematics, nor is it remarkably interesting from a strictly mathematical point of view. But is is a subject in which mathematicians naturally take a good deal of interest. For that reason, I venture to offer some remarks upon it to the Mathematical Society. Some years ago I proposed to send something of the sort to the *American Journal of Mathematics*. Thereupon, the then editor replied that he would insert it if I would say that it was mathematics. I felt the justice of this, and being unable to say so, the matter remained unwritten. But of late several members of the Mathematical Society have kindly expressed a desire that I should submit my ideas concerning multitude to the Society, and I am thus emboldened to do so. It would, however, be very unwise in me to attempt to give my matter a more mathematical form than it has in my own mind. I, of course, depend much upon Cantor, although my own habits of thinking about multitudes were somewhat fixed before I ever made my first acquaintance with Cantor's work in Vol. II of the *Acta Mathematica*, or had so much as heard of Bolzano's celebrated definition. By the time Whitehead's and other works had appeared, I was so engaged in the struggle with my own conceptions that I have preferred to postpone reading those works until my own ideas were in a more satisfactory condition, so that I do not know in how much of what I have to say I may have been anticipated.

It appears to me that the great difficulty under which the doctrine of multitude labors is that its implicit hypotheses have never been explicitly stated; and there is great logical difficulty in stating them. Thus, Cantor has never defined a collection (*Menge*). I know that he thinks he has done so. But he has used a method (a favorite one with Kant and later German philosophers) which, if I am not mistaken, produces a proposition that ought not to be called a definition at all, and which, at any rate, is not a mathematical definition. Namely, the method consists in directing the reader to perform an experiment which is described (it is usually a psychological experiment), and he is informed that the result of that experiment will be the *definitum.* This might be a useful way of describing yttrium or any rare element, though no chemist would call it a definition; because the person who should perform the experiment would obtain the metal in a state fit for further investigation of it. But in case such investigation is to be performed by mathematical reasoning, for which explicit hypotheses are indispensable, such a proceeding is of no use at all. A *definition* ought to state distinctly of what subjects or kinds of subjects the *definitum* can, and of what it cannot, be predicated, and what predicates are true of it, what false. A mathematical definition ought to state explicitly what mathematical hypotheses, applicable as premisses of mathematical reasoning, the predication of the *definitum* of any subject, of the assertion of any predicate of it, to be understood as involving. This Cantor's supposed definition of a *Menge* so utterly fails to do, that he himself does not seem to think it of any mathematical use.

I have seldom met with a conception more difficult of logical analysis than that of a collection in the sense in which the word is wanted in the doctrine of multitude. In a more general sense, a collection is simply an individual object whose being consists in the being of whatever objects there may be of a certain general description, these objects being called its *members*, so that every proposition concerning the collection as subject is equivalent to some relative proposition concerning the members as subjects. But this definition leaves it undetermined what the relation is to be between what is predicated, in any proposition, concerning the collection and the relative predicate concerning the members in the interpreting proposition. If we use the word *plural* to signify a collection of which any predicate is or is not true according as it is or is not true of every object of a certain general character, every such object, and none other, being a *member* of the plural, then it is obvious that a collection in the sense of the doctrine of multitude is not a mere plural, since what we principally predicate of collections in this doctrine is that they have this or that multitude, which is not generally true of the members separately.

Thus the principal question, in finding the definition of a collection is, what is the relative predication concerning the members of the collection which is equivalent to saying that a collection has a certain multitude. We may say, vaguely, that to say that a collection has a given multitude is to affirm that certain schemes of otherness subsist among members of it while other schemes of otherness are denied of all its members. If among the members of one collection there is a complication of otherness that is not to be found among the members of another collection, — as, for example there are among the members of a quartet othernesses corresponding to the edges joining summits of a tetrahedron, while no such scheme is to be found between the members of a triplet, — the collection among the members of which there is a scheme of otherness wanting among the members of another collection may be said to be *more*, or *more multitudinous*, than the latter collection.

But what precisely do we mean by a scheme of otherness? We cannot answer that question without further study. In the collections of the experiences we have had, the scheme of otherness simply consists in an otherness between any two members. This is a somewhat vague statement; and we must hesitate to say that it is adequate to everything that we should call a collection. Further examination of the matter is necessary. Nevertheless, there is one thing that we may say with considerable confidence. There is an important logical distinction between a dyadic relation and a genuine triadic relation. A dyadic relation is a general character of *dyads*, or ordered pairs. A triadic relation is a general relation of *triads*, or ordered triplets. A genuine dyadic relation is one which does not consist in the two members of any dyad between which it subsists having certain qualities. A genuine triadic relation is one which neither consists of dyadic relations nor of monadic characters. Now there is no such thing as a genuine tetradic or *n*-adic relation, where *n* is any finite ordinal number. It must be finite or the relation would be inconceivable to us, unless in some way it were expressible by a finite number of relations of a finite order. . . .

E. ROUGH SKETCH OF SUGGESTED PROLEGOMENA TO YOUR [JAMES MILLS PEIRCE'S] FIRST COURSE IN QUATERNIONS (87)[1]

Mathematics is the business of the mathematician. The mathematician is a person to whom are presented descriptions of real or imaginary states of things, and he is expected to say whether certain propositions will be true in such states of things and to answer other questions concerning them. The description presented is generally confused; and the first part of the mathematician's task is to substitute for that a series of propositions, which he calls his *hypotheses*, which are perfectly definite so far as they need to be in order to answer the question proposed or to ascertain that the description affords no certain answer to it. The hypotheses are in all cases intricate; for were they not so, the problem could be solved without any professional aid. The second part of the mathematician's task is to find a point of view for considering the hypotheses which renders them easily comprehensible. The third part of his task is to employ that mode of viewing the matter so as to solve the problem. But if the point of view solves one problem, it is capable of solving innumerable others; and the development of the methods of

[1] In an undated letter to his brother which Professor Fisch dates as December 1877 or January 1878, Charles writes "I am glad to hear you are making an Introduction to Quaternions." At a much later date (9 Oct. 1901) Jem was to write to Charles: "You know probably that I have from time to time for some years, busied myself in laying the foundations of a Treatise on Quaternions, more satisfactory on the theoretical side than Tait, more limited in scope and more practical than Hamilton. I have lately been working over again the Introductory Chapter, to which, however, I wish to prefix a biographical note on Hamilton and his work. I now send you a typewritten copy of the chapter, in the shape I have now given it. Will you take the trouble to read it and let me know whether you think it worth while for me to go on with the work. I should be glad to finish before I die something that will be a real contribution to Quaternions and will promote its continued and more advanced study. But a treatise is a big job, and though I have a good many materials, the actual licking into shape is sure to lead to much writing and rewriting. It seems to me that what I have here written is both comprehensible and important. Perhaps it is too much attenuated, and perhaps there is too much repetition. Do you think so? It is a book for beginners, not for past masters. I shall be glad of your criticism of whatever kind."

treating a broad class of problems from one general point of view is a *mathematical theory.*

Quaternions is a particular theory of tridimensional space. The description presented to it is that there is such space. But this is confused in the extreme. The questions put will relate to points, lines, surfaces, and solids; as well as to coincidences, curvatures, distances, angles, etc. etc. The hypotheses must define, in respect to their logical form, all classes of relations between these that are impossible. If this is thoroughly accomplished, any relation which does not conflict with some explicit hypotheses is *possible*, and space being regarded as *room* or a mere field of possibility, whatever relation is possible in space exists in space in the only sense logically pertinent. It is usual to add hypotheses defining, in respect to their logical form, all relations between spatial objects which are at once possible and pertinent. This is needless, as far as the logical sequence of conclusion upon premiss goes; but it is an aid in forming a diagram, and all mathematical reasoning relates to some schema of the nature of a diagram, that is, a sign having parts related similarly to the objects denoted, and having letters or other indices to distinguish those parts.

At the outset of our endeavor to draw up geometrical hypotheses, we remark that every *body* has a relation of *occupation* to a determination of space called a *part* of it. And there is a complementary part of space that is *unoccupied*, these two parts having no part in common which any body could occupy. But the occupied and unoccupied parts of space have a part of space in common called their *common surface*. We endeavor to state this more exactly.

1. Every *body precisely occupies* either the whole of space or such a part of space that this body together with other bodies could precisely occupy the whole of space. A part of space that can be precisely occupied by a body is called a *solid*. A part of space of which a part or the whole can be precisely occupied by a body is called a *solid figure*. A solid which is such that the bodies precisely occupying "the rest of space," so as to leave room for no other body, would be n in number is said to have a *periphraxis* of $n-1$. But this definition will be replaced by another below. For according to this, there is no information in saying what the periphraxis of all space is, since it is necessarily -1. Now it is important that periphraxis should be so defined that it may convey information about space to say what its periphraxis is.

2. Given a solid of periphraxis zero, a body precisely occupying it and another body could be such as to leave no room for any third body. Then

those solids have a part in common, which could not be precisely occupied by a body, called *the entire surface of* the body of periphraxis 0. There is a sort of thing called an *image* which *precisely occupies* an entire surface or a part of such a surface of a like nature with an entire surface, called generally a *surface.*

3. Two solids of periphraxis zero do not necessarily have any surface in common. If they together precisely make up a solid of periphraxis zero, the triplet of solids consisting of these two and "all the rest of space" form three pairs, the members of each pair having in common a surface not an entire surface in common. But any two of these three surfaces make up an entire surface. No two have any surface in common, but all three have a part in common called a *non-singular line.* A part of space of the same general kind is called a *line.* There is a class of things called *stresses*, such that every stress precisely occupies a line, and every line may be precisely occupied by a stress.

4. If three solids of periphraxis zero are such that the three pairs of them as well as all three make up solids of periphraxis zero, then if all three have a non-singular line, and only one, in common, this with the three non-singular lines which pairs of them have in common with the rest of space, can at most have two parts in common, each of which is indivisible and is called a *point.* There is a class of things called *particles*, each of which precisely occupies a point, and every point may be precisely occupied by any particle.

This mode of analysis of the conceptions of solid, surface, line, and point, which is substantially given by Plato, is imperfect, in that it affords no satisfactory definition of a surface or line in general. The alternative method is to begin with the point. But in order to pursue this method, it is convenient and, indeed, almost necessary to begin with the analysis of relations of time. This method will now be pursued.

Time is a certain general respect relative to different determinations of which states of things otherwise impossible may be realized. Namely, if P and Q are two logically possible states of things (abstraction being made of time), but are logically incompossible, they may be realized in respect to different determinations of time. To say that a state of things is realized *at* any given time means that it is realized in respect to that determination of time.

A complete, or individual, determination of time is called an *instant*. An incomplete, or general, determination of time may be considered collectively so as to form an individual object whose being consists in the possibility of all the further determinations of that general time. It differs from a collection proper in that the latter's being consists in the actual existence of individuals called its members, so that the truth of whatever is true of the collection consists in the existence of some relation between members of it. A collective time, on the contrary, consists in the *possibility* of all further determinations of the general time, and the truth of what is true of it consists in relations that would subsist between any possible states of things which should be realized "at" further determinations of the general time. (We may thus speak of *mankind*, not meaning the collection of existent men but of possible men.) Such a collective time may be termed a *time-whole*, or a time taken collectively, in opposition to the same time taken distributively. *All time* is the time-whole of Time without further determination.

It may be assumed that there are two instants called the *limits of all time*, the one being A, the *commencement* of all time and the other being Ω, the *completion* of all time. Whether there really are such instants or not we have no obvious means of knowing; nor is it easy to see what "really" in that question means. But it seems to me that if time is to be conceived as forming a collective whole, there either must be such limits or it must return into itself. This is an interesting question. At any rate, it is a help and no inconvenience for the present purpose to assume such limits. It remains to explain in what sense they are called limits. There is a certain triadic relation of *intermediacy* or *betweenness* such that if t, t', t'' are any three different instants whatsoever, there is one of them and only one which is *between* the other two. This one is never one of the limits of all time. Let t' be this one. Then it is the same to say that t' is between t and t'' or to say that it is between t'' and t. Moreover, whatever instants t and t'' may be, if t is between t'' and one of the limits of all time, then t'' is between t and the other limit of all time. Further to say that t' is between t and t'' is the same as to say that t' is between t and a limit of all time and that t'' is also between t and the same limit of all time.

A *lapse* of time is a general determination of time taken collectively, this determination having the following properties. Namely, the lapse has two limits, its commencement and its completion, which are two different instants such that the commencement of the lapse is between the completion and the commencement of all time. And if an instant has the general determination of the lapse it is between the limits of the lapse; and every instant between the limits of the lapse has the general determination of the lapse. Such an instant is said to be *in* the lapse. It is to be observed that an instant does not *exist* merely by virtue of being "in" a lapse. An instant only exists in the sense that some possible state of things is realized "at" it. If all the instants in a lapse existed they would form a collection and this would have a multitude. But a lapse, being a general determination of time, has "in" it instants exceeding all multitude, these instants not existing at all, unless the realization of a state of things constitutes the existence of some of them. They are mere possibilities and, as they are "in" the lapse, abstractedly from states of things, have no individual identity and are therefore not instants.

Every instant of time has a "*whenabouts*," which is an indefinite lapse in which it is, such that between this instant and any other instant whatever, there are instants of the whenabouts beyond all multitude. The instants of the whenabouts of any instant are said to be *near* it.

A possible state of things which is of such a nature that if it be realized it must be realized "at" some determination of time may belong to any one of the following classes: Class 1 is composed of *momentary* states of things. A momentary state of things is a possible state of things which, if it be realized at all, can only be realized "at" some instant; and if t be any instant at which it is realized and t'' be any other instant whatsoever, then there are instants beyond all multitude between t and t'' at which it is not realized. Class 2 consists of *prolonged* states of things. A prolonged state of things is a possible state of things such that if it be realized at all, it can only be realized at instants, and if t' is any instant at which it is realized, there are two instants t and t'' such that t is between them, and the prolonged state is realized at every instant between t and t'', while if t''' be any instant such that one of the instants t and t'' is between the other and t''', then between t''' and that intermediate one of t and t'' there must be two instants, t'^{v} and t^{v} such that the prolonged state is not realized at any instant between t'^{v} and t^{v}. At the limiting instants t and t'' the prolonged state may be indifferently regarded as realized or not, and may be said to be *half-realized.*

Class 3 consists of *gradual* states of things, or *changes*. A gradual state of things is a possible state of things which cannot be realized "at" any instant but may be realized at a lapse of time. This class has two subclasses consisting the one of *states of change*, or *indefinitely gradual* states of things; the other of *entire changes*, or *entire gradual* states of things. A state of change is a gradual state of things such that, if it be realized "at" a lapse whose limits are t_a and t_z and if t_b and t_y be two different instants between t_a and t_z then the state of change is realized "at" the lapse whose limits are t_a and t_z. An *entire change* is a gradual state of things such that, if it be realized at all, there is such a lapse "at" which it is realized that if t_b and t_y be instants between the limits of that lapse, or even if one of them be one of those limits, then the entire change is not realized "at" that lapse whose limits are t_b and t_y.

Class 4 comprises *relational* states of things. A relational state of things is a possible state of things which cannot be realized "at" any instant nor "at" any lapse, but may be realized at a set (or ordered collection) of collectively taken determinations of time.

We now recur to space. *Space*, like Time, is a general respect to whose determinations realizations are relative. Only, in the case of space, the realizations instead of being of states of things signified by propositions are of objects representable by terms of propositions. Namely, if a proposition be so analyzed as to throw all general characters into the predicate, —as when we express 'all men are mortal' as 'whatever exists is either not a man or is mortal,'—then, if the universe of discourse is a collection of objects of a certain kind called *things*, each individual thing denoted by a subject of the proposition (reckoning as 'subjects' not only the subject nominative but the direct, indirect, and prepositional objects) each such individual exists and has such characters as it has, relatively to some determination of space. A determination of space taken collectively is called a *place*. If of two determinations of space one involves the other then, when these are taken collectively, the place corresponding to the former is said to be *contained in* the place corresponding to the latter. If of two different places one is contained in the other, it is said to be a *part* of it. If a thing exists relatively to a determination of space and not to any further determination, it is said to *occupy* the corresponding place and to be *in* every place that contains the place it occupies. Those characters of a thing that are relative to the same determination of space as its existence is may be termed the *spatial characters* of the thing.

Any thing at any instant occupies a place, and of course, one place only; and no place can at any one instant be occupied by two different things. At different instants the same thing can occupy different places; but it can only do so under certain restrictions which I shall develop. Such change of place under those restrictions is called *motion*, if it is a state of change, and *movement* if it is an entire change. Under still further conditions it is called *mere* motion*; but if these conditions are violated it is called *disruptive motion*. (*Generally called continuous, which seems to me to be capable of being otherwise understood.)

An individual determination of space is called a *point.* A thing which at an instant occupies a point is called a *particle.* In a place occupied by a thing each point is occupied by a particle, and such particle is called a particle *of*, or *belonging to*, or a *part of*, the thing occupying the place in which the point lies.

Just as every instant has a whenabouts, so every point has a *whereabouts*, or *neighborhood*, which is an indefinite place such that no movement can take place in which the same particle at one instant occupies one point, *P*, and at another instant occupies a different point, without at each instant of the whenabouts of the instant at which the particle in motion occupies the point, *P*, its occupying points beyond all multitude in the whereabouts of that point, *P*. If one point is in the whereabouts of another, the latter is in the whereabouts of the former. In *mere movement*, two particles which at any instant occupy points in each other's whereabouts do so at every instant of the lapse of the movement. In disruptive motion this is not the case. This defines the distinction.

Any collection of points is called a *point-figure.* A place which contains whatever point is occupied by any particle of a collection of particles at any instant during a possible movement and contains no other point is called a *line-figure.* If the collection of particles is a single particle, the line-figure is a *line.* A thing which at any one instant occupies a line-figure may be termed a *stress-complex.* If the line-figure is a line the stress-complex may be termed a *stress.*

A *departing* motion (and movement) is a mere motion (or movement) of a thing such that in the whenabouts of any instant *t* in the lapse of the motion (or movement) there is no instant at which the place of the thing has more than a multitude of points in common with the place of the thing at the instant, *t*. That is to say, if there be an instant, *t'*, at which the place of the thing has a line in common with the place of the thing at *t*, then there is an intermediate instant at which the place of the thing has no line in common with the place of the thing at *t*.

A place which contains all the points occupied by particles of a stress-complex at instants in the lapse of any departing movement of the stress-complex, and contains no others, is termed a *surface-figure.* If the stress-complex is a stress, the surface-figure is called a *surface.* A thing which at one instant occupies a surface-figure may be termed a *species-figure.* If the surface figure is a surface, the species-figure may be termed a *species.*

A place which contains all the points occupied by particles of a surface-complex at instants in the lapse of any departing movement of the surface-complex, and contains no others, is termed a *solid-figure*. If the surface-complex is a surface the solid-figure is termed a *solid*. A thing occupying at one instant a solid-figure may be called a *body-complex*, or of a *solid*, a *body*.

In three-dimensional space a departing motion of a body is impossible.

If a mere movement is such that no thing which at one instant in the time of the movement belongs to one of the four classes, particles, stresses, species, and bodies, belongs at any other instant in the time of the movement to another of these classes, then, and only then, it is to be called a *strict movement*. For example, if a species occupying the surface of sphere shrinks to a particle, this is a strict movement except at the whenabouts of the last instant when it becomes a particle. Even then it does not cease to be a mere movement.

A *generation* is an action whereby a motion of a thing through a place creates a thing occupying the place moved over. Dimensionality is the number of acts of generation requisite to produce a thing occupying a place from a collection of particles. Strict movement is movement in which the moving thing retains the same generality.

An *ordinary place of a place* is a place in the latter place from which the modes of departing movement are the same as from innumerable other places in its neighborhood in the same place; wherein we reckon two movements as of the same mode if they traverse the same parts of the whereabouts of the place departed from. A *singular place of a place* is a place within the latter place from which the modes of departing movement are fewer or more than from ordinary places of the same dimensionality.

From an ordinary point in a line-figure the modes of departing movement are two, namely along the line in either sense. A singular point of a line-figure is either an *outlying point*, an *extremity*, or a *point of branching*.

From an ordinary point in a surface-figure the mode of departing movement is one, since the neighborhood of a point in a surface is not severed by the point. An *isolated singular point* of a surface figure is either an *outlying point* or a *tack*.

From an ordinary line in a surface-figure the modes of departing movement are two, on the two sides of the line.[2] A *singular line* of a surface is either an *outlying line*, an *edge*, or a *split*. The *singular points of singular lines of a surface-figure* are of three classes, first, those which are extremities or points of branching of such lines; second, those which are common to a

[2] A marginal note is as follows: "Two ways of departing from a line ."

singular line and some other part of the surface figure; third, those which are complications of tacks with other singularities. Of the first class there are *extremities* and *points of branching of outlying lines* (an edge cannot have an absolute extremity), *even way branch points of edges, extremities of even way splits, branch points of splits.* Of the second class are *points of outlying lines that are otherwise ordinary points* of the surface, *points at which an edge is continued into a split, points at which splits alter the number of their sheets, points at which three slits come together, points common to outlying lines and edges, points common to outlying lines and splits,* etc. etc.; of the third class, there are *tacks on outlying lines* (a tack on two edges is equivalent to a branch point of an edge), *a tack between an otherwise ordinary point of one sheet and an edge of another, a tack at the point where three slits come together,* etc.

From an ordinary point in a solid the mode of departing movement is one. In three dimensional space, the only kind of isolated point singularity of a solid is an outlying point.

From an ordinary line in a solid the mode of departing movement is one, and the only kind of isolated line-singularity of a solid possible in three dimensional space is an outlying line.

From an ordinary surface in a solid the modes of departing motion are two. The only surface-singularities of solids possible in three-dimensional space are outlying surfaces and external or limiting surfaces. These may have line singularities which in their turn may have point-singularities.

In order to make the subject thoroughly understood it should be carried into four-dimensional space and into peculiarly shaped non-singular 3-dimensional space. But this has not been done. Especially this should be done for the theory of knots.

The *order* of a simply singular place of a place *in quo* is the excess of the number of independent ways of departing from it in the place *in quo* over that from an ordinary place of the same dimensionality. The lowest order of a singular place is -2.

Two places may be said to be of the same *shape-class* if and only if it is possible for a thing precisely occupying the one to come by a mere movement, strict or otherwise, to precisely occupy the other. *Listing's numbers* are certain quantities measuring the degrees in which any place possesses separativeness of different kinds, where by separativeness is meant something exemplified in the action of any two-sided closed surface in separating all space into two parts, these numbers together with the singularities and the dimensionality sufficing to distinguish all *shape-classes*. For three-dimensional space there are four Listing numbers; the *cho'risy* (Gr. χώρισιζ, a separation), or punctual separativeness exemplified in any point as being distinct from any different point; the *cyclo'sy* (Gr. κύκλωσις, a surrounding), or linear separativeness, exemplified in a selfreturning line; the *periphrax'y* (Gr. περίφραξις, a fencing round), or superficial separativeness, exemplified in a closed surface; and the *apei'ry* (Gr. ἀπειρία, immensity), or solid separativeness. Listing's memoir on the subject is of high importance and of such originality that it is difficult to banish the suspicion that Listing's colleague in Göttingen, Gauss, was in some way a factor in the production of it. At the same time the treatment of the matter by Listing is not altogether satisfactory. The definitions of the Listing numbers by their author are not free from vagueness; and the evaluations of them that he gives as examples are not always unquestionable. He pays no regard to the singularities of places which are essential to their shape-classes. The census theorem which may be regarded as the purpose and end of the numbers is not stated by Listing in the most comprehensive form.

In what follows, I shall not begin by defining the four Listing numbers and go on to deduce the theorem; but I shall, on the contrary, assume the form of the theorem to be true and go back from that to determine precisely how the four quantities ought to be defined and how they may be ascertained.

The following is a correct statement of the census-theorem;

1. There are certain fixed quantities, called the constant census-values, which measure the *units* of chorisy, cyclosy, periphraxy, and apeiry of point-figures, line-figures, surface figures, and solid figures; namely let X_0, X_1, X_2, X_3 denote respectively census-value of a unit of chorisy of a point figure, line-figure, surface figure, and solid figure; K_1, K_2, K_3 denote respectively the census value of a unit of cyclosy of a line figure, surface figure, and solid figure; Π_2, Π_3 denote respectively the census-values of the unit of periphraxy of surfaces and of solids; A_3, the census value of the unit of apeiry of a solid.

There are also fixed census-values of simple singularities, as follows:

U^n_m, the census value of an isolated singular point of a place of dimensionality *m*, *n* being the order of the singularity;

V^n_m, the census-value of an isolated singular line, *m* and *n* having the same significations;

W^n, the census-value of a singular surface, n being the order of the singularity.

$^nI^l_m$, the census-value of a singular point of a singular line of a place of dimensionality m, n being the order of singularity of the line, and l that of the point;

$^nJ^l$, the census-value of an isolated singular point of a singular surface, n being the order of singularity of the surface and l that of the point;

$^nN^l$, the census-value of a singular line of a singular surface, n being the order of singularity of the surface, and l that of the line.

$^n\overset{l}{M}{}^k$, the census-value of a singular point of a singular line of a singular surface, n being the order of singularity of the surface, l that of the line, and k that of the point.

Two or more of these seven symbols may be placed together in parenthesis to denote the census-value of a singular place which unites two characters.

The census-value of a Listing number is the product of that number into the census-value of the unit.

The census-value of a place is the sum of the census-values of its Listing numbers and of its singularities.

The census-value of a place the aggregate of two places is the sum of the census-values of those places, less the census-value of their common place.

The census-value of one point is one. The census-values of two places of the same shape-class is the same.

The above is Listing's census-theorem, as modified by me, so as to take account of singularities. Let us provisionally suppose it to be true and try whether we can determine the fixed census-values and the definitions of the Listing numbers so as to preserve its truth.

We may begin with a point-figure. No mere motion can convert a particle into two particles. We must, therefore, assume that the chorisy of a point-figure is the number of its points, and that $X_0 = 1$. For in this way only (with whole numbers) will the census-figure of one point be one and that of a number of points be equal to their number.

Let us next consider a simple limited line. Since this is of the shape-class of a point (for it can shrink to a point by a mere motion, not strict at its completion), its census number is one. We, therefore, have $X_1 + 2U_1^{-1} = 1$. If the extremities of x simple lines are at one point on another simple line, the result is a line with $x + 2$ extremities and one singularity of order

x. Thus, reckoned by parts, its census-number is

$$(x+1)X_1 + 2(x+1)U_1^{-1} - xX_0$$

while, reckoned as one whole, it is

$$X_1 + (x+2)U_1^{-1} + U_1^x$$

Equating these, we get

$$U_1^x = -xU_1^{-1}$$

This formula holds for $x = -1$; so that it presumably holds for $x = -2$, that is for the outlying point. But since it is doubtful what the chorisy of a line with any outlying point is, let us assume it to be 4, and determine x so as to make the above formula hold for $y = -2$. Let a simple line pass through an outlying point of a line-figure consisting of a simple line and an outlying point. Then the census value of the result, calculated by parts, is

$$(yX_1 + 2U_1^{-1} + U_1^{-2}) + (X_1 + 2U_1^{-1}) - X_0$$

while, for the result, it is

$$2X_1 + 4U_1^{-1}$$

Equating these and assuming $U_1^{-2} = 2U_1^{-1}$, we get $y = 2$. Therefore the chorisy of a line figure is the number of separate pieces.

A non-singular line is self returning and has cyclosy 1. It may be conceived as composed of any number, l, of simple terminated lines joined at l points. The census value, calculated by parts, is

$$l(X_1 + 2U_1^{-1}) - lX_0 = 0$$

while the census value of the whole is

$$X_1 + K_1$$

Hence $K_1 = -X_1$.

The junction of two ordinary points of a line figure by a simple line may diminish the number of parts by one. If it does not do so, it will increase the number of enclosures by one. Let us suppose that in the former case, it increases the chorisy by $-x$, and in the latter case increases the cyclosy by y. Then its effect on the census-value, calculated by parts is to add to it

$$X_1 + 2U_1^{-1} - 2X_0$$

While calculated from the whole the same addition is, in one case,

$$-xX_1+2U_1^1$$

Equating, we have $xX_1=X_1$ or $x=1$.
In the other case, the effect on the whole is

$$yK_1+2U_1^1$$

and since $K_1=-X_1$, this gives $y=1$. Hence the cyclosy is the number of smallest circuits.

There is nothing whatever to determine the value of X_1, from which those of K_1, and U_1^n will follow; nor is it necessary for calculating the census-value of the line figure. We may make it zero, when $K_1=0$, and $U_1^{-1}=\frac{1}{2}$, and $U_1^n=-\frac{n}{2}$, so that the census value of the figure will be minus half the sum of the orders of the singularities. Or we may make $X_1=1$, when $U_1^n=0$ and $K_1=-1$ and the census-value will be the excess of the chorisy over the cyclosis. It follows that this is equal to minus half the sum of the orders of the singularities.

We now come to surface figures. A simple surface with one boundary has chorisy 1, and the other Listing numbers zero. Its census value is, therefore,

$$X_2+V_2^{-1}$$

But such a surface may be formed by putting together two surfaces of the same topical shape. Hence, calculated by parts, its census-value is

$$2X_2+2V_2^{-1}-(X_1+2U_1^{-1})=2X_2+2V_2^{-1}-1$$

Equating the two computations we have

$$X_2+V_2^{-1}=1. \tag{1}$$

If in such a surface h holes be made, there is a cyclosy h; so that the census value is

$$X_2+hK_2+(h+1)V_2^{-1}$$

If one of the holes is plugged by a surface, the result is that there are only $(h-1)$ holes; and the census-value is

$$X_2+(h-1)K_2+hV_2^{-1}$$

or calculated by parts

$$\begin{aligned}&2X_2+hK_2+(h+2)V_2^{-1}-(X_1+K_1)\\&=2X_2+hK_2+(h+2)V_2^{-1}\end{aligned}$$

Equating the two computations

$$X_2 + K_2 + 2V_2^{-1} = 0$$

or by (1) $K_2 + V_2^{-1} = -1$

The census value of the surface is consequently

$$1 - h.$$

A cap placed on a disk gives a closed surface. This is the type of periphraxy one. The periphraxy is in the surface itself not in the space in which it is. Therefore, it does not consist in the enclosure of space by the surface; for it would not enclose three-dimensional space if it were in a space of four dimensions. The periphraxy must consist in the incapacity of the species occupying it to shrink to a stress by a mere movement in the surface; or must consist in something like this in relating to what is possible in the surface itself. The surface has no *cyclosy* because a self returning stress in it can shrink to a particle by a mere movement in the surface. Its chorisy is one because it is all one piece,—that is, a particle occupying any point in it can by mere movement in the surface come to occupy any other point. Its census-value, according to this, is

$$X_2 + \pi_2.$$

But, calculated by parts, it is

$$2X_2 + 2V_2^{-1} - (X_1 + K_1) = 2X_2 + 2V_2^{-1}$$

Equating the computations

(2) $X_2 - \pi_2 + 2V_2^{-1} = 0$

or by (1) $\pi_2 - V_2^{-1} = 1$ or $X_2 + \pi_2 = 2$

If c additional caps are put on, all meeting closed surface in the same ring, the periphraxy gets an addition of one for each; but no cyclosy is developed. Hence the census-value is

$$X_2 + (c+1)\pi_2 + V_2^c$$

or, calculated by parts, is

$$(c+1)X_2 + \pi_2 + cV_2^{-1} - c(X_1 + K_1) = (c+1)X_2 + \pi_2 + cV_2^{-1}$$

Equating the two computations

$$c(X_2 - \pi_2 + V_2^{-1}) = V_2^c$$

or by (2) $V_2^c = -\ cV_2^{-1}$

The census-value of the surface is, therefore,

$$2+c$$

which holds for c equal to -1 or to -2. The equation for V_2^c shows that it makes no difference where the caps are put on.

We thus have the convenient expressions

$$\begin{aligned} X_2 &= 1+V_2^1 \\ K_2 &= -1+V_2^1 \\ \pi_2 &= 1-V_2^1 \\ V_2^c &= cV_2^1 \\ U_2^1 &= V_2^1 \end{aligned}$$

If two simple closed surfaces are tacked together, the census-value is

$$X_2+2\pi_2+U_2^1$$

But, calculated by parts, it is

$$-2X_2+2\pi_2-X_0=2X_2+2\pi_2-1$$

Equating the computes

$$U_2^1=X_2-1=V_2^1$$

The census-value of the surface is thus 3.

Since a terminated line has the census-value of a point, it is evident that a surface formed of two simply closed surfaces having a terminated line in common has the same census-value, 3, as if they had but a point in common. Equating, then, the census-value to 3, we have

$$X_2+2\pi_2+V_2^2+2\,{}^2I_2^{-1}=3$$

We then compute the value of ${}^2I_2^{-1}$ as follows;

$$\begin{array}{lll} {}^2I_2^{-1} = 1\tfrac{1}{2} & & = 1\tfrac{1}{2} \\ \quad -\tfrac{1}{2}X_2 & & -\tfrac{1}{2}-\tfrac{1}{2}V_2^1 \\ \quad -\pi_2 & & -1+1V_2^1 \\ \quad -\tfrac{1}{2}V_2^2 & & -1V_2^1 \\ & = & -\tfrac{1}{2}V_2^1. \end{array}$$

If we join to such a surface s additional simple closed surfaces, so that one of the first pair has a line in common with each of the additional surfaces, these lines having no point common to two except one point common to all at an ordinary point of the line joining the first pair of surfaces, nor any of the surfaces having any other place common to two of them, the census value, from characters of the whole, is

$$X_2 + (s+2)\pi_2 + V_2^2 + (s+2)^2 I_2^{-1} + {}^2I_2^s.$$

But, computed from parts, the same (remembering that each of the $s+1$ lines is on two surfaces, and that the census-value of it must therefore be once subtracted from the sum of the census-values of the surfaces, while the common-point is on $s+2$ surfaces and must be subtracted $s+1$ times, which is just the number of times it is subtracted in subtracting the $s+1$ lines) is

$$(s+2)X_2 + (s+2)\pi_2 - (s+1)(X_1 + 2U_1^{-1}) = (s+2)(X_2 + \pi_2) - (s+1)$$

$$\begin{array}{llll}
\text{Hence } {}^2I_2^s = & (s+1)X_2 & = s+1 & +(s+1)V_2^1 \\
& -V_2^2 & & -2V_2^1 \\
& -(s+2)^2 I_2^{-1} & & (+\tfrac{1}{2}s+1)V_2^1 \\
& -(s+1) & -s-1 & \\
& & = & \tfrac{3}{2}sV_2^1
\end{array}$$

The census-value of the surface is $3+s$ and it makes no difference how the lines of junction of the s surfaces are distributed among the two first surfaces.

If m simple closed surfaces are all joined together in pairs along n lines which have one extremity in common to all, and neither surfaces nor lines have any other places in common, the census-value from the description of the whole is

$$X_2 + m\pi_2 + V_2^2 + n^2 I_2^{-1} + {}^2I_2^{n-2}$$

By computation from the parts it is

$$mX_2 + m\pi_2 - n(X_1 + U_1^{-1}) + (n-m+1)X_0$$
$$= mX_2 + m\pi_2 - m + 1 = m + 1$$

F. NOTES ON THE THEORY OF MULTITUDE (s-1)

I make these notes with the aid of the Report of Arthur Schoenfliess on "Die Entwickelung der Lehre von dem Punktmannigfaltigkeiten," which composes the second Heft of Vol. VIII of the Jahresbericht der Deutschen Mathematiker-Vereinigung, published 1900 Sep. 13.

It seems that the catholic theologian Bernhard Bolzano, — a persecuted man, — in his posthumous work "Paradoxieen des Unendlichen" [1850] first showed that a part of an infinite collection may be equal to the whole.

PROPOSITIONS CONCERNING THE DENUMERABLE

I The collection [of] Rational Numbers is denumerable. Proved by Cantor Journ. F. Math. LXXXIV. I prove it by the theorem that $\frac{a+x}{b+y}$ has a value intermediate between $\frac{a}{b}$ and $\frac{x}{y}$.

II The rational numbers of any dimensional v Space form a denumerable collection

III The collection of all algebraic numbers is denumerable

IV Every collection of regions in space excluding one another is denumerable

(V I should think that all values of any definite form of function must be denumerable.)

Schoenfliess says it is not proved that two collections cannot be each greater than the other "und somit fehlt es an dieser stelle der Theorie an dem nötigen Fundament." But it is simply that this is owing to the fact that *all possible* collections cannot be otherwise defined than by assuming this.

To say that the *a*s are more multitudinous than the *b*s is to say that whatever group of pairs each of an *a* and a *b* you may take, some *a*, a_x, can be found such that whatever *b*, say b_u, be taken either this group does not contain the pair $a_x b_u$ or else it contains a pair $a_y b_u$ where a_y is not a_x.

But the former alternative may be dismissed for if the group of pairs taken does not include $a_x b_u$ there certainly is a group of pairs such that whatever *a* a_x may be there is some *b* say b_u such that the pair $a_x b_u$ is included.

Therefore we are reduced to the other alternative that whatever group of pairs you take in which every *a* has been paired with a *b*, you find that some *b* has been paired to two different *a*s.

But we can reduce the group of pairs by not admitting the same *b* into two pairs.

Then it comes to this that there are groups of pairs in which every *b* is paired with at least one *a* to which no other *b* is paired.

Here is a Discovery, though I said something like it but more vague in the Monist in 1893:

Objects that exist by Secondness are necessarily finite in multitude.

Mr. Frankland said *everything* was finite except time. But this was too vague.

Proof. For whatever collection is infinite contains whatever there can be of some general description, and thus is *by law* and not by blind secondness alone.

Moreover, what exists by secondness only exists so far as it is *done*,—so far as the act of becoming is done. Consequently, the whole collection is done.

To be sure, it may be said that it is only so far as we conceive them that this is so. But we must give "existence" some meaning involving the idea of law, that is, that *whatever* the law demands exists, in order to assert that anything not actually experienced exists. And unless this law is true of the objects in their being, it is not true. We cannot therefore mean anything by existence except that it is in the nature of the existing objects to be experienced.

The question needs to be considered at the very outset whether the individuals of any collection whatever are not necessarily in some linear relation.

Every object has an indvidual relation to every other, that is, forms with it an ordered pair.

A collection of these pairs is a relation.

If we define a *connection* so that if *A* has a given relation, ρ, to *B* then

A is connected by ρ with B and if C is connected by any relation, r, to D, and D is connected by r to E then C is connected by r with E, but nothing is connected with another that is not so necessitated to be connected, it follows that there are only a denumerable collection of grades of connection.

Here is the thing I was after!

Postulate 1. Given any general quality whatsoever, there are general qualities narrower than it.

This is an established postulate of logic.

There is therefore an endless series of general qualities narrower and narrower.

Postulate 2. This suffices to discriminate any individual from all others [and] is (or follows from) that clause of the principle of indiscernibles which has never been disputed.

[Restatement]

Postulate 1. Any general character whatever is universally predictable of some general characters more determinate than it.

This postulate, universally admitted, is what Kant (Critik d.r.V. page 656) calls the "Logische Gesetz der Specification." Though he calls it a regulative, not a constitutive, principle, this limitation only means that it cannot be used to infer the actual existence of species. Characters are pure affairs of logical possibility. Although no sound logician will call this in question, I must show that it is a necessary postulate of logic.

Postulate 2. In every possible universe of individual objects of any category every individual is distinguished from all others by general characters.

This is also generally received. It must be distinguished from two other doctrines. One is that as a matter of natural fact all individuals are unlike. The other is the doctrine of the identity of indiscernibles, that things that did not differ in any general respect would be identical. The true doctrine, stated by Cardinal Cusa is that "all things must of necessity differ from one another." (Du Docta Ignorantia, iii, 1)

1903 July 20

I now return to Schoenfliess 1st Abschnitt Chap. 5.

Cantor connects the idea of *ordinal number* with that of a collection; so that he says that a collection arranged in a particular way *has* an ordinal number. It would be better to consider the ordinal numbers simply as points of a row of points.

(p. 28) He speaks of the *Ordnungsgesetz* meaning the character of being arranged in one type. But he does not say whether two different finite numbers have the same Ordnungsgesetz or not. He says finite number has but one. An *Ordnungstypus* seems to be the same thing schematically exhibited.

1. A collection is *geordnet* or more precisely einfach geordnet in case there is a recognized transitive relation among its members.

Two collections are called *similar* (ähnlich) if they are equal [and] *geordnet*.

(p. 29) A *Typus* or Ordnungstypus is the Allgemeinbegriff (therefore *not* a schema) which is got by leaving out of view the Beschaffenheit of the elements and attending to their Rangordnung.

It would therefore seem that different finite numbers are of different types.

The Type of the positive integers is called ω

Its reverse is $^*\omega$

ν denotes a finite ordinal number.

However directly below it appears as the *type*.

(p. 30) The sum of two types $\mu + \nu$ is the type resulting from making the entire type μ *precede* (it ought to be the other way) the entire type ν

$$\lambda + \omega = \omega$$

But $\omega + \lambda$ is a different type.

The obvious law is that a type with an end preceding a type with a beginning welds with it

$^*\omega + \omega$ is a simple type

while $\omega + {}^*\omega$ is a double type

The product $\mu \cdot \nu$ of two types is obtained by substituting the type μ for every element of ν. (It ought to be the other way.)

Thus $\nu\omega = \omega$

But $\omega(\mu + \nu) = \omega\mu + \omega\nu$

(p. 31) $\omega\omega = 1_1, 1_2, 1_3, 1_4,$ etc. $2_1, 2_2, 2_3,$ etc. etc.

$\omega\omega\omega = 1_1, 1_2, 1_3,$ etc. $2_1, 2_2, 2_3,$ etc. etc.

$1'_1, 1'_2, 1'_3,$ etc. $2'_1, 2'_2, 2'_3,$ etc. etc.

$1''_1, 1''_2, 1''_3$

etc.

These are supposed to be different. I suppose they are.

In any geordnet collection any partial collection of the type ω (for the same way of ordering) is called an *ascending fundamental series* steigende Fundamentalreihe while of the type $*\omega$ it is called a *descending fundamental series* fallende. If an ascending fundamental series has a *limiting* element Grenz-element, a_ω,—that is a_a is higher in the series than every element of the ascending fundamental series but so that no element of the whole collection that is lower than the limiting element is higher than every element of the fundamental series then this limiting element of the series is called a capital element (Hauptelement) of the whole collection and so is any limiting element of a descending fundamental series.

(p. 32) A type (or a collection) is called secluded or concluded abgeschlossen if every fundamental series in it has its limit within it.

It is called *close* (in sich dicht) if every element is a capital element.

It is called *perfect* (perfect) if it is both secluded and close. (That is, if you take away an element from a perfect type it will be *close* just the same but will cease to be *concluded.* But if you add an element it will still be *concluded* but will cease to be *close.* Hence we may translate *in sich dicht* by *dense* or *condensed* or *concise but as a whole,* and *abgeschlossen* by *complete.*)

An interval (Intervall) is a part of a type including two elements and all the intervening elements. Those two are the *terminal elements* Endelemente while any between them is an *interior inneres* element.

If a type is everywhere überall condensed, that is, condensed in every interval, it is also *condensed as a whole.* But it may be condensed as a whole without being everywhere condensed, that is an interval may have terminal elements which are not limits of any series in that interval. On the other hand it may be not condensed as a whole without being nowhere condensed, but if it is nowhere condensed, it is not condensed as a whole. If it is complete as a whole it is everywhere complete and conversely if it is not complete as a whole it need not be nowhere complete although if it is nowhere complete it must be incomplete as a whole.

Chap. 6

The theory of wohlgeordnet collections was given by Cantor in 1897 Math Ann. xlix 207. But a good deal had been given by him before. The foundations for it were laid in 1882. (My paper was 1881.)

Cantor's "transfinite" ordinal numbers correspond to the following

2^0 2^1 2^2 2^3 2^4 . . . . and so on indefinitely

His numbers 1 2 3 4 5 . . . . .

$2^0\cdot 3 \quad 2^1\cdot 3 \quad 2^2\cdot 3$ . . . . and so on indefinitely

$\omega \quad \omega+1 \quad \omega+2$ . . . .

$2^0\cdot 3^2 \quad 2^1 3^2 \quad 2^2 3^2$ . . . . and so on indefinitely

$\omega\cdot 2 \quad \omega\cdot 2+1$ . . . .

$2^0\cdot 3^3$

ω^3

$2^0\cdot 3^4$

ω^4

and so on indefinitely

$2^0\cdot 5 \quad 2^1\cdot 5$

$\omega^2 \quad \omega^2+1$

$2^0\cdot 3\cdot 5 \quad 2^1\cdot 3\cdot 5$

$\omega^2+\omega \quad \omega^2+\omega+1$

And so he has a number corresponding to every integer.

Consequently there is a denumeral collection of such numbers. I do not know whether he has ω^ω which would in the above system correspond to the denumerable collection. If he had done so he would have had sufficient without resorting to $\omega^\omega\cdot 2$ to give a number to every multitude.

The best I can do is this:

If any thing is above anything

Then whatever character q may be

There is something having the character q that is not above anything that has the character q.

The character μ is such that whatever character q may be there is something, i, such that if anything, j, possesses the characters μ and q then i possesses those characters and whatever k may be either k does not possess μ or j is above k or i is not above k.

(p. 36) Three corollaries from the def.

1. Every part of a Cantorian collection is Cantorian.
2. Every element is either the last or has a next following.
3. There is no series of elements (adjacent or not) which has not a first.

A *segment* "Abschnitt" is a part of a Cantorized collection which consists of all the elements preceding a given element.

(p. 37) After taking away a section, what remains is the *balance* (rest).

1. A section of a section is a section.

2. Two sections cannot each be greater [than] the other.
3. The sections and the elements of a Cantorized collection are in one-to-one correspondence. The last sections to the last elements.
4. Every part of the collection of sections has a smallest.
5. The necessary [and] sufficient condition of two Cantorized collections being similar in that to every section of either corresponds a section of the other.

Two elements are called *near* (benachbart) to one another or *remote* (getrennt) from one another according as there are a finite or infinite number of elements between them.

(p. 39) I. Two Cantorian collections are either similar or one is a segment of the other.

(p. 40) II. Any part of a Cantorian collection is similar either to that collection or to a segment of it.

III. No two ordinal numbers can be each greater than the other.

(p. 41) IV. No collection of ordinal numbers is of the type $^*\omega$. On the contrary it has a smallest member.

V. Every collection of ordinal numbers is a Cantorian collection in respect to magnitude.

VI. The sum or product of two ordinal numbers is itself an ordinal number. In the case of the sum even an infinitely continued sum is so.

VII. If any Cantorian collection be substituted for every member of a Cantorian collection the result will be a Cantorian collection.

(p. 42) VIII. Every increasing series of ordinal numbers of any possible type has a next greater number.

(p. 43) Such a number is called by Cantor a *limes*-number. In my opinion a *limen* number would be more appropriate.

It is to be observed that the only limes-numbers are those whose regular expression ends in ω or $\omega \cdot n$

They are denoted by Q_ω

(p. 44) *Chap. 7*

He calls the denumeral the *first* multitude.

He calls that segment of the whole Cantorian body of ordinal numbers which first transcends the finite,—what I call

$$2^0\ 2^1\ 2^2\ 2^3\ 2^4\ \ldots.$$

the *first class* of ordinal numbers.

They are the finite ordinal numbers.

The *second class* are all those which begin with ω and are below ω^{ω}— corresponding in my arrangement to finite integers.

The numbers of the first class are derived by a single *principle of production*, which he calls the *first*. This principle is that there is a number next following any given number.

The numbers of the second class bring in a *Second Principle of Production* which is that to any denumeral increasing series of numbers there is a *limes-number*.

(p. 46) I. Every ordinal of the second class is either a limes or is a limes plus a finite.

II. The first and second principles of production suffice to give all possible types of denumeral collections.

The multitude of all numbers of the second class is called by Cantor $\aleph_1$. He says it is the "second" multitude. But it is evidently (by my diagram) only of the first.

(p. 46) Schönfliess

Cantor's Blunder. Not a blunder

Zunächst beweisen wir, dass das zweite Zahlklasse nicht abzählbar ist. . . .

Wäre sie nämlich abzählbar, so könnte man sie in die Form eine Reihe vom Typus ω setzen, die wir durch

$$\{\Phi_{\nu}\} = \Phi_1\, \Phi_2\, \Phi_3 \, . \, . \, . \, .$$

bezeichnen. (No doubt.)

Nun ist zunächst klar, dass es under den Zahlen der zweiten Zahlklasse keine grösste giebt. (So clear that the proof may be skipped.)

He goes on thus:

Es giebt daher notwendig Zahlen, die grösser sind als Φ_1; von ihnen sei Φ_λ die erste Zahl der Reihe (He means the first of the series $\{\Phi_\nu\}$). Ebenso sei Φ_μ die erste Zahl, so dass $\Phi_\mu > \Phi_\lambda$; so weiter schliesend gelangen wir wieder zu einer unendlichen Reihe von Ordnungszahlen

$$\Phi_1 < \Phi_\lambda < \Phi_\mu$$

so dass zugleich

$$1 < \lambda < \mu < \, . \, . \, . \, .$$

ist. Diese Reihe definirt wieder eine Limeszahl Φ_ω, so dass $\Phi_\omega > \Phi_\rho$ fur *jedes* ρ ist; diese Zahl kann daher in der obigen Reihe nicht enthaltensein. . . .

G. [PLAN FOR SIXTY LECTURES ON LOGIC] (745)

Lecture I

Definition of Logic. A practical science. What is to be expected from the study of it.

Lecture II

Physiological and psychological basis of logic. Thinking, as cerebration, subject to the laws of nervous action. Five properties of nerves. 1. Irritability. 2. Conveyance of irritation. 3. Spreading of irritation. 4. Fatigue. 5. Habit. Hypothesis to account, by these five principles, for the direction of action toward an end. Illustration by cards.

Direction of discharge. Essential. Intricacy of connections. Inhibition. Spontaneous explosions. Visceral excitation.

Imagination. Not sensation. Action in an unforeseen emergency. Fancied action with inhibition of external volition.

Attention and self-direction.

Lecture III

The fixation of belief. Four methods: 1, Tenacity; 2, Authority; 3, Natural Inclination; 4, Scientific Investigation. Close connection between Logic and Ethics.

Nature of reality. Three grades of clearness of apprehension: 1, Familiarity; 2, Formal distinctness; 3, Apprehension of the practical or sensible issue. Rule for attaining the third grade. Illustrations.

Lecture IV

Continuation of the same subject. Reality defined. Historical sketch of different forms of idealism.

Lecture V

The flow of time and inference. The parts of inference:
1, Premise; 2, Conclusion; 3, Leading principle

P
∴ C.

The universe of discourse; definition of possibility. Nature of logical principles.

Propositions are either, 1, *Universal*, affirming leading principles, asserting non-existence, or 2, *Particular*, denying leading principles, asserting existence.

$$A \prec B \qquad A \precneq B$$
$$\Pi_i u_i \qquad \Sigma_i u_i$$

Distinction of categoricals and hypotheticals, unimportant. Syllogism and dialogism.

Three properties of the copula:

1st, If $A \prec B$ and $B \prec C$, then $A \prec C$.
2nd, $A \prec A$.
3rd, If $A \prec B$ and $B \prec A$, then $A = B$.

Lecture VI

Syllogistic, or the algebra of the copula.

$$a \prec (b \prec c) = b \prec (a \prec c).$$

Indirect syllogism.

Properties of the negative. Principles of contraposition, contradiction, and excluded middle. Syllogism depends on principle of contradiction, dialogism on principle of excluded middle.

Canons of syllogism. The middle term must be distributed once and only once. The order of particularity of the conclusion is the sum of those of the premises. Spurious propositions, of the second and higher orders of particularity. Antispurious propositions, of negative orders of particularity.

Lecture VII

Introduction to the Boolian algebra. Let a variable's coming to one fixed value, v, mean that a certain proposition is true, and its coming to another fixed value, f, mean that the same proposition is false. The principle of contradiction expressed by the fundamental equation

$$(x - f)(v - x) = 0.$$

The proposition 'if X then Y' is expressed by

$$(x - f)(v - y) = 0.$$

The negative of x is expressed by any function of x which equals f when $x = v$ and equals v, when $x = f$. The simplest such function is

$$f + v - x.$$

The logical aggregate of x, y, z, is expressed most simply by

$$v - (v - f)\frac{(v - x)(v - y)(v - z)}{(v - f)(v - f)(v - f)}.$$

The logical compound of x, y, z, is most simply expressed by

$$f + (v - f)\frac{(x - f)(y - f)(z - f)}{(v - f)(v - f)(v - f)}.$$

Different values for v and f.

The two terms of second intention, *being* and *nothing*. Definitions of logical addition and multiplication.

$$\left\{\begin{array}{l}\text{If } a \prec x \text{ and } b \prec x \text{ then } a + b \prec x.\\ \text{If } x \prec a \text{ and } x \prec b \text{ then } x \prec a \times b.\end{array}\right\}$$

$$\left\{\begin{array}{l}\text{If } a + b \prec x, \text{ then } a \prec x \text{ and } b \prec x.\\ \text{If } x \prec a \times b, \text{ then } x \prec a \text{ and } x \prec b.\end{array}\right\}$$

$$\left\{\begin{array}{l}(a + b) \times c \prec (a \times c) + (b \times c)\\ (a + c) \times (b + c) \prec (a \times b) + c\end{array}\right\}$$

Other formulae.

Lecture VIII

Solution of problems in non-relative deductive logic. Various methods. Rules of lecturer's method.

First process. Analysis of the premises.

$$\{(a+b) \prec c\} \prec (a \prec c) \times (b \prec c)$$
$$\{(a \prec (b \times c)\} \prec (a \prec b) \times (a \prec c)$$

$$\{(a+b) \precsim c\} \prec (a \precsim c) + (b \precsim c)$$
$$\{a \precsim (b \times c)\} \prec (a \precsim b) + (a \precsim c).$$

Second process. Elimination of middle terms by means of the principles of syllogistic, combined with the following formulae of transposition:

$$(x \prec y+z) = (x \times \overline{y} \prec z)$$
$$(x \times y \prec z) = (x \prec \overline{y} + z).$$

Third process. Recomposition of results to find antecedents or consequents of any term.

Examples.

Lecture IX

Application of the Boolian algebra to the inversion of the order of integration and summation.

Lecture X

Introduction to the logic of relatives. Individuals. Nominalism and realism. The principle of individuation. Historical sketch of the controversy. Application to this question of the principle of continuity.

Lecture XI

The logic of relatives, continued. The proposition that something exists considered as a principle of logic. Universal and particular propositions represented by the two types

$$\Pi_i u_i \qquad \Sigma_i u_i$$

Conception of a relative term. Limited universe in relative logic. The universal block.

A:A	A:B	A:C	etc.
B:A	B:B	B:C	etc.
C:A	C:B	C:C	etc.
etc.	etc.	etc.	etc.

Representation of a relative in the form

$$l = \Sigma_i \Sigma_j l_{ij} \,(\mathrm{I}:\mathrm{J}).$$

By means of the numerical coëfficients, l_{ij}, every problem in the logic of relatives is reduced to a problem in the Boolian Algebra.

In dual relatives, there are four species of propositions:

$$\Pi_i \Pi_j \mu_{ij} \quad \Sigma_i \Pi_j \mu_{ij} \quad \Pi_j \Sigma_i \mu_{ij} \quad \Sigma_i \Sigma_j \mu_{ij}$$

We have, in general,

$$\Pi_i \Pi_j = \Pi_j \Pi_i$$
$$\Sigma_i \Sigma_j = \Sigma_j \Sigma_i$$
$$\Sigma_i \Pi_j \prec \Pi_j \Sigma_i$$

The last is the most important formula of logic. But we have

$$\Sigma_i \Pi_j \mu_i \nu_j = \Pi_j \Sigma_i \mu_i \nu_j$$
$$\Sigma_i \Pi_j (\mu_i + \nu_j) = \Pi_j \Sigma_i (\mu_i + \nu_j)$$

Lecture XII

General formulae of the logic of relatives. Two types of individual dual relatives.

A:A A:B

Classification of relatives according to their including all, some, or none of the individuals of either of these types.

		Of the type A:A		
		All	Some	None
Of the type A:B	All	∞	Negatives of equiparants	11
	Some	Negatives alio-relatives	Free	Alio-relatives
	None	1	Equiparants	0

Converse and negative.

$$\breve{l}_{ij} = l_{ji}$$
$$\breve{\breve{l}} = l \qquad \bar{\bar{l}} = l$$
$$\breve{\bar{l}} = \bar{\breve{l}}$$
$$(l \prec b) = (\bar{b} \prec \bar{l}) \qquad (l \prec b) = (\breve{l} \prec \breve{b})$$

Relative addition and multiplication defined by the formulae

$$(lb)_{ij} = \Sigma_x l_{ix} b_{xj}$$
$$(l \dagger b)_{ij} = \Pi_x (l_{ix} + b_{xj})$$

Multiplication a *particular*, addition a *universal* mode of combination. The parts *undistributed* in both. Associative principle:

$$l \dagger (b \dagger s) = (l \dagger b) \dagger s$$
$$l(bs) = (lb)s.$$

Two important formulae:

$$l(b \dagger s) \prec lb \dagger s$$
$$(l \dagger b)s \prec l \dagger bs.$$

Formulae of transposition:

$$(l \prec b \dagger s) = (\breve{\bar{b}} \prec s \dagger \breve{\bar{l}}) = (\breve{\bar{s}} \prec \breve{\bar{l}} \dagger b)$$
$$(bb \prec s) = (b\breve{\bar{s}} \prec \breve{\bar{l}}) = (\breve{\bar{s}}l \prec \breve{\bar{b}}).$$

Distributive principle:

$$(l + b)s = ls + bs$$
$$s(l + b) = sl + sb$$
$$l{,}b \dagger s = (l \dagger s){,}(b \dagger s)$$
$$s \dagger l{,}b = (s \dagger l){,}(s \dagger b).$$

Negatives of combinations:

$$\overline{l + b} = \bar{l}{,}\bar{b} \qquad \overline{l{,}b} = \bar{l} + \bar{b}$$
$$\overline{l \dagger b} = \bar{l}\bar{b} \qquad \overline{lb} = \bar{l} \dagger \bar{b}.$$

Converses of combinations:

$$\overset{\smile}{l + b} = \breve{b} + \breve{l} \qquad \overset{\smile}{l{,}b} = \breve{b}{,}\breve{l}$$
$$\overset{\smile}{l \dagger b} = \breve{b} \dagger \breve{l} \qquad \overset{\smile}{lb} = \breve{b}\breve{l}.$$

Formulae relating to the relatives of second intention, ∞, 0, 1, n.

$$0 \prec x \qquad x \prec \infty$$
$$x + 0 = x \qquad x, \infty = x$$
$$x + \infty = \infty \quad x, 0 = 0.$$

$$x \dagger \infty = \infty \qquad x0 = 0$$
$$\infty \dagger x = \infty \qquad 0x = 0.$$

$$x \dagger n = x \qquad x1 = x$$
$$n \dagger x = x \qquad 1x = x.$$

$$x + \bar{x} = \infty \qquad x, \bar{x} = 0.$$
$$1 \prec x \dagger \check{\bar{x}} \qquad x\check{\bar{x}} \prec n.$$
$$1 + n = \infty \qquad 1, n = 0.$$

Method of using these formulae illustrated by examples.

Lecture XIII

Logical extension and comprehension. Germs of the doctrine among the Stoics, and in Porphyry. John of Salisbury. Durandus and Scotus. The Port Royal Logic. Kant and his followers.

Various ways in which it has been understood. Objections which have been made to it.

Proof that the doctrine should be symmetrical. Difficulty of rendering it perfectly so.

Application of the logic of relatives and in particular the principles of the last lecture solves every difficulty. Limited and unlimited universes of marks. New theory of names springing from this method of treatment. Names are *universal* or *particular*, like propositions. Perfected theory of categorical propositions.

Lecture XIV

General Method with the Logic of Relatives. The premises are first multiplied together and the operators Σ and Π are all brought to the left of all the operands by the formulae

$$\Pi_i u_i \cdot \Pi_j v_j = \Pi_i \Pi_j u_i v_j.$$
$$\Pi_i u_i \cdot \Sigma_j v_j = \Pi_i \Sigma_j u_i v_j.$$
$$\Sigma_i u_i \cdot \Sigma_j v_j = \Sigma_i \Sigma_j u_i v_j.$$

To these formulae may be added the following:

$$\Pi_i u_i + \Pi_j v_j = \Pi_i \Pi_j (u_i + v_j).$$
$$\Pi_i u_i + \Sigma_j v_j = \Pi_i \Sigma_i (u_i + v_j).$$
$$\Sigma_i u_i + \Sigma_j v_j = \Sigma_i \Sigma_j (u_i + v_j).$$

We may also often simplify the expression by means of the following.

$$\Pi_i \Pi_j u_i v_j = \Pi_k u_k v_k.$$
$$\Sigma_i \Sigma_j (u_i + v_k) = \Sigma_k (u_k + v_k).$$

It will then be found that the number to be eliminated occurs in the two forms u_i and $\overline{u}_j$, and supposing the operator relative to i to be to the left of that relative to j, it becomes necessary to substitute j for i.

The most perfect method of doing this, consists in finding a system of relatives

w, w′, w″, w‴, etc.

such that

$$\Pi_i \Sigma_j w_{ij} \; \Pi_i \Sigma_j w'_{ij} \; \Pi_i \Sigma_j w''_{ij} \text{ etc.}$$
$$\Pi_i \Pi_j \{\overline{w_{ij}} + \overline{(wn)}_{ij}\} \; \Pi_i \Pi_j \{\overline{w'_{ij}} + \overline{(w'n)}_{ij}\} \text{ etc.}$$
$$\Pi_i \Pi_j (w_{ij} + w'_{ij} + w''_{ij} + \text{etc.})$$

Then, we have

$$\Sigma_j u_j = u_i + (wu)_i + (w'u)_i + (w''u)_i + (w'''u)_i + \text{etc.}$$
$$\Pi_j u_j = u_j \times (wu)_j \times (w'u)_j \times (w''u)_j \times (w'''u)_j \times \text{etc.}$$

After this, we shall use the process of elimination of the Boolian algebra.

Required, for instance, to eliminate s from the premises

$$\Sigma_h \Pi_i \Sigma_j \Pi_k (\alpha_{hik} + s_{jk} l_{ji})$$
$$\Sigma_u \Sigma_v \Pi_x \Pi_y (\varepsilon_{uyx} + \overline{s}_{yv} b_{vx}).$$

Multiplying, we have

$$\Sigma_u \Sigma_v \Pi_x \Pi_y \Sigma_h \Pi_i \Sigma_j \Pi_k \, (\alpha_{hik} + s_{jk} l_{ji}) \, (\varepsilon_{uyx} + \overline{s}_{yv} b_{vx}).$$

This is the same as

$$\Sigma_u \Sigma_v \Pi_x \Pi_y \Sigma_j \Pi_k \, (\alpha_{uxk} + s_{jk} l_{jx}) \, (\varepsilon_{uyx} + \overline{s}_{yv} b_{vx}).$$

Lecture XV

Logic of arithmetic; its true nature made manifest by the application of the logic of relatives. The reason of the important part played by relations of correspondence in general and by counting in particular, made plain from the principles of the last lecture. Conception of quantity, in general. Let g be 'as great as.' Then,

$$1 \prec g$$

$$gg \prec g.$$

In the universe of quantity

$$g,\breve{g} \prec 1.$$

In linear quantity.

$$\infty \prec g + \breve{g}.$$

In continuous quantity

$$g \dagger g \prec g;$$

in discrete quantity, we have the conception of the *next*.

$$l,(\bar{l} \dagger \bar{l}).$$

Conception of finite and super-infinite quantity. Quantity may be *limited*, in the sense of having an absolute maximum without being finite. The logic of relatives first shows us this, which we are then able to make intuitively evident. Unlimited quantity is defined by the formula

$$g \dagger 0 \prec 0.$$

Finite quantity, how distinguished from the super-infinite. Infinite quantity is that which is at once unlimited and finite. Mathematical induction generally fallacious; when valid.

Counting is a relative of correspondence. Defining formulae of such a relative

$$\breve{c}c \prec 1 \qquad c\breve{c} \prec 1.$$

If c be a number used in counting, an s an object of the lot counted; we have

$$s \prec \breve{c}$$
$$\breve{g}c \prec c.$$

If the lot counted is a finite and limited lot there is a maximum number reached in the counting. Calling this q, we have

$$q \prec c$$
$$\bar{g}q \prec \bar{c}.$$

Nature of mathematical reasoning shown by the application of the principles of the last lecture. Syllogisms of transposed quantity generally fallacious; when valid.

Lecture XVI

Logic of the differential calculus. Application of the rule for attaining clear ideas to the conception of continuous quantity. The doctrine of limits. The value at the limit is the sum of an infinite convergent series, and this was explained in the last lecture. The conception of continuity involves the principles of the logic of probability. Evaluation of indeterminate expressions. Cases of ambiguity. The solution sometimes depends on the number of dimensions. Real indeterminacy in the path of a moving body beyond a certain point shows that the time occupied in reaching that point is infinite.

Lecture XVII

Continuity continued. The sophism of Achilles and the tortoise and its congeners. The part played by these sophisms in philosophy. Herbart and Hegel. Analysis by means of the logic of relatives. The principle of continuity.

Lecture XVIII

The so-called absurd quantities. Application of the rule for attaining clear ideas to the question how far such quantities are real and how far fictitious. The square-root of the negative. Multiple algebra. Sketch of the history of these conceptions. The non-Euclidean geometry in its two aspects. Every system of finite or super-infinite quantity reducible to a system of simply infinite quantity. The doctrine of infinitesimals. Intuitional representation by means of a spiral. In regard to the *zero* and the *infinite* orders we have still to distinguish the order of the orders, and so on *ad infinitum.* Nothing false or really absurd about infinitesimals.

Lecture XIX

Enumeration and logical examination of the methods of geometry. Nature of geometrical axioms.

Lecture XX

Logic of Mechanics. Methods of establishing the first principles. The chief principles of mechanics. Methods of mechanics and of geometry compared. Imaginaries in mechanics.

Lecture XXI

Traditional Logic. General sketch of its history. The doctrine of the *predictables.*

The Categories. Trendelenburg's view. Kantian categories. Other lists, including the lecturer's.

Lecture XXII

The *traditional syllogistic* set forth. Aristotle. Ancient development of hypotheticals. Petrus Hispanus. Sketch of the doctrine of modals. Lambert's *Neues Organon.* DeMorgan. Criticism in the light of the logic of relatives.

Lecture XXIII

The *Posterior Analytics.* Principal contents of the books.

Lecture XXIV

The *Topics.* Aristotle. Later developments. The *ars magna* of Raymund Lully. Possible future of this part of logic.

Lecture XXV

Fallacies. The traditional list. Different classifications.

Utility of the doctrine. But is there, after all, any fallacious thinking? Harmfulness of logic, too narrowly studied. Drink deep, etc.

Lecture XXVI

The *parva logicalia.* Suppositiones. Distributio. Exponibilia. Insolubilia. Obligationes.

The formalitates of Scotus. Terministic views. Paulus Venetus and his *Sophismata aurea.*

Traditional scraps of logic. Ass of Buridanus. Argumentum ad hominem. *Exceptio probat regulam.*

Lecture XXVII

Introduction to the *Theory of probabilities.* Crude conception of probability as a reduced mode of existence. Conception of probability as the measure of just belief unduly emphasizes a secondary circumstance. Application of the rule for attaining clear ideas. Probability attaches primarily to modes of inference and denotes the frequency with which they carry truth with them. Probability of an event an abridged expression. Probability a matter of fact. The feeling of probability.

Lecture XXVIII

Relative numbers, in general. Rules for combining them. Independent relatives. Probabilities a species of relative numbers.

Connection with logical algebra, as especially with the Boolian calculus. Effect of assigning different values to *f* and *v*. Probabilities and odds. The feeling of probability and the psychophysical law of Fechner.

Lecture XXIX

Elementary problems in probabilities. The binomial development.

Boole's method in probabilities reformed, with examples. Extension of this method to problems involving other relative numbers.

Lecture XXX

The law of high numbers. Important consequences of certain numbers being large in different branches of science; such as political economy, theory of gases, physiology, doctrine of natural selection, and wherever there is a tendency toward an end.

Lecture XXXI

The law of error. Natural classes. Mr. Galton's methods.

The principles of least squares. Ferrero's theory.

Lecture XXXII

Examples in *least squares*.

Lectures XXXIII–XXXVI

Equations of finite differences, treated mainly as in Boole.

A function in finite differences has not generally a differential coëfficient.

The algebraic properties of the symbol E.

Linear equations, simultaneous equations, partial equations, applied to problems in probability.

Equations of higher degrees.

The problem of the duration of play, applied to the theory of natural selection, and to philosophy.

Lectures XXXVII to XLVI

The theory of *Induction*, treated as in the lecturer's essay in the "Studies in Logic."

Lecture XLVII

Pure *induction* illustrated by chemical theories. Mendelejeff's [Mendeleev's] law.

Lecture XLVIII

Hypothetic reasoning illustrated by the attempts to discover the identity of Junius.

Lecture XLIX

The *à priori* element of science illustrated by Galileo's dialogue, and other dynamical speculations.

Lectures L to LV

The history of *Astronomy*. Ptolemaic astronomy, expounded. Copernicus. Tycho Brahe. History of Kepler's work. The Newtonian discovery and its consequences. Hegel a pretended rival of Newton.

Lectures LVI–LVIII

The *Kinetical theory of gases* and speculations on the constitution of matter. Stallo's objections.

Lecture LIX

Logical principles of political economy.

Lecture LX

Anthropomorphic science. Judgment of men. Physiognomy. Art. Natural theology. One-sidedness of physical science.

H. [REMARKS ON CANTOR'S BEITRÄGE] (821)

Mathematische Annalen 1895. Bd XLVI. p. 48[1].
Beiträge zur Begründung der Transfiniten Mengenlehre.
By Georg Cantor, Dated Halle. 1895 March
His motives are curious

"Hypotheses non fingo." [Newton]

"Neque enim leges intellectui aut rebus damus ad arbitrium nostrum, sed tanquam scribae fideles ab ipsius naturae voce latas et prolatas excipimus et describimus."

"Veniet tempus, quo ista quae nunc latent in lucem dies extrahat et longioris aevi diligentia."

§1 DER MÄCHTIGKEITSBEGRIFF ODER DIE KARDINALZAHL

He defines a "Menge," that is, a Collection, as follows: "jede Zusammenfassung M von bestimmten wohlunterschiedenen Objecten *m* unsrer Anschauung oder unseres Denkens zu einem Ganzen."

This is remarkably accurate. No need of the subjectivism of "unsrer Anschauung oder unseres Denkens." He notes that the *members* which he calls "Elemente" are *definite*, "bestimmten" but not that they are also *individual*, and that they are *independent*, "wohlunterschiedenen." But he does not analyse that character, and say in what it consists. He implies that the relation of the collection to a member is that of inclusion but this he does not analyse.
He denotes this thus

(1) $M = \{m\}$

The aggregate (Vereinigung) of several collections M, N, P, *which have no common members* he writes

(2) $(M,N,P,\ldots)$

The members of this collection are the members of the collections M, N, P, etc.
Part "Theil" or "Theilmenge" is a *different* (andere) collection whose members are all members of the whole.
A part of a part is a part.
Every collection has a "*Multitude*" "Mächigkeit" which he also calls *Cardinalzahl*. He ought not to confound them.

He defines multitude in the German fashion by negations. I think this is idle business.

"Mächtigkeit oder Cardinalzahl von M nennen wir den Allgemeinbegriff welcher mit Hülfe unseres activen Denkvermögens (what has this to do with the subject in hand?) dadurch aus der Menge M hervorgeht, dass von der Beschaffenheit ihrer verschiedenen Elemente m und von der Ordnung ihres Gegebenseins abstrahirt, wird."

This is bad. He ought first to define the relation of greater and less and show if possible that of any two collections both cannot be greater than the other.

p. 482 He denotes the multitude by

(3) $\overline{\overline{M}}$

He speaks of the mental "Abbild oder Projection" of a collection.

He defines the *equality* "äquivalent" of two collections as consisting in the possibility of putting them into one-to-one correspondence; thus leaving his definition of multitude behind as the idle performance it is. He denotes equality thus

(4) $M \sim N$

Then every part of M has its equal part of N. Transpositions are possible

(5) $M \sim M$

(6) If $M \sim P$ and $N \sim P$ then $M \sim N$.

Two collections have the same multitude if and only if they are equal. Here is a little better definition of multitude. But not yet good.

(7) if $M \sim N$ $\quad \overline{\overline{M}} = \overline{\overline{N}}$

(8) if $\overline{\overline{M}} = \overline{\overline{N}}$ $\quad M \sim N$

Why he calls one *equivalence* and the other *equality* I don't see. It is a distinction without a difference. By equality he probably means identity. p. 483 He vaguely deduces all this from his definition

(9) $M \sim \overline{\overline{M}}$

If $M \sim M'$ $N \sim N'$ $P \sim P' \ldots$
then

$$(M, N, P, \ldots) = (M', N', P' \ldots)$$

§2 GREATER AND LESS

He defines as follows:

If No part of M is equal to N
But a part N_1 of N is equal to M

the same thing is true of $M' \sim M$ and $N' \sim N$. This, therefore, is a relation of multitude.

Then M and N are not equal, for since then $M \sim N$ and $N' \sim M$ it follows the $N' \sim N$; and since $M \sim N$ there is a part of M say M' which is $M' \sim N'$ and consequently $M' \sim N$ contrary to hypothesis.

He writes

(1) $a = \overline{\overline{M}} \quad b = \overline{\overline{N}} \quad a < b \quad b > a$

and it is easily proved that

(2) if $a < b$, $b < c$ then $a < c$

And if P_1 is part of P and $a < \overline{\overline{P}}_1$ then $a < \overline{\overline{P}}$ and if $\overline{\overline{P}} < b$ then $\overline{\overline{P}}_1 < b$ He thus proves that the three relations

$$a = b \quad a < b \quad a > b$$

are mutually exclusive.

But that one or other always holds is in this place "Kaum zu beweisen."

I define as follows:

If there is no relation, ρ, such that every M is ρ to an N to which no other M is ρ, then, and then only $M > N$.

If M is not greater than N and N not greater than M then M is equal to N.

From this point of view the difficulty is to prove that it cannot be that $M > N$ [and] $N > M$. It is evident that either $a > b$ or $b > a$ (or both) or $a = b$.

As to a *part* identity is the relation in which every member of it is in to a member of the whole to which no other is in the same relation. Therefore the part is either fewer than or equal to the whole.

If in my sense $N > M$ then it is not evident that M is not also greater than N and therefore is not evident that $N > M$ in his sense. My $N > M$ is the same as either his $N > M$ or his *neither*.

But if in his sense $N > M$ so that a part of $N_1 = M$ there is a relation ρ in which every M is ρ to an N to which no other M is ρ and therefore M is not greater than N and therefore either $M = N$ or $N > M$ in my sense. But if $M = N$ then a part of M, say M_1 would be equal to N_1 and therefore $M_1 = N_1 = M = N$ and a part of M would be equal to N contrary to the first clause of his def. Hence if $N > M$ in his sense it is so also in my sense. In short of his four relations $a = b$, $a > b$, $a < b$, a neither to b, our equals are the same and my greater means his greater or neither. His $a > b$ is my $a > b$ and $b \ngtr a$.

He now gives 5 propositions of which the latter 4 are he says corollaries from the 1st

A. If a and b are multitudes either $a = b$ or $a > b$ or $b > a$

B. M_1 part of M. N_1 part of N. If $M \sim N_1$, $N \sim M_1$ then $M \sim N$

I remark that it is evident that a part is either equal to or less than its whole. Therefore

Either $M = N$ or $M < N$
Either $M = N$ or $M > N$

Hence $M = N$ without using A at all

C. M_1 part of M, M_2 part of M_1. If $M_2 \sim M$ then $M_1 \sim M$

The same remark applies.

D. If N is not equal to M nor to any part of M, then a part of N_1 of N is $N_1 \sim M$. If in his sense $N < M$, $N = M_1$ part of M.

Hence if this is not the case in *my* sense $N > M$. But that $N_1 = M$ requires A

E. If $\overline{M \sim N}$ (M not $\sim N$) but $N_1 \sim M$, $\overline{M_1 \sim N}$

I. OUR SENSES AS REASONING MACHINES (1101)

The new psychology which now has its laboratories in every university and is acknowledged by scientific men generally to have, at last, taken its place among the progressive sciences, dates from about 1860. Among its first fruits was a crop of experiments showing that the senses of ordinary sane persons, no matter how little imaginative, furnished something more than plain, unvarnished facts of the outer world, and that no amount of direct scrutiny could enable us to say what part of that which we seem to see or hear is due to stimulations of the nerve-terminals of our eyes and ears and what part is a quasi-inferential interpolation of our own minds.

Nothing that the psychologists have since discovered has been so surprising or so instructive as those phenomena; and their significance has not yet been exhausted. I propose to consider here only one of their lessons, a practical one for us all, by showing that instinct and reason shade into one another by imperceptible gradations.

Let us first examine a very different phenomenon. The reader is doubtless aware that logical machines have been constructed. You feed one of these machines, say for example that of Prof. Allan Marquand, of Princeton, with premises, by moving certain parts until those premises appear upon the face of the machine, expressed in a conventional system of signs, and then you turn a crank. Immediately, the necessary conclusion from those premises,—by no means, always a conclusion so simple as that of the syllogisms or sorites to which the ordinary logic-books restrict their studies,—appears expressed upon the face of the machine. Such apparatus as has hitherto been constructed in this line has had a very modest aim. It will do for us all the necessary ratiocination that the text-books teach, and much more; but the circumstance that all such machines propose to exhibit *the* conclusion from given premises shows that they are not adequate to mathematical reasonings. Thus, the theory of numbers from a few simple premises deduces not one conclusion, but hundreds of marvellous theorems. There is absolutely no end to the series of conclusions that might be deduced. But while no machine has been constructed that will deduce more than one conclusion; yet it has been shown that all possible general conclusions can be arranged in serial order and as soon as anybody wishes to defray the not extravagant cost, the specifications will be ready for a machine that will actually turn out new theorems from a given set of premises, one after another, as long as they continue to have any interest. But though a machine could do all that, and thus accomplish all that many an eminent mathematician accomplishes, it still cannot properly be called a reasoning-machine, any more than the sort of man I have in view can be called a reasoner. It does not reason; it only proceeds by a rule of thumb.

It is necessary to arrest our attention upon this point, in order that we may be led to see how much more like reasoning the operations of our senses are than the performances of any mere machine can be. A really accurate explanation would carry us into modern exact logic, which is fine ground for a man that is spoiling for a tough intellectual wrestle, but not at all what a tired reader takes up a magazine to find. However, it is not difficult to see that one cannot go far in reasoning.

J. MULTITUDE (GER. *MÄCHTIGKEIT*, *CARDINALZAHL*; FR. *PUISSANCE*) (1147)

That character of a collection by virtue of which it is greater than some collections and less than others. A *collection* (Ger. *Menge*; Fr. *ensemble* . . .) is defined by Dr. Georg Cantor, who is the chief authority upon the doctrine of multitude, as a "Zusammenfassung von bestimmten wohlverschiedenen Objecte;" but this appears to be a *circulus in definiendo*, since a "Zusammenfassung" seems absolutely synonymous with "Menge." At least, things cannot be "zusammengefasst" unless they are "Objecte" and "wohlunterschieden," whether they must be "bestimmt," or not. The following definition may perhaps be found unobjectionable: A collection is an individual (in the logical sense) which possesses no character whatever which is not conceived to consist in the possession of certain corresponding characters by all of certain other individuals, called its *members* (Ger. *Elemente*) or *ineunts*, which are distinct from and independent of one another; and the being of the collection consists in the possibility of considering those members as thus determining a derivative individual, without being drawn into absurdity. By saying that *A* is "distinct from" *B*, is here meant that *A* has some character, a relative one, if not an absolute one, which *B* has not. By saying that *A* is "independent of" *B*, is here meant that the possession or non-possession of any character by *B* does not logically prevent *A* from possessing or not possessing any character, unless there is such a relation, direct or indirect, between the characters themselves, that the same thing would be true of any two subjects whatever. Thus, all shades of red do not make up a collection, because shades of red are not individual; all the males and negroes in the U.S. do not make a collection, because they are not distinct; a spherical shape and its surface do not make a collection because they are not independent.

A collection, as the *A*s, is said to be "greater" than another, the *B*s, and the latter to be "less" than the former, according to the definition given by Cantor in 1873 (*Journal von Rein. u. angew. Math.* of Borchardt.

LXXVII. 258), (here modified in an unessential particular) if taking any relation, ρ, whatsoever, there is some A which either does not stand in the relation ρ to any B at all, or only to a B to which some other A stands in the relation ρ. That is to say, whatever the relation may be, some A fails to be the *sole* A that stands in that relation to any B. It is then a theorem, and a very difficult one to prove, that no two collections are each greater than the other.

Cantor makes "multitude" (*Mächtigkeit*) and "cardinal number" (*Cardinalzahl*) perfect synonyms (*Math. Ann.* XLVI, 481). But "multitude" is an abstract noun, while any particular cardinal number is either a concrete adjective, or is not properly a noun of any kind but merely a vocable used as an instrument for performing the experiment of counting a collection. For that reason, it seems somewhat objectionable to call an adjective of infinite multitude a number.

A "finite" multitude is substantially defined by C. S. Peirce (*Am. Jour. Math.* 1881, IV, 93) as that of a collection for which the syllogism of transposed quantity is valid. Dedekind (*Was sind und was sollen die Zahlen*, 1888, ¶ 64) defines a finite collection (*System*) as one which is not equal to any true (*echte*) part of itself. By a "true" part as contradistinguished from the whole, Schröder has easily shown that the two definitions are virtually equivalent.[1]

Cantor proves that the smallest infinite multitude, which he calls "denumerable," is that of all the finite whole numbers. He proves that this is equal to the multitude of rational fractions, and to that of the imaginary quantities each of which satisfies [an] algebraic equation with rational coefficients, and to that of any collection of which every member can be described so as to distinguish it from all others. He denotes this multitude by $\aleph_0$ and calls it by the name "Aleph-zero."

Cantor further states (and there is no room to question the assertion) that there is an endless sequence of multitudes, greater than aleph-zero, each having another next greater than it, with no possible intervening number. These he denotes by $\aleph_1$, $\aleph_2$, etc.

The multitude $\aleph_1$ is that of the possible collections of finite whole

[1] The missing contents of an empty parenthesis were to have identified the reference to Schröder. Peirce wrote in the margin "Where? Will Mrs. Franklin know?" This definition was written for *Baldwin's Dictionary* and Mrs. Ladd-Franklin was helpful in editing Peirce materials.

numbers. Now if we consider that any of our real quantities contemplated by the calculus could be designated by a number, if this could be carried out into an endless series of decimal places; and if we consider that this could equally well be expressed in the binary system of numerical notation, and further that such a quantity could equally well be expressed by the ordinal numbers of the fractional places occupied by 1 and not by 0 (the only two digits of the binary system), we see that analytical real quantities constitute a collection corresponding, one-to-one, to all possible combinations of whole numbers.

K. [PART OF A LOWELL LECTURE] (from 450)

The first kind of reasoning to be studied is *Deduction.* Deduction is that kind of inference in which the fact expressed in the conclusion is inferred from the facts expressed in the premisses, regardless of the manner in which these facts have come to the reasoner's notice. Deduction is either necessary or probable. *Necessary deduction* is that sort of inference in which the fact concluded is conceived to be involved in the facts premised. It is the reasoning of mathematical demonstration.

It has taken two generations to work out the explanation of mathematical reasoning. This delay has been partly due to many writers entirely missing the point and directing their energies to ascertain the sequence of mental phenomena in reasoning instead of the logical sequence of the argument, which need not be closely related to the psychological sequence. The delay has also been due in part to the circumstance that some students attempted to divest thought of its garment of expression and to get at the naked thought itself, an attempt analogous to that to remove the peel from an onion so as to get at the naked onion itself. Reasoning is nothing but the discourse of the mind to itself. Divest thought of signs and it ceases to be thought, and becomes, at best, direct perception.

What is requisite is to take really typical mathematical demonstrations, and state each of them in full, with perfect accuracy, so as not to skip any step, and then to state the principle of each step so as perfectly to define it, yet making this principle as general as possible. For routine demonstrations there is no particular difficulty; but for the major theorems there is much. If we attempt to make the statement in ordinary language, success is practically impossible. Our syntax was not made with a view to such propositions; and it sometimes defies ingenuity to express them, I do not say clearly, but accurately with whatever intricacies of expression in words. At all times, the burden of language is felt severely, and leaves the mind no energy for its main work. Yet this is the least of the disadvantages of ordinary speech. After we have weeded out its ambiguities, it presents so many forms whose precise difference of meaning we are not prepared to define that we are very apt to pass over important steps of reasoning as mere grammatical transformations; and in many cases we see pretty clearly that an inference holds good but are provided with no sure way of stating its principle in general terms. Mathematicians have found themselves obliged to resort to algebraical arrays of letters in order to express themselves. But their algebras were devised for a purpose quite inconsistent with that of logical analysis, and are of no material help. It is necessary to devise a system of expression for the purpose which shall be competent to express any proposition whatever without being embarrassed by its complexity, which shall be absolutely free from ambiguity, perfectly regular in its syntax, free from all disturbing suggestions, and come as near to a clear skeleton diagram of that element of the fact which is pertinent to the reasoning as possible. I am going to devote this lecture to a brief description of a system of expression which, if it does not quite satisfy my ideal of what such a system ought to be, is at any rate the best I have been able to devise during forty years study of the problem. If you will learn this system and will then train yourselves to the use of it, I can promise that it will help you much to unravel tangles of thought.

Only, let not its aim be mistaken. I wish to declare distinctly and once for all that it is *not* intended to furnish a speedy or ready way by which to pass from premisses to conclusion. It aims in the diametrically opposite direction, namely, to break up reasoning into the greatest possible number of distinct steps, so that the constitution of reasonings may be studied. If we wished to obtain speedy passage from premisses to conclusion, we should, on the contrary, seek to make the steps as few and as large as we could. In short this system is meant not as an aid in reasoning but as an aid in the minute analysis of reasonings. Practice with it, however, will make thought clearer, and will so conduce indirectly to skill in reaching conclusions.

This system is a system of *diagrams*. A diagram has the advantage of appealing to the eye, and to that adds others due to the prominence it gives to *conventional signs*. Every conventional sign or other symbol is employed over and over again. The word *the* will occur several times on every page of English print; and it is everywhere one and the same word. Thus, it is the general type that constitutes the self-identity of the symbol; that is, it is its being formed in conformity to certain general precepts. But every one of those single embodiments of the word *the* which we find on a page;

what are we to call them? A word is wanted for the purpose. I will call the single embodiments of a symbol, whether conventional or natural, its *replicas*.

The special system of diagrams that I am about to describe is called the *Method of Existential Graphs*. This system will repose upon 14 conventions or agreements into which you and I shall have to engage as to the significations of parts of our graphs. When I say there are 14 conventions, there is one that I do not count for the reason that it does not belong to this particular system any more than it does to any other system of conventions relating to no matter what subject. I will therefore number this Convention Number Zero. It reads as follows. . . .

L. ON THE CLASSIFICATION OF THE SCIENCES (1345)

Every systematic philosopher must provide himself with a classification of the sciences. Comte first proposed to arrange the sciences in a series of steps, each leading on to another. This general idea may be adopted; and we may adapt our phraseology to the image of the well of truth with flights of stairs leading down into it.

We divide the whole into three great parts,

I. *Mathematics*, the study of ideal constructions without reference to their real existence,
II. *Empirics*, the study of phenomena with the purpose of identifying their forms with those mathematics has studied,
III. *Pragmatics*, the study of how we ought to behave in the light of the truths of empirics.

Mathematics is divided into

1. *Geometry*, the study of continua;
2. *Arithmetic*, the study of an infinite discrete series, to which the Fermatian inference applies; and
3. *The theory of finite groups*, the study of manifolds to which the syllogism of transposed quantity applies.

This division of mathematics will seem very strange. In the first place, it will be objected that I overlook the important division into pure and applied mathematics. I fully admit the importance of the division. Namely, pure mathematics is the study of ideal constructions interesting in themselves, while applied mathematics is the study of ideal constructions less interesting in themselves, but selected for study on account of their analogy with real systems of interest. But the reason these constructions of applied mathematics are less interesting in themselves is that pure mathematics has not reached that stage of development in which it can with the utmost advantage take up the study of them. They are destined ultimately to be absorbed in pure mathematics. Hence, I think this division far less important *for the purposes of philosophy* than the division of pure mathematics.

M. [GENERAL REMARKS ON PROBABILITY] (245)

There is a certain branch of logic to which, in nice propriety, the appellation "the logic of science *propter excellentiam*" might be exclusively restricted; seeing that it is an instrument of reasoning that has been created by modern science. It is actually applied, too, to the purposes of science, and indeed, quite constantly so in the more precise of those disciplines that are known as "the exact sciences," which reap substantial profit from the application. I see that I am making an enigma of my characterization; and so far, you might well guess that the Differential Calculus was meant. But that I will exclude by capping all the merits of the epithet, with what for our present purpose will be its crowning excellence — the instrument I refer to affords quite the most advantageous standpoint for the review, comparison, criticism, and organization of the different types of scientific reasoning; so that it is not merely an instrument of the reasoning, but is furthermore an organon of the logic, of modern science. What else, now, can this describe but the Doctrine of Chances, or, as sticklers, — I will not say for the "high-piping Pehlevi" of science, — but for verbal habiliments of irreproachable dignity for all things scientific, have rechristened "the calculus of probabilities"?

This doctrine is the sole logical instrument that modern science has invented for its own use and in its exacter researches constantly uses to advantage.

That people which, in point of native intellect, far outclassed any other that has trod this planet never had the first notion of it; and to one who remembers how it was not until the middle of the proud XIXth century that modern geometers succeeded in placing themselves in every respect in advance or abreast of Euclid, who knows how late it was before modern architects possessed themselves of the secret of the Parthenon's beauty, and who in reading Plato, — quite apart from his poetic flights, — has here and there been struck breathless at the involution of the thought even in the simplest sentences, at the subtlety of the syntax, the opulence of the vocabulary, is amazed when he comes to reflect that in the huge lexicon of Greek the principal ingredient of the meaning modern science attaches to the word "probability" is neither expressed nor implied by any noun, verb, adjective, or adverb.

But some reader may ask why not. "The Greeks had dice of two kinds, *κύβοι* and *ἀστράγαλοι*, and two kinds of dice box, *φιμός* and *πύργος*, and some variety of backgammon-board, *ἄβαξ*. They bet upon dice too; for we read of men being ruined by dicing. Now in betting they must sometimes have given and taken odds; and they would wish to determine what the odds ought to be. Therefore, one might confidently expect that they would calculate the chances." Very good! Thus, the fact that they never did so puts in a strong light the failure of the Greeks to consider subjects quantitatively. They never measured anything but multitudes, space, time, weight, prices, and verses. "Which argues that they were dull." Not at all! It only illustrates what the whole history of science and judging from my own, the experience of every man go[es] to emphasize, that the generalization of concepts is an operation of thought which is performed by all men that perform it at all only with difficulty and with timidity. Thus the Egyptian arithmeticians never got a clear notion of any fraction having any other numerator than 1, with the single exception of 2/3; and even Ptolemy, a mathematician strong enough to work out spherical trigonometry for himself, always expressed 5/6 as 1/2 and 1/3, and would have denoted 2/5 as 1/3 and 1/15 and 2/7 as 1/4 and 1/28 etc. So the early modern algebraists boggled over negative numbers, and still more over negative and fractional exponents; and so on.

N. BOOLIAN ALGEBRA (s-38)

The algebra of logic was invented by the celebrated English mathematician, George Boole, and has subsequently been improved by the labors of a number of writers in England, France, Germany and America. The deficiency of pronouns in English, as in every other tongue, begins to be felt as soon as there is occasion to discourse of the relations of more than two objects, and forces the lawyer of today in speaking of parties, as it did Euclid of old in treating of the relative situations of many points, to designate them as *A*, *B*, *C*, etc. This device is already a long stride toward an algebraical notation. Two other kinds of signs, however, must be introduced at once. The first embraces the parentheses and brackets which are the punctuation marks of algebra. The imperfection of the ordinary system of punctuation is notorious; and it is too stale a joke to fill up the corner of a newspaper to show a phrase may be ambiguous when written from which the pause of speech would exclude all uncertainty. In our algebraical notation, we simply enclose an expression within a parenthesis to show that it is to be taken together as a unit. We thus easily distinguish the "black (lady's veil)," from the "(black lady)'s veil." In case it becomes necessary to enclose one parenthesis within another, we resort to square brackets [ ] for the outer one. The other signs of which we shall have immediate need are +, −, =, [and] are of the nature of abbreviations. The sign +, now read plus, can be historically traced back, through successive insensible modifications, to the ancient word ET, and. Without stopping to explain the origin of − and =, I merely remark that these signs are, virtually at least, mere phonograms of minus and est. Everybody knows how much abbreviations may lighten the labour of thought. In our ordinary Arabic notation for numbers, we have two kinds of signs, first, the ten figures, and second, the decimal places. Important as the figures are, they are not nearly so much so as the decimal places. Of the two conceivable imperfect systems of notation which should discard one and the other of these two classes of signs, we should find that one the more useful which should write for 123456 one, two, three, four, five, six, rather than that which should write 1 hundred and 2-ty, 3 thousand 4 hundred and 5-ty, 6. For what we need to aid our reasoning is a sign the parts of which stand to one another in relations analogous to those on which our reasoning is to hinge, so that we may just think of the signs themselves that are before our eyes, and not have to think of the things signified, which we could only do after all by calling up some mental image or sign which might answer the purpose of reasoning better than those that would be written down. Because the Arabic figures fulfill this condition to a certain extent, we are able to rattle off a long multiplication, thinking only of the figures and not of the numbers; and because we possess no notation for numbers which fulfill the condition perfectly, we find a great difficulty in reasoning about the divisibility of numbers and such like problems. A similar quasi diagrammatical power is what gives the algebraical signs $+$, $-$, $=$, their great utility.

In that particular modification of the Boolian algebra to which I shall first introduce you, and which I shall chiefly use, the letters of the alphabet are used to signify statements. The special statement which each letter signifies will depend on the convenience of the moment. The statement signified by a letter may be one that we believe or one that we disbelieve; it may be very simple or it may be indefinitely complex. We may, if we like, use a simple letter to signify the entire contents of a book, or the sum total of omniscience, or a falsehood as such. To use the consecrated term of logic, which Appuleius, in the second century of our era, already speaks of as familiar, the letters of the alphabet are to be PROPOSITIONS. The final letters x, y, z, will be specially appropriated to the expression of formulae which hold good whatever statements these letters may signify. Of the special signs of invariable significance, the first consists in the writing down of a proposition by itself; and this has the effect of asserting it. This sign will receive a further development further on.

Equality and the cognate words, as well as the sign $=$, are used in such a sense that $x = y$ (no matter what statements x and y may signify) means that x and y are equally true, that is, are either both true or both false. Thus, let d signify that the democrats will carry the next election and R that the republicans will lose it; then $d = R$ means that either the democrats will carry the next election while the republicans will lose it, or the democrats will not carry it nor the republicans lose it. The exact meaning of the sign of equality, then, may be summed up in the following propositions, which I mark L, M, N, for convenience of future reference.

L. If $x = y$, then either x is true or y is false. (I never use the locution "either . . . or . . ." to exclude the case of both members being true.)

M. If $x = y$, then either x is false or y is true.

N. If x and y are either both true or false, then $x = y$.

From this definition of the sign of equality, it follows that in this algebra it is subject to precisely the same rules as in ordinary algebra. (Note that the rules of algebra are "rules" in rather a peculiar sense. They do not compel us to do anything, but only permit us to perform certain transformations.) These rules are as follows:

Rule 1. $x = x$.
Rule 2. If $x = y$, then $y = x$.
Rule 3. If $x = y$, and $y = z$, then $x = z$.

I proceed to give formal proofs of these rules; for though they are evidently true, it may not be quite evident that their truth follows necessarily, or how it does so, from the propositions L, M, N. At any rate the proofs will be valuable as examples of demonstration carried to the last pitch of formalism.

Rule 1. Any proposition, x, is either true or false. Call this statement E. In N, write x in place both of x and of y. From N, so stated, together with E, we conclude $x = x$.

Rule 2. Suppose $x = y$, which statement we may refer to as P. Then, all we have to prove is that $y = x$. From L and P, it follows that either x is false or y is true. Call these alternatives A and A' respectively. We examine first the alternative A. By M and P, either x is true or y is false. Call this statement (having two alternatives) B. But no proposition, x, is both true and false. Call this statement C. From B and C, we conclude that y is false. Thus, the first alternative, A, is that x is false and y is false. Next, we examine the other alternative, A'. From M and P, we conclude B, as before. But no proposition, y, is both true and false. Call this statement C'. From B and C', we conclude that x is true. Then the second alternative is that y is true and x is true. Thus, there are but two alternatives, either that x and y are both true or that they are both false. Call this compound statement D. In the statement of N, substitute x for y and y for x. Then from N as stated, together with D, we conclude that $y = x$, which is all we had to prove.

Rule 3. Any proposition, y, is either true or false. Call these two alternatives A and A'. We first examine the alternatives A. No proposition, y, is both true and false. Call this statement C. By M, A, and C, if $x = y$, then x is true. Call this conditional proposition B. In the statement of L substitute y for x and z for y. Then, from L, so stated, A, and C, we conclude that if $y = z$ then z is true. Call this conditional proposition D. In the statement of N, substitute z for y. Then, from N, so stated, from A, D, it follows that if $x = y$ and $y = z$, then $x = z$. Next, I examine the other alternative A'. From L, A', and C, it follows that if $x = y$, then x is false. Call this statement B'. From M, stated as before, A', and C, we conclude that if $y + z$, then z is false. Call this statement D'. Then from N, stated as before A', and B', and D', we conclude that if $x = y$ and $y = z$, then $x = z$. This being the case under both alternatives, we conclude it unconditionally.

Addition and multiplication, and their cognate words and algebraical signs, are used in such sense that $x + y$ means that either x or y is true (without excluding the possibility of both being so), while xy means that both x and y are true. More explicitly, the meanings of the sum and product are summed up in the following propositions, which are lettered A, B, C, X, Y, Z, for convenience of reference.

A. Either x is false or $x + y$ is true.
B. Either y is false or $x + y$ is true.
C. Either $x + y$ is false or y is true.
X. Either xy is false or x is true.
Y. Either xy is false or y is true.
Z. Either x or y is false or xy is true.

From these definitions it follows that in this algebra, all the ordinary rules of addition and multiplication hold good, together with some other rules besides. The rules common to logical and arithmetical algebra are the following.

Rule 4. The associative principle of addition. $(x + y) + z = x + (y + z)$.
Rule 5. The associative principle of multiplication. $(xy)z = x(yz)$.
Rule 6. The commutative principle of addition. $x + y = y + x$.
Rule 7. The commutative principle of multiplication. $xy = yx$.
Rule 8. The distributive principle of multiplication with reference to addition. $x(y + z) = xy + xz$.

The rules peculiar to logical algebra may be stated as follows:

Rule 9. $x + x = xx$.
Rule 10. If $x + y = xz$, then $x + y = x$.

Exercise 1. Prove the above rules, from propositions $L, M, N, A, B, C, X, Y, Z$.

The above rules are made to conform as much as possible to those of ordinary algebra, and suppose that we are dealing with equations. But as a general rule, we shall not have any equations, but having written down a statement, the problem before us will be to ascertain what follows from it. In that case, it will be better to work by the following system of rules, which for the sake of distinction, I shall term principles.

Principle 1. The commutative principle. The order of factors and additive terms is indifferent, that is, $x + y = y + x$ and $xy = yx$.

Principle 2. The principle of erasing parentheses. We always have a right to erase a parenthesis in any asserted proposition. This includes the associative principle, and also permits us to infer $x + yz$ from $(x + y)z$.

Principle 3. From any part of an asserted proposition, we have the right to erase any factor; and to any part we have a right to logically add anything we like. Thus, from xy we can infer $x + z$.

Principle 4. We have a right to repeat any factor, and to drop any additive term that is equal to another such term. Thus, from x we can infer xx, and from $x + x$ we can infer x.

Exercise 2. Prove the above four principles from proposition L, M, and N, together with Rules 4 to 10.

Exercise 3. 1. By means of the ten rules alone, prove that addition is distributive with respect to multiplication; that is, that $x + yz = (x + y)(x + z)$.

2. By means of the four principles alone, show that from $x(y + z)$ we can infer $xy + xz$.

3. A chemist having a substance for examination, finds by one test that it contains either silver or lead, by a second test that it contains either silver or mercury, and by a third test that it contains either lead or mercury. Show by the four principles that it contains either silver and lead or silver and mercury or lead and mercury.

4. Show the same thing by means of the ten rules.

5. A substance known to be simple salt is shown by one test to be either a potassic or a sodic salt, by a second test to be either a potassic salt or a sulphate, by a third test to be either a sodic salt or a nitrate, and by a fourth test to be either a sulphate or a nitrate. Show by the four principles that it is either potassic nitrate or sodic sulphate. Show the same thing by the ten rules.

6. A simple salt is shown by one test to be either a salt of calcium,

strontium, or barium; by a second test to be salt of calcium or strontium or an iodide; by a third test to be either a salt of strontium or barium or a chloride; by a fourth test to be either a salt of barium or calcium or a bromide; by a fifth test to be either a salt of calcium or a bromide or iodide; by a sixth test to be either a salt of strontium or a chloride or iodide; by a seventh to be either a salt of barium or a chloride or bromide; and by an eighth test to be either a chloride or bromide or iodide. Prove that it is either the chloride or calcium or the bromide of strontium or the iodide of barium.

O. SIGNIFICS AND LOGIC (part of 641)[1]

In this paper I propose for the Reader's assent or dissent some analyses of relations between Semeiotic, or the general physiology of Signs, and Logic considered as the theory of reasoning. I can do little more, herein, than state what at present seems to me tolerably certain; for the arguments which have led me to these opinions could not, with any justice to them, be compressed into a few pages. I am striving with all my might so to set them forth in a book so that they may be critically examined; but whether my powers hold out for so great a task is dubious. Perhaps, meantime, it may be found of interest, and even of value to some future thinker who may pass over somewhat the same line of research as I have done, to see what conclusions I have reached.

An argument is, of course, a Sign; namely, a Sign of the truth of its conclusion. For it is a Sign in the first place, of the truth of its premisses; and then this truth being a Sign of that of the conclusion, the Argument itself is a Sign of that same. I do not say that every Sign of another Sign is, *ipso facto*, a Sign of that of which the latter is a Sign; but it is so in this case.

It evidently follows that there is a somewhat close affinity between Significs, in so far as it coincides with or is a branch of Semeiotic* (*The words I "capitalize," such as "Significs" and "Semiotic" in this sentence, are so marked in order to call the Reader's attention to the fact that the precise, or approximately precise, sense in which each of them is used throughout this paper is somewhere explained in it.)[2] and Logic, in the sense just assigned to this word. It is often used in other senses, in consequence, I suppose, of the opinion often expressed in books in which it is taken in those other senses, that the theory of reasoning was perfected by Aristotle, or by Apelt (!), or by some unspecified writers. This notion is as false as possible. In Germany, false views of reasoning are still so general

[1] Peirce's dating of this paper ranges from 3–18 November 1909.
[2] Peirce uses both *semiotic* and *semeiotic* in *The Century Dictionary*.

and so important as to lead to strengths being attributed to many arguments which they do not come near to having, while on the other hand obvious arguments of great weight are entirely overlooked from the same cause. The logic of relations, which strikes so deep as to put quite another face upon deductive reasoning, has been really studied but by a very small group, and even by them often with little suspicion of its true significance. The Doctrine of Chances, which is incomparably the most important contribution that has been made to logic since the era of modern science set in, has been utterly neglected by logicians, who have abandoned it to the mathematicians. The mathematicians at first did, in their usual admirable manner, what it lay in them to do for it. But mathematicians are apt to imagine that all problems can be treated in the ways which have been successful in their proper field; and when they take up a question of a different nature, great mathematicians, no less than small ones, are apt to go ridiculously wrong; and so they have done about the "calculus of probabilities," and this has, no doubt, contributed to prevent the recognition of the true nature of non-mathematical scientific reasoning.

Logic naturally could not advance beyond a certain point during the middle ages, when there was no scientific inquiry at all, and therefore no prominent examples of scientific reasoning to be explained. For just as thermodynamics had to await the steam-engine before it could amount to anything, so the theory of reasoning could not amount to much until reasoning could exhibit some genuine triumphs. Still, considering the state of science during those ages, it ought to be acknowledged that their logic was all that could be expected, if not considerably more. It was better than modern logic would have been but for the work of mathematicians; and if logical studies could have been continued with the same earnestness after the advent of modern science as they had been previously, I for my part have no doubt that they would have done great things for our scientific ages. Unfortunately, the whole Christian church set its face against modern science, and largely against its forerunner, the revival of learning. The universities were in church hands, and opposed the new learning. The Constantinopolitans and their disciples knew nothing of logic or metaphysics; but finding the obscurantists of the universities were Dunces, — that is, followers of Duns, men whose subtile reasonings they were utterly unable to cope with, — they naturally struck hands with the only adversaries of that great analyst who dared to enter the lists of controversy with his disciples; and these were the Nominalists. The promoters of the revival accordingly adopted the Nominalist opinion,

—in words; for they never gave thought enough to the subject to know what it was to which they were assenting; and having given their assent, they dropped the whole distasteful subject; for to these mere enjoyers of literature severe reasoning, especially upon abstract subjects, was a pain that they lacked the fortitude to. . . .

I will now, Reader, set before you the truth of the matter and will do what I can to help you to see that it *is* the truth. I am going to show you, by a process that I call "Logical Analysis" that there are just three kinds of elements in any individual case of cognition, and no more; and although two of these three kinds can be further subdivided, yet the sub-kinds do not at all stand on the same footing. You will find this requires much meditation, even with the help of what I say. I myself, with such aid as I derived from Kant and other writers, had no great difficulty in seeing that there are three elements, and approximately what their natures are. Three years, during which I had the matter almost incessantly before me, day and night, more than sufficed. But to render my ideas accurate took me near three times as long, though I only went over the subject occasionally, and purposely dismissed it twice for a year and once for three years; and after all this, the process of satisfying myself that the whole thing was not a delusion but was assuredly a pretty close approximation required far more effort than the rest and occupied probably full double the whole time I had previously given to the subject. For although I am willing, if unhappily I find the situation to require it, to maintain my own opinion unshared against the protests of millions and should have little respect for the human intellect that would not do so, yet I will not utter a word that may be denied until I fully appreciate all that can be said against it and am quite sure that I have gone *much* deeper into the subject *from every side* than any of those who take the other side of the question; and that is the reason I have written so little. I don't doubt that my writings are full of minor blunders; but I have reason to believe that their main points have been subjected to a far severer inquisition at my own hands than they have been or will be likely to receive for a generation or so. All this is put forward, not to serve as a *succedaneum* for rational argumentation, but to persuade the Reader that in taking the trouble to examine carefully into the justice of what I shall urge he will not be wasting his pains upon the fancies of an idle dreamer; and I may urge that I have made contributions to several well-developed sciences in which critical judgments are sure; and if none of these,—in undertaking which my principal motive was always to improve and test my powers of reasoning,—was of very high importance, they have generally been recognized as sound contributions. When, at a gathering in Paris in 1875 of all the leading geodesists of Europe, to which I was invited, I was publicly asked by the President, General Ibañez, to give my opinion of the determinations of the absolute acceleration of gravity that were then getting made, I found myself compelled to reply that I believed them to be subject to a constant error about a hundred times greater than that which had been deemed probable, owing to the elastic swaying of the brass tripod with the pendulum. The idea being novel to my auditors, little was said at the time; for men of science like them do not declaim without exact observation to support their opinions. But the next year, at a meeting in Brussels, three of the leading men described experiments they had made which had convinced them that "our American *confrère*" had—well, had found a mare's nest. I received the report of these utterances just as there was about to [be] a third meeting of the body,—the *Europäische Gradmessung* or association of the governmental surveys of the continent of Europe, was about to be held in Stuttgart, and at once applied to my superior officer, the Superintendent of the Coast and Geodetic Survey, for leave to be present at that meeting and there defend what had before been little more than a well-grounded opinion but which further mathematical analysis and measurements had converted into certain knowledge. But the Superintendent was so over-awed by the great authority of those who had declared against my thesis, that I should not have been able to appear at the meeting if I had not resorted to a *ruse*, by which I got a paragraph inserted in the editorial page of that New York daily that was most influential in Washington, strongly urging that *some* Coast Survey officer be dispatched to the meeting. The next day I received no permission, but telegraphic *orders* to be present there; and after I had detailed my analysis and my experiments and exhibited my records, my three former opponents, one after another, rose and fully acknowledged that I was in the right; for they were Men of Science, and their desire was for the truth and not personal glory.

I am forced [to] confess that there are investigators of science who cannot be called "Men of Science," if by that term be understood men who prefer scientific Truth to their own glory. For in 1882 I gave a proof that all my father's "Linear Associative Algebras" could be put into the form of Matrices.[3] To be sure, I did not use the word "Matrix," since, not being a

[3] Charles had urged his father to produce this work which appeared in lithographed

professional mathematician, I was entirely unacquainted with the doctrine of Matrices, which had been set forth in 1858 in a memoir, famous among mathematicians, by Cayley. But I had been led to the very same idea in my studies of the Logic of Relations; and I expressed the proposition I proved by saying that every Linear Associative Algebra could be "put into Relative Form." My proof was of the extremest simplicity and directness, and I never heard of anybody's questioning its correctness except the illustrious algebraist Jacob Joseph Sylvester, who probably thought it made a certain discovery of his of whose importance he had talked much appear as of very limited truth. He always declared that he could not see that I had proved my proposition at all; and the difficulty was not that he did not see that my "Relative Forms" were simply Matrices; for he once asked me what they were, and when I explained, he indignantly exclaimed, as if I had been attempting a plagiarism, "Why, they are nothing but Matrices." If he had said that I had not proved and could not prove my proposition because it was a truism, there would have been some justice in the criticism; and in fact when it was first printed from a letter of mine written to my father, it was set forth as requiring but a few words of explanation. Yet Sylvester never could be brought to assent to its truth (while he remained in America: and I have no reason to suspect that his position on the subject was ever at all modified). The proof is so easy, —every bit as easy as the *Pons asinorum,*—that I will repeat it here; for the Reader must, I feel sure, be quite competent to judge of it. I must, however, first explain what a Linear Associative Algebra is, what an Algebraical Matrix is, and what my "Relative Forms" were.

A Linear Associative Algebra will best be defined after some of its belongings have been described. It has in every case more than one "Unit"; it may certainly have any finite Multitude of Units, and I do not know how many making more. Though these are all Units, no two of them, in the same Algebra, are *Equal,* any more than a dollar is Equal to an hour or a pound avoirdupois, or any more than two differently shaped triangles, though they be of Equal *area,* are called Equal in geometry. For each has some other character besides that of being a Unit. The characters by

form in 1870. He continued to be interested in the subject and finally edited his father's work for publication in the *American Journal of Mathematics* in 1881. For example a note to his father from the U.S. Coast Survey Office in Washington dated 20 March 1873 opens as follows: "Suppose an algebra put into one of your forms. I can prove that every vid is capable of expression as a sum of fractions $A:B$ such that

$$(A:B)(B:C) = A:C$$
$$(A:B)(C:D) = 0."$$

which one is distinguished from another are however entirely Indefinite, beyond the fact that they are different, and beyond their behaviours as Multipliers and Multiplicands, which I shall explain presently. As to my word "Indefinite," I use it primarily to signify a certain way in which a Sign may refer to, or indicate, the Object for which it stands, or which it Represents; but in addition, I apply the adjective "Indefinite" to express a particular Mode of Being of some Objects, considered in themselves, regardless of any Relation to Other Objects, excepting the Relation of being Other than them. I call a Sign "Indefinite" if it does not distinguish each Single Object it Denotes from every one which it does not Denote or Denotes differently. Thus, any Proposition is a Sign; and the Proposition "Cain killed his brother Abel," or its precise equivalent "Abel was killed by his brother Cain" is a Definite Sign; but the Proposition "One of the brothers Cain and Abel killed the other" is Indefinite, though all three denote the same pair of Objects, Cain and Abel. Every non-relative Sign Denotes a Single Object; every Relative Sign denotes a set of Single Objects: that is, such is my form of stating the matter. The Proposition "Some peas are green" is precisely equivalent to "Some green pea is other than Some green pea;" for "some" implies Existence at some time. It is also equivalent to "There is a pea that resembles another pea in being green."

KEY TO GREEK TERMS

Professor Ralph L. Ward of Hunter College of City University of New York undertook the task of transcription and transliteration for the production of this appendix which has been compiled by him. He notes that there are numerous etymologies of Latin, Greek and English words in Peirce's work which are faulty from the present day standpoint but which have been left unaltered in this edition. Moreover no attempt has been made in this edition to insert the usual asterisk before purely reconstructed words adduced by Peirce; nor has there been an attempt to insert macrons over vowels of Latin words cited in the text.

ἀπειρία	apeiria	p. 76
ὁρισμός	horismos	p. 97
λόχος	lochos	
εὔλογον	eulogon	p. 143
πᾶς	pas	p. 167
ἐπαγωγή.	epagōgē	p. 183
ἐπαγωγή	epagōgē	
ἐπί	epi	
ἄγω	agō	
ἐπάγειν	epagein	
ἐπαγωγή	epagōgē	p. 190
ἐπαγωγή	epagōgē	p. 192
ἐπαγωγή	epagōgē	p. 193
ἡ γὰρ ἐπαγωγὴ διὰ πάντων	hē gar epagōgē dia pantōn	p. 200
σημεῖον, ἐπιφάνεια, γωνία, κύκλος	sēmeion, epiphaneia, gōnia, kyklos	
ὅρος, αἴτημα, πρότασις, ἔκθεσις, κατασκευή, ἀπόδειξις	horos, aitēma, protasis, ekthesis, kataskeuē, apodeixis	

ὅπερ ἔδει δεῖξαι	hoper edei deixai	p. 233
πρῶτον ψεῦδος	prōton pseudos	p. 400
μηδέν	mēden	p. 425
οἱ πολλοί	hoi polloi	p. 432
ὅρος	horos	p. 570
κύκλος	kyklos	p. 619
ἀριθμῷ	arithmōi	p. 747
ἔπεα ἔπτερόεντα	epea apteroenta	
ἔπεα πτερόεντα	epea pteroenta	p. 762
πρώτη οὐσία	prōtē ousia	p. 773
κύριον ὄνομα	kyrion onoma	
κῦρος	kyros	p. 842
ῥόδεος	rhodeos	
κυανοῦς	kyanous	p. 849
παρασκευή	paraskeuē	p. 890
πολλά	polla	
πλῆθος πόστον	plēthos poston	p. 915
ἀφαίρεσις	aphairesis	p. 917
βίος	bios	
βίος	bios	p. 935
Διόφαντος	Diophantos	p. 937
πρᾶγμα	pragma	p. 946
χώρισις	chōrisis	
κύκλωσις	kyklōsis	
περίφραξις	periphraxis	
ἀπειρία	apeiria	p. 1081
κύβοι	kyboi	
ἀστράγαλοι	astragaloi	
φιμός	phimos	
πύργος	pyrgos	
ἄβαξ	abax	p. 1125

INDEX OF NAMES

INDEX OF SUBJECTS

(numbers refer to pages)